Integration of Distributed Resources in Smart Grids for Demand Response and Transactive Energy

Meng Song · Ciwei Gao

Integration of Distributed Resources in Smart Grids for Demand Response and Transactive Energy

A Case Study of TCLs

Springer

Meng Song
School of Electrical Engineering
Southeast University
Nanjing, China

Ciwei Gao
School of Electrical Engineering
Southeast University
Nanjing, China

ISBN 978-981-16-7169-2 ISBN 978-981-16-7170-8 (eBook)
https://doi.org/10.1007/978-981-16-7170-8

This Springer imprint is published by the registered company Springer Nature Singapore Pte Ltd.
The registered company address is: 152 Beach Road, #21-01/04 Gateway East, Singapore 189721, Singapore

To rest on the absolute perfection.

—The school motto of Southeast University

Preface

The grids are embracing more renewable energy, but its inherent variability also poses enormous challenges to the power systems. Traditional solutions from the supply side generally cause irreversible damage to generators and increased economic expense. In contrast, demand response (DR) and transactive energy can address the same problem from the demand side at less cost. Distributed resources, the typical participants in DR and transactive energy programmes, contribute significantly to addressing the imbalance between supply and demand because of the constant improvement in the penetration level of renewable energy.

Thermostatically controlled loads (TCLs), i.e. air conditioners, water heaters and refrigerators, represent an example of distributed resources, accounting for up to 30–40% of total power consumption in developed countries. They offer tremendous potential in adjustable power consumption and attract considerable interest in DR and transactive energy opportunities, which highlights their advantages in responding to uncertainties in power systems.

This book provides an insight into TCLs as typical distributed resources in smart grids for DR and transactive energy to address the imbalance between supply and demand. The key research on analysis of uncertainty parameters, modelling of cold load pickup, aggregated control, battery modelling, parameter identification, hierarchical scheduling and transactive control of TCLs in DR and transactive energy is all included. This book is written to offer the researchers, the staff in the power grid and college students a better understanding of the distributed resources integrated with smart grid for DR and transactive energy. Also, it is expected to help address the problems in practice.

The book is divided into ten chapters.

Chapter 1 provides the overview of TCLs in smart grids, including fundamental models, response modes, control strategies, modelling and control issues, system models, dispatch strategies, discussions and future research of TCLs.

Chapter 2 analyses the impacts of uncertain parameters on TCL power capacity. Because the parameters of TCLs are heterogeneous, the identification of those parameters that have a greater impact on their operational behaviour is of major importance. It helps to save the efforts on data collection and processing.

Chapter 3 models the most important characteristic of TCLs after an outage, namely cold load pickup. When TCLs are regulated in DR or transactive energy programmes, the cold load pickup phenomenon cannot be ignored and should be carefully considered.

Chapters 4 and 5 present two methods to model and control the aggregated TCLs at the aggregator or device level. Given the large number of TCLs, which are difficult to be regulated directly, aggregation is required in DR programmes.

Chapter 6 models TCLs as thermal batteries, which is quite different from other flexible loads. When TCLs are treated as batteries, they can be scheduled in a convenient manner.

Chapter 7 compares the thermal batteries of TCLs at the power system level under different control methods. TCLs are generally regulated by temperature set-point adjustment and switch control in DR, which pose favourable characteristics for capacity-type and power-type batteries, respectively.

Chapter 8 models the operational behaviours of TCLs in multi-time scales. A parameter identification method is proposed due to the large number of TCLs and their varying parameters, which is a key point for scheduling TCLs at the power system level.

Chapter 9 presents the hierarchical scheduling method of TCLs for transactive energy based on the aggregated models in Chaps. 6–8. At the lower level, TCLs are regulated by aggregators via a virtual market. At the higher level, aggregator represents TCLs to trade with other aggregators.

Chapter 10 provides a multi-time scale transactive scheduling of TCLs to smooth microgrid tie flow fluctuations. TCLs are regulated by temperature set-point and frequency adjustment in hour and minute time scale, respectively. Moreover, the model parameters of TCLs in minute time scale are determined by the control behaviours of TCLs in hour time scale.

Nanjing, China
August 2021

Meng Song
Ciwei Gao

Acknowledgements

This book summarizes our research on the applications of TCLs in demand response and transactive energy from 2015 to 2020. These studies were carried out at four different universities/labs, including Southeast University, Lawrence Berkeley National Lab, Illinois Institute of Technology and University of Central Florida. Many people contributed to this book. We offer our tonnes of thanks to Prof. Mohammad Shahidehpour and Dr. Mingyu Yan from Illinois Institute of Technology, Prof. Wei Sun and Dr. Reza Roofegari nejad from University of Central Florida, Prof. Zhiyi Li from Zhejiang University and Prof.Yifei Wang from Southeast University. We also thank Ms. Qin Xiao, Mr.Yunfeng Cai, Mr.Jing Meng, Mr. Xudong Wang, Ms. Gujing Lin, Mr. Kang Chen, Ms. Yifan Huang, Mr.Jiacheng Cao, Mr. Wenhan Zhang, Mr. Shengbing Guan, Mr. Wei Wei, Mr. Congzhong Shu and Ms. Yan Chen for their careful examination of the entire book. Finally, we appreciate the assistance and help of the staff at Springer.

This book is supported in part by National Natural Science Foundation (Grant No. 52007030), in part by the National Natural Science Foundation of China (Grant No. 51577029) and in part by the National High Technology Research and Development Program of China (Grant No. 2015AA050401). We do appreciate their supports.

Meng Song
Ciwei Gao

Contents

Nomenclature

Chapter 2

T_{out}	Ambient temperature (°C)
T_i	Indoor temperature (°C)
C_{a}	Equivalent thermal capacity (kJ/°C)
R	Equivalent thermal resistance (°C/kW)
Q	Cooling power of TCL (kW)
$Q*$	Power for heating, illumination and so on (kW)
$S(t)$	Switch state at time t. $S(t)$ is 1 if TCL is on. Otherwise, $S(t)$ is 0
H	Energy efficiency ratio of TCL
P	Rated electric power of TCL (kW)
δ	Temperature dead bandwidth (°C)
T_{set}	Temperature set-point (°C)
$\Delta t'$	A small-time increment (s)
T_i^S	Normalized indoor temperature
Δt	Dispatch period (h)
$T_i(t + \Delta t)$	Indoor temperature at time $t + \Delta t$ (°C)
f_0	Constant zeroth-order function
$f_i(x_i)$	First-order function denoting the effect of x_{i} acting independently upon $f(x)$
$f_{ij}(x_i, x_j)$	Second-order function describing the cooperative effect of x_i and x_j upon $f(x)$
$f_{1,2,\ldots,\text{n}}(x)$	Cooperative effect of variables upon $f(x)$
k, l, m	Highest order of orthonormal polynomials
α, β	Constant coefficients which need to be identified
r, p, q	Order of orthonormal polynomials
$x_{\min}, x_{\max}$	Minimum and maximum input, respectively
x_i, x_i^*	Actual and normalized input, respectively
$x_{\text{ind}}, x_{\text{HDMR}}$	Power capacity obtained by individual simulations and HDMR (MW), respectively

N	Number of dispatch periods
N_s	Number of samples
N_a	Actual number of TCLs
C	Actual TCL power capacity (MW)
Nt	Number of TCLs in simulations (10,000)
C_t	Capacity when the number of TCLs is N_t (MW)
C_l, C_u	Capacity when the parameter has lognormal and uniform distribution (MW)
μ, σ	Mean and standard derivation of normal distribution, respectively
a, b	Lower and upper limits of uniform distribution, respectively
R_μ	Mean value of R (°C/kW)
C_{a}_μ	Mean value of C_a (kJ/°C)
Q'_μ	Mean value of Q' (kW)
P_μ	Mean value of P (kW)
T_{set}_μ	Mean value of T_{set} (°C)
δ_μ	Mean value of δ (°C)
η_μ	Mean value of η
R_σ	Standard derivation of R (°C/kW)
C_{a}_σ	Standard derivation of C_a (kJ/°C)
Q'_σ	Standard derivation of Q' (kW)
P_σ	Standard derivation of P (kW)
T_{set}_σ	Standard derivation of T_{set} (°C)
δ_σ	Standard derivation of δ (°C)
η_σ	Standard derivation of η
T_a	Critical temperature to identify TCL with perfect parameters (°C)
n	Number of input variables
$\phi(x)$	Orthonormal polynomials
D, D_i	Total, first-order partial and second order
D_{ij}	Partial variances, respectively
S_i, S_{ij}	First- and second-order sensitivity indices
$P(P_0, P_i, P_{ij})$	Projection operators
N_p	Number of dispatch periods

Chapter 3

t	Time step
L	Set of nodes
T	Set of DSR time steps
i, j	Index of nodes
m	Sampling step
K	Set of lines
k	Index of line
S	Set of substation nodes

r	The exact time step when the load is restored
ϕ	Set of three-phase
G	Set of distributed generators
C	Set of capacitors
$K\ (i,.)$	Set of lines starting from bus i
$T_{\text{min}}, T_{\text{max}}$	Small range of TCL internal temperature around the set-point (°C)
T_{out}	Outdoor temperature (°C)
ω_i	Priority weight of load
$P_{i,t}^D$	Active base power of load (kW)
$P_i^{DG\_\min}$	Minimum active power of DG (kW)
$P_i^{DG\_\max}$	Maximum active power of DG (kW)
$Q_i^{DG\_\min}$	Minimum reactive power of DG (kVar)
$Q_i^{DG\_\max}$	Maximum reactive power of DG (kVar)
Δt	Time step of DSR (min)
P_i^{ramp}	Active power ramp rate of DG (kW)
$P_k^{L\_\max}$	Active line power limit (kW)
$Q_k^{L\_\max}$	Reactive line power limit (kVar)
N	Number of sampling intervals (time steps)
$P_{i,t}^{\text{TD\_ baseline}}$	Active baseline power of time-dependent CLPU (kW)
$Q_{i,t}^{\text{TD\_ baseline}}$	Reactive baseline power of time-dependent CLPU (kVar)
$Q_{i,t}^D$	Reactive base power of load (kVar)
$\sigma_{i,t}$	Penetration level of TCLs
Z^{D0_et}	Baseline power of extended CLPU (kW)
Z^{D1_et}	Peak power of extended CLPU (kW)
α_i	Decay rate of CLPU for the load at node i
M	A large positive number
$V^{\min}, V^{\max}$	Voltage limits (kV)
$Q_i^{\text{cap\_ max}}$	Reactive power limit of capacitor bank (kVar)
Z	Power of CLPU (kW)
$s_{i,t}^D$	Binary variable indicating the status of load
$s_{i,t}^{DG}$	Binary variable indicating the status of DG
$P_{i,t}^{DG}$	Active power of DG (kW) qDG
$Q_{i,t}^{DG}$	Reactive power of DG (kVar)
$\Delta Z(m)$	Incremental power of CLPU (kW)
$\Delta P_i^{\text{TD}}(m)$	Incremental active power of time-dependent CLPU (kW)
$\Delta Q_i^{\text{TD}}(m)$	Incremental reactive power of time-dependent CLPU (kVar)
$P_{i,t}^{\text{TD}}, Q_{i,t}^{\text{TD}}$	Active and reactive power of time-dependent CLPU (kW, kVar)
$\Delta Z^{et}(m)$	Incremental power of extended CLPU (kW)
$Z^{et}(m)$	Power of extended CLPU (kW)
$\Delta P_i^{\text{TD}-\text{et}}(m)$	Incremental active power of extended CLPU (kW)
$\Delta Q_i^{\text{TD}-\text{et}}(m)$	Incremental reactive power of extended CLPU (kVar)
$P_i^{\text{TD\_ et}}(m)$	Active power of extended CLPU (kW)
$Q_i^{\text{TD}-\text{et}}(m)$	Reactive power of extended CLPU (kVar)

$P_{k,t}^{L}$	Active line power (kW)
$Q_{k,t}^{L}$	Reactive line power (kVar)
μ	Fluctuation extent of load
$V_{i,t}$	Nodal voltage (V)
P_t^{sub}, Q_t^{sub}	Total three-phase active and reactive restorative power by transmission network (kW, kVar)
$Q_{i,t}^{\text{cap}}$	Reactive power of capacitor bank (kVar)
$\tilde{P}_{1,t}^{L}$, $\tilde{Q}_{1,t}^{L}$	Total three-phase active and reactive power of the line connected to substation (kW, kVar)

Chapter 4

T_{o}	Outdoor temperature (°C)
T_i	Indoor temperature (°C)
C_{a}	Equivalent thermal capacity (kJ/°C)
R	Equivalent thermal resistance (°C/kW)
P	Rated TCL electric power (kW)
$S(t)$	Switch state at time t. If TCL is on, $S(t)$ is 1. Otherwise, $S(t)$ is 0
η	Coefficient of TCL performance
t_{on}, t_{off}	Turn-on/off time of TCL in operating cycle (h)
T_{set}	Temperature set-point (°C)
δ	Temperature dead bandwidth (°C)
P_{baseline}	Power baseline of aggregated TCLs (kW)
N	Number of TCLs
P_i	Rated power of ith TCL (kW)
t_{on}^{i}, t_{off}^{i}	On and off duration of the ith TCL (h)
Δt_{dq}	TCL transition from state bin q to state bin d
T_d^m, T_q^m	Intermediate temperature of state bins d and q
Δt	Dispatch period (h)
T_q, T_{q+1}	Start/end temperatures of TCLs in state bin q (°C)
$T_q^{'}$, $T_{q+1}^{'}$	New start/end temperatures of TCL in state bin q after Δt (°C)
a_q^{d1}, a_q^{d2}, a_q^{da}	Entry $(d1, q)$, $(d2, q)$ and (da, q) in A
ΔT	Temperature interval of each state bin (°C)
T_{d1}, T_{d1+1}	Start/end temperatures of state bin $d1$ (°C)
$T_1, T_2, \ldots, T_{2N}$	Intermediate temperature of state bin 1, 2, …, $2N$(°C)
Δt_1	TCL transition time from T_{q+1} to $T_{\max}$ (h)
Δt_2	TCL transition time from T_q to $T_{\max}$ (h)
t_{lock}	Locked time of TCL compressor (h)
n_p^L	Number of locked TCLs that will be unlocked after nL-p+1 dispatch periods
w	Weighting factor

Chapter 5

C_{a}	Equivalent thermal capacity, kJ/°C
T_{o}	Outdoor temperature, °C
T_{i}	Indoor temperature, °C
R	Equivalent thermal resistance, °C/kW
Q	Cooling power of TCL unit, kW
Q'	The heat generated by the people, lighting and other equipment, kW
$S(t)$	The switch state at the time t. If the TCL is on, $S(t)$ is 1. Otherwise, $S(t)$ is 0
η	Energy efficiency ratio of TCL
P	The rated power of the TCL unit, kW
T_{set}	The temperature set-point
δ	Temperature dead bandwidth,°C
Δt	A small-time period, s
t_{on}, t_{off}	The turn-on and turn-off time of an TCL in an operating cycle, s
N	Number of TCLs
P_0^i	The average electric power of the ith TCL, kW
$n(t)$	The number of TCLs at the time t

Chapter 6

T_{o}	Outdoor temperature (°C)
T_i	Indoor temperature (°C)
C_a	Equivalent thermal capacity (kJ/°C)
R	Equivalent thermal resistance (°C/kW)
Q	Cooling power of inverter TCL unit (kW)
Q'	The heat power of people, illumination, equipment and so on (kW)
P	Electric power of the inverter TCL (kW)
f	Frequency of the compressor (Hz)
k_1, l_1, k_2, l_2	Constant coefficients of the inverter TCL system
$x(k)$	Charge/discharge power of the lithium-ion battery in the kth period (kW)
x^d, x^c	Maximum discharge and charge power of the lithium-ion battery (kW)
$s(k)$, $s(k+1)$	Initial state of charge of the lithium-ion battery at the kth and $(k+1)$th period, respectively.
$o(k)$	The stored energy at the kth period (kJ)
$o^{\max}$	Capacity of the battery (kJ)
$\eta(k)$	Charging/discharging energy conversion efficiency of the lithium-ion battery at the kth period

Δt	The duration time of a dispatch period (h)
η^{in}, η^{out}	Charging and discharging energy conversion efficiency of the lithium-ion battery
s_{min}, s_{max}	Minimum and maximum values of the state of charge of the lithium-ion battery
T_{min}, T_{max}	Minimum and maximum values of the indoor temperature to keep users' comfort (°C)
$O_a(k)$	The stored thermal energy of the TB at the kth dispatch period (kJ)
$T_i(k)$	The initial indoor temperature at the kth dispatch period (°C)
O	Capacity of the TB (kJ)
$Q_{baseline}$	Baseline of the cooling power (kW)
$P_{baseline}$	Baseline of the electric power (kW)
$P(k)$	Electric power of the TCL at the kth period (kW)
P^d, P^c	Discharge and charge power limit of the TB (kW)
$soc(k)$, $soc(k+1)$	Initial state of charge of the TB at the kth and $(k+1)$th period, respectively
$Q(t)$	Cooling power of the inverter TCL at the time t (kW)
$Q(k)$	Cooling power of the inverter TCL at the kth dispatch period (kW)
T_{set0}	Initial temperature set-point of the TCL (°C)
α, β, γ	Parameters of the TB model
$f(k)$	Frequency of the compressor at the kth period (Hz)
$soc_i(k)$, $soc_i(k+1)$	The initial state of charge of the ith TB at the kth and $(k+1)$th period, respectively
α_i, β_i, γ_i	Parameters of the ith TB
P_i^c, P_i^d	Maximum charge and discharge power of the ith TB (kW)
$P_{ch/dis\_i}(k)$	Charge/discharge power of the ith TB at the kth dispatch period (kW)
$P_{agg}(k)$	Electric power of the aggregated TB model at the kth period (kW)
$SOC_{agg}(k)$, $SOC_{agg}(k+1)$	The initial state of charge of the aggregated TB model at the kth and $(k+1)$th period, respectively
ξ, γ_{agg}	Parameters of the aggregated TB model
τ	Maximum value of the state of charge of the aggregated TB model
P^d_{agg}, P^c_{agg}	Maximum discharge and charge power of the aggregated TB model (kW)
$B(k)$	Charge/discharge power of the energy storage device at the kth dispatch period (kW)

$L(k)$, $G(k)$, $p_{rt}(k)$	The forecast of load, renewable energy generation and electricity price at the kth dispatch period (kW, kW, \$/MW)
w	The receding optimization horizon periods

Chapter 7

T_o	Outdoor temperature, °C
T_i	Indoor temperature, °C
C_a	The equivalent thermal capacity, kJ/°C
R	Equivalent resistance, °C/kW
Q	The cooling power of the inverter TCL, kW
Q'	The internal heat generation power, kW
k_1, l_1, k_2, l_2	Constant coefficients
U	The voltage of the grid, V
I	The electric current, A
$SOC(k+1)$, $SOC(k)$	*SOC* of PTBM at the (k+1)th and kth period
$Q(k)$	Cooling power of TCL at the kth period, kW
$P(k)$	Electric power of the TCL at the kth period, kW
$P_{cd}(k)$	Charging/discharging power of *PTBM* at the kth period, kW
P_{min}, P_{max}	Minimum and maximum electric power of the inverter TCL, kW, respectively
m	The number of PTBMs
$P_{agg}(k)$	Aggregated charging/discharging power of the PTBMs at the kth period, kW
P_h^d, P_h^c	The maximal discharging/charging power of the hth PTBM, kW, respectively
T_s, T_c	The desired and the current temperature, °C
ΔT_{max}, ΔT_{min}	Maximum and minimum temperature errors, °C
$Q(k)$	The scheduled power of the aggregator in a day-ahead market, MW
$G(k)$, $L(k)$, $p_{ad}(k)$	The predicted wind generation, load and electricity price, MW, MW, dollar/MW, respectively
w	The receding optimization period

Chapter 8

T_{out}	Ambient temperature (°C)
$\Delta t'$	A small-time increment (s)
$T_i(t)$, $T_i(t+\Delta t')$	Indoor temperature at time t and $t+\Delta t'$ (°C)
C_a	Equivalent thermal capacity (kJ/°C)

R	Equivalent thermal resistance (°C/kW)
$Q(t)$	Cooling power of TCL at time t (kW)
$S(t)$, $S(t+\Delta t')$	Switch state at the time t and $t+\Delta t'$. If FTCL is on, $S(t)$ is 1. Otherwise, $S(t)$ is 0
η	Energy efficiency ratio of FTCL
P	Electric power of FTCL (kW)
δ	Temperature dead bandwidth of FTCL (°C)
T_{set}	Temperature set-point (°C)
$T_{\min}$, $T_{\max}$	Range of FTCL indoor temperature at T_{set} (°C)
$T_{\text{set}}^{\text{new}}$	New temperature set-point (°C)
$T_{\min}^{\text{new}}$, $T_{\max}^{\text{new}}$	Range of FTCL indoor temperature at $T_{\text{set}}^{\text{new}}$ (°C)
t_{on}, t_{off}	On and off time in an FTCL operating cycle (h)
P_{baseline}	Power baseline of FTCLs (kW)
$P_a(k)$	Power valley/peakof FTCLs at kth dispatch period (kW)
$\Delta T(k)$	Temperature adjustment at kth period (°C)
$\Delta T_{\min}$, $\Delta T_{\max}$	Minimum and maximum temperature adjustment (°C)
k_1, l_1, k_2, l_2	Constant coefficients of ITCL
$P_i(t)$	Electric power of ITCL at time t (kW)
$P_{\min}$,$P_{\max}$	Minimum and maximum power of the ITCL (kW)
ΔT_c	Temperature adjustment range of ITCL (°C)
n	Number of TCLs
P^k	Electric power of continuous FTCL power model (kW)
P_o	Power baseline of continuous power model (kW)
$P_h(k)$	Power output of virtual generator at kth dispatch period (kW)
$f(\Delta T(k))$	Power change function (kW)
$P_h^{\min}$, $P_h^{\max}$	Minimum and maximum power adjustment (kW) of virtual generator
ξ, γ, τ	Parameters of virtual battery on minute time scale
$soc(0)$	Initial state of charge of a virtual battery
$P_m(k)$	Charge/discharge power of minute time scale virtual battery (kW)
$P_m^{\min}(k)$, $P_m^{\max}(k)$	Minimum and maximum power of minute time scale virtual battery (kW)
$SOC_m(k)$, $SOC_{m+1}(k)$	State of charge of the minute time scale virtual battery at kth and $(k+1)$th dispatch period
$SOC(k)$, $SOC(k+1)$	State of charge of a second time scale virtual battery at kth and $(k+1)$th dispatch period
$P_{cd}(k)$	Charge/discharge power of a second time scale virtual battery at kth dispatch period (kW)
σ, υ, ψ	Parameters of a second time scale virtual battery
$SOC_s(k)$, $SOC_s(k+1)$	State of charge of aggregated second time scale virtual battery at kth and $(k+1)$th dispatch period
$P_s(k)$	Charge/discharge power of the aggregated second time scale virtual battery at the kth period (kW)

$P_s^{\min}$, $P_s^{\max}$	Minimum and maximum charge/discharge power of aggregated second time scale virtual battery (kW)
ν, κ	Parameters of aggregated second time scale virtual battery
U	Number of input variables
g, l, m	Highest order of orthonormal polynomials
x^*, x	Normalized and actual value of input variable x
$x_{\min}$, $x_{\max}$	Minimum and maximum of input variable x
$(\cdot)_i$	$\cdot$ of the ith TCL

Chapter 9

c, d	Indices for charging and discharging
k	Index for time slot
$i(v)$, j	Indices for aggregator (AGG) and TCL
t	Index for time
g, G	Indices for generator
C_a	Equivalent thermal capacity (kJ/°C)
COP	Coefficient of performance for TCLs
T_{out}	External temperature (°C)
P_{rated}	Rated electric power of a TCL (kW)
R	Equivalent thermal resistance (°C/kW)
T_{set}	Temperature set-point (°C)
δ	Temperature dead bandwidth (°C)
$T_{\min}$, $T_{\max}$	Minimum and maximum internal temperatures of TCL (°C)
κ, γ	Coefficient of TCL virtual battery
α	Coefficient of virtual battery for aggregated TCLs
$P_{\min}$, $P_{\max}$	Minimum and maximum charging/discharging power of virtual battery for aggregated TCLs (kW)
$SOC_{\min}$, $SOC_{\max}$	Minimum and maximum state of charge (SOC) of virtual battery
n, N_k, N_{agg}	Number of TCLs, time slots and aggregators, respectively
ω_1, ω_2	Coefficients of bid price
a, b, c	Coefficients of generator generation cost function
d_{iv}	Equivalent electrical distance between aggregator i and v
h, l	Parameters for Lagrangian multiplier update
P^c, P^d	Maximum charging and discharging power of conventional battery (MW)
η^c, η^d	Charging and discharging efficiency of conventional battery
o	Capacity of conventional battery (MWh)
$soc_{\min}$, $soc_{\max}$	Minimum and maximum state of charge of conventional battery
soc_0	Initial state of charge of conventional battery
T_{in}	Internal temperature (°C)
S	TCL status: S is 1 if TCL is on; otherwise, S is 0
P_{cont}	Electric power of continuous TCL power model (kW)

P_{baseline}	Power baseline of TCL (kW)
P_{cd}	Charging/discharging power of virtual battery for individual TCL (kW)
x	State of charge of virtual battery for individual TCLs
Δt	Time slot duration (s)
SOC	State of charge of the virtual battery model for aggregated TCLs
P	Charging/discharging power of the virtual battery for aggregated TCLs (kW)
N_t	Number of total control signals
N_r	Number of TCL enforced switching
P_{desired}	Desired power of TCLs (kW)
P_{actual}	Actual power of TCLs (kW)
ΔP	Power adjustment of TCLs (kW)
ρ_1, ρ_2	Bid price components
ρ_{pid}	Bid price for the virtual battery ($/kW)
$\rho_{\text{pid}}^{\text{cl}}$	Market clearing price ($/kW)
ρ	Contract price of AGG ($/MW)
$P_{G,i}$	Power of generator owned by AGG i (MW)
$P_{\text{pur},i\text{SO}}$	Power purchased by AGG i from SO (MW)
$P_{\text{pur},iv}$	Power purchased by AGG i from AGG v (MW)
$P_{\text{sel},i}$	Power sold by AGG i to other AGGs (MW)
$P_{\text{battery},i}$	Charging/discharging power of virtual battery owned by AGG i (MW)
$P_{\text{load},i}$	Load of AGG i (MW)
m	Number of Lagrangian multiplier iterations
λ	Lagrangian multiplier

Chapter 10

i	Index of inverter TCL
t	Index of time
k	Index of time slot
ν, κ	Index of current time slot and prediction-horizon time slots in minute time scale
ω	Index of wind generation realization scenarios
N	Set of TCLs
T	Set of time slots in hour and minute time scale
V, E	Set of current and prediction-horizon time slots in minute time scale
Ψ, Π	Set of wind generation realization scenarios in hour and minute time scale
a, sch, f	Index of current, scheduled and predicted variables
ra, rm	Index of a_g and m_g segment in hour time scale

R_1, R_2	Set of a_g and m_g segment in hour time scale
H_1, H_2	Set of a_b and m_b segment in minute time scale
ha, hm	Index of a_b and m_b segment in minute time scale
C_a	Equivalent thermal capacity (kJ/°C)
R	Equivalent thermal resistance (°C/kW)
s, l, k_p, l_p, k_q, l_q	Constant parameters of the inverter TCL
ΔT	Temperature incremental (°C)
$T_{\min}, T_{\max}$	Minimum and maximum indoor temperatures of TCLs (°C)
T_{set0}	Initial (preferred) temperature set-point (°C)
$P_g^{\max}, P_g^{\min}$	Maximum and minimum of P_g (kW)
α, β, γ	Coefficient of virtual battery model
$P_b^{\min}, P_b^{\max}$	Minimum and maximum of P_b
$SOC^{\min}, SOC^{\max}$	Minimum and maximum of SOC
$\Delta t_h, \Delta t_m$	Time interval in hour and minute time scale
SOC_0	Initial SOC of virtual batteries in minute time scale
$\Delta SOC^{\min}, \Delta SOC^{\max}$	Minimum and maximum of SOC incremental
ΔSOC^m	Maximum SOC incremental decided by the MO
μ, υ	Coefficient of bidding curve (kW^2/dollar)
ε	Power conversion coefficient (kW)
π_ω	Probability in scenario ω
λ_0, P_{g0}, Z_0	Numerical value of λ, P_g and Z when λ is the boundary point of the piecewise linear segments
r_0	Number of λ segments in hour time scale
h_0	Number of ξ segments in minute time scale
$\xi_0, G_0, \Delta SOC_0$	Numerical value of ξ, G and ΔSOC when ξ is the boundary point of the piecewise linear segments
Q, P	Cooling and electric power of the inverter TCL (kW)
φ	Compressor frequency of inverter TCL (Hz)
P_{baseline}	Power baseline of the inverter TCL (kW)
$T_{\text{in}}, T_{\text{out}}$	Indoor and outdoor temperature (°C)
P_g	Power of virtual generator (kW)
P_b	Charging/discharging power of virtual battery (kW)
SOC	State of charge of virtual battery
T_{in}^*	TCL temperature set-point determined in hour time scale (°C)
λ, ξ	Hour and minute time scale clearing price (dollar/kW)
ΔSOC	SOC incremental of virtual battery
R_h, R_m	Adjustment fee of inverter TCL in hour and minute time scale (dollar)
P_{load}, P_w	Load and wind generation power (kW)
P_{da}	Day-ahead plan power (kW)
ρ	Penalty factor (dollar/kW)
Y	Absolute value of the tie flow errors (kW)

Z	Adjustment fee of virtual generators
a_g, m_g	Introduced variable to determine λ
G	Adjustment fee of virtual batteries (dollar)
a_b, m_b	Introduced variable to determine ξ

Chapter 1
Overview of TCLs in Smart Grids

Abstract More flexibility is desirable with the proliferation of variable renewable energy for balancing supply and demand in power systems. Thermostatically controlled loads (TCLs) attract tremendous attention because of their specific thermal inertia capability in smart grids. To effectively manage numerous and distributed TCLs, intermediate coordinators, e.g., aggregators, as a bridge between end-users and dispatch operators are required to model and control TCLs for serving the grid. Specifically, intermediate coordinators get access to fundamental models and response modes of TCLs, make control strategies, and distribute control signals to TCLs according to the requirements of grid operators. Moreover, intermediate coordinators provide dispatch models that present the external characteristics of TCLs to dispatch operators for scheduling different resources. The bottom-up key technologies of TCLs in demand response (DR) programs based on the current research are reviewed and the discussions are provided. The promising research field of TCLs in transactive energy is stated at the end of this chapter.

Keywords Thermostatically controlled loads (TCLs) · Demand response (DR) · Transactive energy · Renewable energy · Smart grids

1.1 Introduction

Even though the proliferation of renewable energy enhances the sustainability of power systems, their inherent variability poses great challenges to the planning and operation of large power grids. The corresponding electric power deficiencies can be compensated by fast ramping generators and energy storage devices [1, 2]. However, frequent ramp up/down power adjustments can increase the operation and the maintenance cost of generators. Moreover, storage devices are regarded as costly alternatives. Demand response (DR) can address this problem owing to its attractive and versatile capability in balancing the supply–demand, improving energy efficiency, and enhancing system resilience. The flexible loads can be adjusted rather quickly with no negative impacts on participating end-users by providing financial opportunities to compensate for power deficiencies caused by variable renewable energy.

M. Song and C. Gao, *Integration of Distributed Resources in Smart Grids for Demand Response and Transactive Energy*, https://doi.org/10.1007/978-981-16-7170-8_1

Thermostatically controlled loads (TCLs) (i.e., air conditioners (ACs), water heaters, and refrigerators) is an example of flexible loads, which account for up to 30–40% of total power consumption in developed countries [3, 4]. TCL has aroused significant interest in the DR opportunity due to its huge potential in adjustable power consumption. It has highlighted the advantages of TCLs in responding to uncertainties in power systems.

Due to the small capacity and large quantity of TCLs, individual TCL can not significantly affect the power systems, which makes it necessary to manipulate TCLs in a population. Accordingly, load aggregators which serve as intermediate coordinators will manage TCL operating behaviours and determine strategies to effectively control TCL populations based on the requirements of dispatch operators. On the other hand, load aggregators should be responsible for characterizing the performance and providing the aggregated TCL characteristics to dispatch operators. Dispatch operators will schedule aggregated TCL operations with other resources, e.g., renewable energy [5] for satisfying operational objectives.

This chapter presents a state-of-the-art review of TCLs for DR in smart grids, including fundamental models and response modes of individual TCLs, control strategies, system models, and dispatch strategies of TCLs. Specifically, fundamental models describe the basic characteristics of TCLs, which bridge the electricity and the surroundings constrained by the requirements of customers. TCLs is known to be adjustable in different ways, including on/off switches, temperature set-point adjustment, and the combination way which are characterized by response modes. TCLs perform differently in various response modes, which determine the types of ancillary services provided by TCLs. As described above, load aggregators as intermediate coordinators not only provide aggregated characteristics of TCL populations at a higher level but also make control strategies for TCLs in terms of dispatch signals. In this chapter, system models are utilized to describe the collective performances of TCLs at the level of power systems.

1.2 Fundamental Models of TCLs

The physical model of TCL includes the energy conversion process which converts electric power to cooling/heating, and the energy exchange process which corresponds to the internal TCL temperature with cooling/heating power. These two processes enable TCLs to adjust their power consumption while keeping the internal temperature within a certain range as illustrated in Fig. 1.1.

1.2.1 Energy Conversion

For residential TCLs, the main power consumption is the compressor. In literature, the highlights are TCLs with compressors of fixed speed. In other words, the ratio of

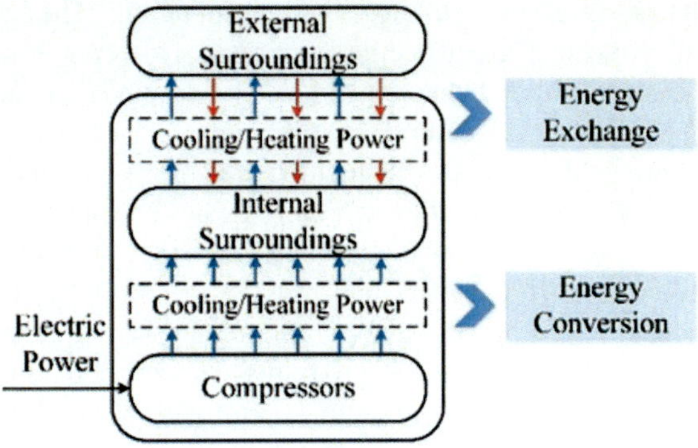

Fig. 1.1 Diagram of TCL fundamental models (the blue lines indicate the cooling power, and the red lines the heat power)

electric power to cooling/heating power is fixed [6]. However, TCL compressors are evolving from fixed speed to variable speed for higher efficiency. The ratio of electric power to cooling/heating power is thus not fixed and depends on their compressor frequency [7, 8]. For large TCLs, e.g., the commercial air conditioning system, power consumption refers to several components, including fan, chiller, etc. [9], which makes the energy conversion more complicated.

1.2.2 Energy Exchange

Two methods are used to model the energy exchange between the internal surroundings of TCLs and the external surroundings, as summarized below.

1. The equivalent electric circuit method [10] utilizes common electrical components, e.g., electric resistance, capacitance, and power source, to model the external and internal surrounding and cooling/heating power of TCLs, as well as the heat transfer process [10–12]. A third-order thermodynamic air conditioning model is proposed in Ref.[12]considering the temperature differences among internal air, and external and internal walls of dwellings. Although higher-order thermodynamic models provide better accuracy, complicated structures with heavy computational burdens limit the application. Therefore, the simplified second-order [13] and first-order [14] thermodynamic models are usually employed for TCL aggregation and control.
2. The cooling load calculation method [11] derives the thermodynamic model of TCLs according to the energy conservation law, i.e., the energy increment of objects equals the difference between the cooling/heating power and the energy obtained from the surrounding environment. Many object parameters, e.g., size and material, need to be measured for the calculations of energy increment.

The comparisons of different methods are summarized as follows [11]. The second and higher-order equivalent electric circuit method has a high modelling accuracy but it has complicated structures and computations. The structure of the first-order equivalent electric circuit method is simple. However, it cannot accurately describe the actual operation behaviours of TCLs. Moreover, parameter identification is required for the models captured by the equivalent electric circuit method. Although the cooling load calculation method free from parameter identification, complicated cooling load calculation is required whose accuracy is low.

The energy conversion and exchange models characterize the operating behaviours of TCLs, which provide the basis for TCL aggregation, control, and scheduling. The physical parameters in TCL fundamental models are critical to modelling TCLs in DR programs. However, parameter heterogeneity, including device types, object structures, external surroundings, and customers preferences poses a great challenge. Although physical parameters can be calculated in terms of individual measurements of each TCL, it may result in privacy disclosure of end-users as well as huge cost. Thus, the evaluation of physical parameters of TCL populations while protecting the information of customers and reducing measurement investment remains a pivotal challenge.

1.3 Response Modes of TCLs

Response modes are defined as the action of TCLs upon receiving control signals. Based on the response modes, aggregators will determine control strategies and generate control signals to end users according to system operating requirements.

Response modes are closely related to the types of TCL compressors. In literature, the highlight of TCLs is fixed-speed, which is regulated by changing the switch status [15] and the temperature preferences [3]. As the power consumption of fixed-speed TCLs is restricted to the constant frequency of compressors, TCLs only have two operation statuses, including the rated and zero power output. This results in periodic changes of internal temperature within a certain range $[T_{\min}, T_{\max}]$ [16]. If the internal temperature is out of the range $[T_{\min}, T_{\max}]$, TCLs will work at rated power or turn off to either reduce or increase the internal temperature, respectively. Based on specific characteristics, TCLs have different operating characteristics on response speed and duration at different control signals.

It is shown that changing switch status can instantaneously modify TCL consumptions for adjusting power output, but the adjustment cannot be kept for a long time due to the temperature preference constraints of customers [17]. Changing switches are usually employed to quickly modify the power consumption of TCLs, such as critical peak shaving [18] and regulation service [19]. On the contrary, the TCL power consumption can be regulated for a longer period by modifying the thermostat set-point, but with a delayed effect. The adjustments to temperature set-points in TCL energy management are applied to minimize energy costs [6] and provide load-following service [20]. Moreover, on/off switches can be combined with temperature set-point adjustment to provide fast response speed, large capacity, and long duration in different applications [17, 18], such as providing power pulses [16]. The

Table 1.1 Comparisons of on/off switches, temperature set-point adjustment and their combination

Methods	Response speed	Response time	Typical application scenario
On/off switches	Fast	Short	Regulation service
Temperature set-point adjustment	Slow	Long	Load following
Combination	Fast	Long	Power pulses

comparisons of switches, temperature set-point adjustment, and their combination are summarized in Table 1.1 [11].

Compared to fixed-frequency TCLs, inverter TCLs can be controlled by temperature set-point change, as well as continuously manipulated by frequency adjustment [7, 21]. Therefore, inverter TCLs can be regulated with more flexibility in response time, capacity, and duration, which are generally applied in minute-based DR programs [8, 22].

1.4 Control Strategies of TCLs

1.4.1 Centralized Control

Aggregated control models for homogeneous TCLs: When residential TCLs are regulated in a centralized way, a centralized controller needs to characterize the TCL aggregated characteristics and generate control signals for each TCL. To guide the centralized controller, it is critical to developaggregated control models that characterize the distribution of TCL internal temperatures and switch states. Initially, TCL dynamic behaviours are characterized by Fokker–Planck equations [23], which are solved in a non-closed-form analytic way with heavy complexity. To reduce the complexity of aggregated TCL models, some simplifications are developed by using discrete temperature bins and time settings [24].

The common aggregated control models of TCLs include the state queuing model [17], the transport model [25], the Markov chain [26], and the state space/bin transition model [27]. As shown in Fig. 1.2, different state bins are employed to describe the collective behaviours of TCLs according to their internal temperature and switch states. It is shown that more states with a shorter discretization time slot [25], state bin with non-uniform temperature ranges [28], two-dimensional state-bin model based on the second-order thermodynamic model of TCLs [29] are helpful to improve the modelling accuracy.

Aggregated control models for heterogeneous TCLs: TCL parameters could also be heterogeneous as a result of different object structures, comfort requirements of end-users, device types, and geographic locations. Most aggregated control models are initially developed for homogeneous TCLs and then modified to heterogeneous ones [28, 30]. The heterogeneous TCLs can be classified into several clusters using

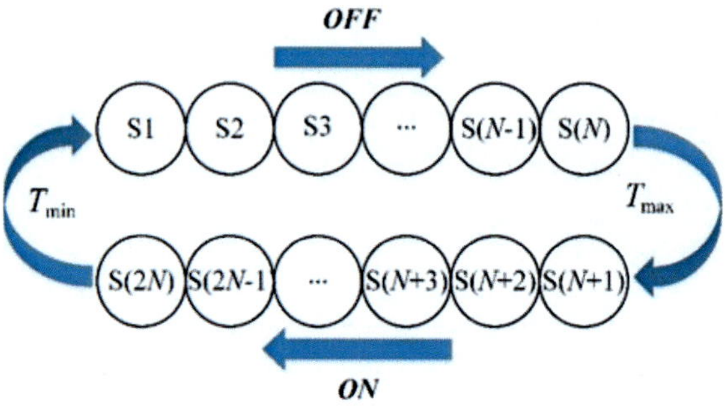

Fig. 1.2 Diagram of TCL state transition (S means state and N is the number of TCL operating states when their switch status is on/off)

k-means algorithms [28]. Within each cluster, TCLs are regarded as homogeneous, which can be aggregated using aggregation techniques, e.g., state-space/bin transition models. Then the sum of different aggregated models of homogeneous TCLs is seen as the model of heterogeneous TCLs.

There are other methods for heterogeneous TCLs. An improved aggregated model with the lockout time is developed in Ref. [31] to extract the collective dynamics of heterogeneous TCLs by analytical derivations. A similar modelling methodology is proposed in Ref. [32] to describe the behaviours of aggregated TCLs.

Control methods: Control signals are generated by the centralized controller to regulate TCLs according to the reference signals from dispatch requirements. Priority stack [28, 33] and probability control [27] are often utilized to determine the control signals. Besides, some other methods are also applied to control TCLs [30, 34]. For example, the model predictive control with on/off switching is developed [3] for the sequential temperature set-point adjustment for power adjustment. A hierarchical centralized control algorithm is presented in Ref. [30] to provide load-following without affecting the comfort of end-users.

One major disadvantage of these model-based control approaches is that aggregated models of TCLs are required to forecast and optimize the behaviour of TCLs. However, it is difficult to obtain accurate models of aggregated TCLs as a result of parameter heterogeneity and an extremely large number of TCLs [13]. Therefore, model-free control methodologies have been developed recently to regulate TCLs without aggregated models. Advanced solution algorithms are developed to make sequential decisions according to TCL dynamics. The queueing method is utilized in Ref. [35] to continuously select proper TCLs for regulation services based on the power and energy limits of TCLs. A three-step approach is proposed in Ref. [13] to regulate both steady-state and transient thermal dynamics of TCLs by identifying tracer devices. Reinforcement learning is studied in Ref. [36] to improve the scalability of TCL centralized control.

1.4.2 Decentralized/Distributed Control

The decentralized/distributed control enables end-users to optimize their behaviours for providing services to the grid, based on local information or the exchanged information from other end-users [35]. For example, the stochastic switching of refrigerators for frequency control is developed in Ref. [37], considering the lockout constraints of devices and the incremental power consumption. A stochastic control approach is developed in Ref. [24] to accurately model the collective power for TCL control. To eliminate the two-way communication, a decentralized approach similar to the droop control of generators is suggested in Ref. [38] for frequency regulation. A decentralized cooperative algorithm is proposed in Ref. [39] to make the plan of energy usage based on the hybrid system model of TCLs and electric vehicles. Recently, the transactive control strategy [19] has attracted much attention, in which self-interested entities are encouraged using market mechanisms to serve the grid. The advantage is that the price and power consumption quantity are the only required information, which works well to protect the privacy of entities. For example, a transactive control method is developed in Ref. [19] to regulate commercial buildings for peak load shaving.

Comparing the two control methods, the centralized control has good controllability but with a high communication burden and time delay. In contrast, decentralized control can respond faster but with less predictability. The characteristics of both methods have been compared in the literature. A hierarchical control scheme is developed in Ref. [40], in which the TCLs in a community are centrally controlled by an aggregator and different aggregators are controlled by decentralized controllers.

1.5 Modelling and Control Issues of TCLs

1.5.1 Cold Load Pickup

Mechanism: When controlled by on–off switches or temperature set-point adjustment, the TCL operating state heterogeneity is destroyed and becomes homogeneous. The aggregated power may experience oscillation [18] and power system operations appear more variability. For example, the evolution of TCL collective operating states in the operation of raising temperature set-point is depicted in Fig. 1.3. In Fig. 1.3a, TCLs are working at temperature set-points and their operating states are uniformly distributed in terms of temperature, resulting in steady power. When the temperature set-point increases by ΔT (i.e., the internal temperature of TCLs varies between $[T_{\min} + \Delta T, T_{\max} + \Delta T]$), as demonstrated in Fig. 1.3b, the TCLs with the switch state of on and the internal temperature lower than $T_{\min} + \Delta T$ will be immediately turned off while the other TCLs with the switch state of on will be turned off once their internal temperature reaches $T_{\min} + \Delta T$. For TCLs with the switch state of off, they will remain off until their internal temperature reaches $T_{\max} + \Delta T$. Gradually,

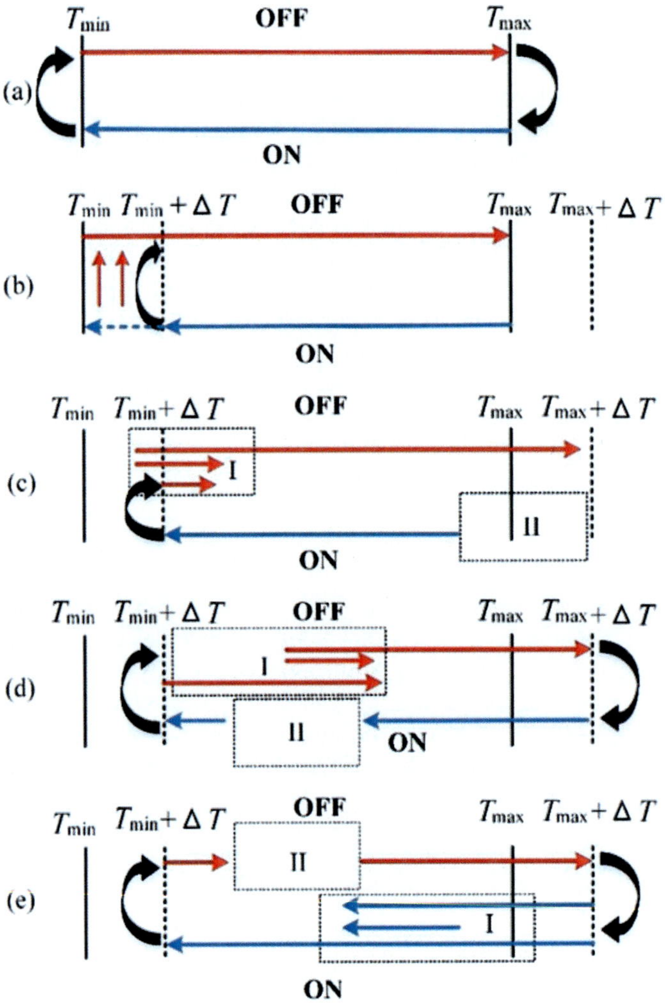

Fig. 1.3 State evolution of TCLs with an incremental temperature

the number of TCLs in Box I keeps increasing and there are no TCLs in Box II, as exhibited in Fig. 1.3c. In other words, the aggregated power consumption of TCLs decreases to cause a power valley. With more TCLs being turned off, as displayed in Fig. 1.3d, the collective TCL power continues decreasing to achieve the desired load shedding. However, moving forward, more TCLs will be turned on while fewer TCLs are turned off, as described in Fig. 1.3e. At this stage, the collective power of TCLs presents a peak that is larger than the original one.

In summary, as the temperature set-point is adjusted, TCL collective power will be oscillatory with the evolution of TCL operating states, as a result of the loss of operating state heterogeneity. This is the cold load pickup phenomenon, which is usually characterized by the peak duration and magnitude [41].

Solutions: The cold load pickup phenomenon is influenced by many variables. A sensitivity analysis to identify the important factors that have great impacts on the aggregated dynamics of TCLs is proposed in Ref. [42], which is helpful to facilitate the control design for restraining the cold load pickup. A methodology is provided in Ref. [34] to estimate the peak magnitude of the cold load pickup, considering the outdoor temperature, the TCL concentration level and the outage duration. Moreover, the cold load pickup of TCLs can cause undesired power fluctuations. To address this issue, it is shown that the operating state heterogeneity of TCLs should be restored as soon as possible [18, 22]. For example, several safety protocols are proposed in Ref. [18] to restrain the undesired power oscillations by embedding instructions and memory to the TCL controllers, which relieve the communication burden and protect the privacy of end-users.

1.5.2 Impact Factors of TCL Control Strategies

Among different impact factors on the operating behaviours of TCLs, two main factors, i.e., the lockout phenomenon and the outdoor temperature have great impacts on the control strategies.

Lockout phenomenon: The durability of the TCL compressor can be destroyed by too frequent switches [3, 27]. There are two main methods to address this problem. The first one is to utilize a minimum time constraint when TCLs are off [31]. In other words, TCL compressors are kept off for a certain period, during which TCLs will be locked and not respond to any external signals. When the predefined time is passed, TCLs will be unlocked and begin to react in terms of the external control signals. The second one is the priority-stack method [33, 35], in which TCLs are ranked and given different priorities to change their switch states. When selecting TCLs to be turned on, those TCLs whose switch status is off will be sorted in descending order according to their internal temperatures. Those TCLs whose internal temperatures are closer to $T_{\max}$ will be turned on with higher priorities. Similarity, when selecting TCLs to be turned off, the TCLs whose internal temperatures are closer to $T_{\min}$ will be turned off with higher priorities. In summary, TCLs close to the end of their operating cycles will be given higher priorities to change their switch states for reducing the forced switch times.

External temperature: The power consumption of TCLs is sensitive to external temperature. It is shown that a lower external temperature will cause more power consumption due to the longer and more frequent operation of heat pumps to maintain the internal temperature at the set-point [43]. Similarly, as demonstrated in Ref. [8], a higher outdoor temperature will increase the discharge power limit, which can reduce the charge power limit of thermal batteries for inverter air conditioners. Therefore, the time-varying external temperature is a critical factor in the design of TCL control strategies. In Ref. [33], a forecast-and-update method is developed to consider the changing outdoor temperature in TCL control algorithms, in which the external temperature is first fixed during a time slot and then updated by measurements. The impact of external temperature on the aggregated behaviours of TCLs is characterized in a mathematical formulation for TCL control in DR programs [44]. Real-time outdoor temperature deviations, the ramping rate of temperature set-points, and the related deviation are considered in Ref. [30] as three inputs of the aggregated models, which provide the frequency capacity and ramping rate of TCLs under time-varying surroundings.

1.6 System Models of TCLs

Aggregation is critical to integrate TCLs into power system operations. The majority of studies conducted on aggregated TCL models are focused on characterizing the detailed evolutions of internal temperatures and switch states at control signals of TCLs from the perspective of control [22].

To dispatch TCLs in power system operations, a concise aggregated model at the system level is required to accurately characterize TCL primary behaviours. TCL system models can be formulated as common components (e.g., batteries), which offer additional compatibility for power system operation [8]. In this way, TCL populations with complicated structures and alien parameters are invisible. In other words, TCLs are modelled as common batteries or generators connected with the grid from the perspective of operation and dispatch.

It is shown that TCLs have different response performances, in terms of response speed and duration, using diverse control methods. Accordingly, there are three categories for the system models of TCLs in terms of response time scales, which will provide various ancillary services for the grid.

Hour-time scale: When scheduled on an hour-time scale, TCLs are usually utilized for peak load shaving. Thus, TCLs should have a longer duration but with less need for a fast response speed. The temperature adjustment is more suitable to regulate TCL power profiles for the hour-time scale dispatch. A virtual generator model is developed in Ref. [22] to describe the TCL response capacity with temperature control, considering a control strategy for a cold load pickup. In Ref. [45], the virtual power plant that consists of a large TCL population is developed to manage distributed energy resources.

Minute-time scale: At this time scale, TCLs are generally employed to provide load-following service. Thus, TCLs are required to respond faster but with less need for long-duration time compared to that in hour-time scale. It is widely known that TCLs can make thermal energy stored in related objects (e.g. buildings). For example, a leak storage unit is proposed in [46] for security-constrained economic dispatch and multi-service allocation. A thermal battery model of inverter air conditioners is developed in Ref. [8, 47] to reduce the economic loss caused by renewable energy uncertainty in the real-time market. The capacity of TCLs at the 10-min time scale is estimated in Ref. [48] to dispatch generators in terms of the day-ahead temperature forecast of the market.

Second-time scale: TCLs can provide the second-time scale frequency regulation service to the grid. Accordingly, TCLs should act very quickly and tend to be controlled by on–off switches. Battery models are the most common and popular system models for TCLs [22, 49]. A battery model that describes the external characteristics of TCLs under dispatch signals is modelled in Ref. [22]. A stochastic battery model with dissipation is developed in Ref. [35] by aggregating the flexibility of TCL populations, considering the corresponding power limits and energy capacity. The optimization-based algorithms to build a virtual battery model are proposed in Ref. [49] for representing the TCL flexibility.

In summary, system models of TCLs are closely related to control methods, in order to provide different ancillary services with various response characteristics. When scheduling aggregated TCLs with other resources, accurate system model parameters are critical. However, there are only a few parameter estimation methods for TCL system models. In Ref. [22] a high dimensional model representative method is developed to identify the aggregated parameters of multi-time-scale TCL system models, in which the probability distribution of the TCL parameters is utilized. An alternative approach is presented in [50] to identify the aggregated parameters without the information of individual TCL parameters.

1.7 Dispatch Strategies of TCLs

It is an optimization problem when TCLs are dispatched in power system operations. The objective functions are to minimize load [45, 51], cost [9, 50, 52], or the uncomfortableness of end-users [53]. Decision variables can be power adjustment [8, 9, 54] and off-shift time [55]. Constrains are related to system operating states and TCL adjustment capacity, including power balancing [52], power capacity limit [53, 54], and load payback [53]. Solution algorithms are usually based on linear programming [9, 45, 51] and dynamic programming [53].

As TCLs can be aggregated as common system components (e.g. batteries), the dispatch strategies of these components can be applied to the integration of TCL aggregated models into system operation. For example, there has been extensive literature focusing on the dispatch strategies of unit commitment [56], peak load shifting [57], output smoothing of wind generation [58], and economic optimization in the electricity market [59]. Aggregated battery models of TCLs can substitute batteries to provide similar grid services, which shows the compatibility of TCL battery models with current dispatch strategies. Besides similarity between the battery and aggregated TCLs models, there are also some differences in operating behaviours [8]. The optimal coordination of virtual storage (i.e. TCL battery models) and real storage (e.g. conventional batteries) is an interesting topic [9].

1.8 Discussions of TCLs in DR Programs

1.8.1 Non-invasive Parameter Estimation

Non-invasive parameter estimation is a method that employs the external information of power, current or voltage to analyze and obtain the related internal parameters without installing the devices into the objects. This method is effective in protecting the privacy of objects without causing widespread concerns. For example, a non-invasive method is developed in Ref. [60] to identify the parameters of turbine, generator, and exciter using real network data and the stochastic optimization algorithm. However, there are few studies of non-invasive parameter estimation in the field of TCLs.

TCL physical parameters are relevant to structures, locations, device types, and behaviours of customers. Moreover, TCLs have large a quantity and are widely distributed. Thus, it is difficult to obtain TCL parameters because of the limited data available, measurement errors, and costly device installations. TCL parameters are generally assumed to be distributed in the form of normal [61], lognormal [62], or uniform distribution [46]. Few studies have been conducted on the parameter estimation of TCLs, considering privacy protection of end-users for control and dispatch. Besides, the actual parameter distributions of TCL populations or accurate physical parameters of individual TCL remain a problem for the practical applications of TCLs. The identification of TCL parameters in a non-invasive way is an important topic.

1.8.2 Reward Allocation Mechanism

TCL parameters are heterogeneous with different operating characteristics. In other words, heterogeneous TCLs have diverse contributions in service provision with different monetary rewards. To encourage the participation of end-users in DR programs, a fair reward allocation mechanism is required to evaluate the response performances and distribute the compensation to participating TCLs, in terms of their contributions in ancillary services. However, few studies have been conducted on the fair reward allocation mechanism. Only the authors in Ref. [63] develop a specific index to evaluate the diverse capacity of TCLs and make a priority list to manage them accordingly. Besides, the compensation for TCLs is also determined according to a rewarding curve. When TCLs participate in energy service, the index of incremental average power is used to quantify the contribution of TCLs during the given time interval. As the time scale of the ancillary service becomes shorter (e.g., load following), a minute-to-minute power measurement will be given to monitoring the TCL capability. The power adjustment capability will affect the reward allocation, and intermediate coordinators may even receive a penalty if the actual power deviates from the required amount. When the regulation service is provided, the on/off times of TCLs are important factors to be considered in contribution evaluation. It can be seen that not only the system requirement but also the specific performances of TCLs in different control methods need to be taken into account in the reward allocation mechanism. The design of a fair and reasonable reward allocation mechanism is important for TCLs to actively participate in DR programs.

1.8.3 Optimal Coordination of Virtual and Real Energy Storage

Energy storage devices are critical to smooth out the output of renewable generation and keep the supply–demand balance in real-time. Energy storage devices have been utilized in unit commitment [56], shifting peak [57], smoothing wind generation [58], and economic optimization in the electricity market [59]. TCLs can store thermal energy in related objects, with similar energy storage characteristics as conventional energy storage devices. Thus, TCLs can be aggregated as virtual energy storage models to be compatible with existing dispatch strategies. The coordination of TCLs and real energy storage devices are helpful to improve power system operation. Virtual and real energy storage are optimally coordinated by the algorithms proposed in Ref. [9] for both grid and customers' services. Moreover, the locations and scheduling costs between virtual and real energy storage devices should also be taken into account in developing coordination strategies, which could be another challenging topic.

1.9 Promising Research Field—Transactive Control and Scheduling of TCLs

TCLs as important DR resources usually participate in the wholesale market in a hierarchical way under load-serving entities or aggregators. Therefore, TCLs have no access to market prices, which makes them insensitive to market signals and cannot be scheduled with full capability. Recently, the transactive energy market has been proposed with intensive studies. The authors in Ref. [64] present a transactive framework in the day-ahead market. To optimize the market operation, transactive markets of both the transmission and distribution levels are simultaneously cleared. Reference [65] also studies the trading mechanism and clearing process of the transactive energy market using Lagrangian relaxation. When the transactive energy market is utilized to manage TCLs, it will allow participants to interact with each other in a distributed scheduling way without a central controller. Then, the market information can be fully employed to effectively manage distributed resources with optimized objectives.

Moreover, the transactive energy market will provide a distributed control method to manipulate TCL populations by market prices. This can not only protect the privacy of customers but also relieve the communication burden between aggregators and customers. A transactive market to control electric vehicles is developed in Ref. [66], considering the cost of utilizing the charging flexibility of electric vehicles. The authors in Ref. [14] propose a transactive control method to manipulate commercial buildings for the ancillary service.

In summary, the transactive energy market mechanism is a powerful tool to manage TCLs effectively and economically, which will attract more attention from researchers. Compared to DR, transactive control and scheduling of TCLs that operate based on the transactive energy market has attractive advantages.

Generally, DR is mainly to set prices or directly control customers' devices for shaping their electricity profiles. Thus, it can be stated that customers follow the price/control signals from the power systems to change their electricity usage behaviours when the grid needs power adjustment in DR programs. The grid is the leader while customers work as followers. Price/control information flows from the grid to customers while customers cannot provide feedback in terms of their will or capacity information to the grid. This will greatly lower the customers' initiatives to explore their flexibility, which cannot help to balance renewable energy generation fluctuations at the level of distribution systems.

Transactive energy changes the one-way leader–follower pattern between the grid and customers. It emphasizes the two-way peer-to-peer (P2P) patterns between customers while the grid is to show the operating states of power systems via price information. As customers can trade with other peers in terms of their operating behaviours, their power adjustment flexibility can be employed to the most extent. Accordingly, renewable energy generation is promoted to be locally consumed, which is helpful to keep the real-time balance between supply and demand.

The comparisons between DR and transactive energy are summarized in Table 1.2

1.10 Conclusions

TCLs as important distributed resources have been intensively studied in smart grids. This chapter summarizes current research on fundamental models, response modes, control strategies, system models, and dispatch strategies of TCLs in DR programs from a bottom-up perspective. Comparisons of different research methods in each aspect are presented. Besides abundant studies conducted on the modelling and control of TCLs, there are still great potentials of TCLs in smart grids. Three future research topics of TCLs on DR are discussed, including non-invasive parameter estimation, reward allocation mechanism, and optimization of virtual and real energy storage. Compared to DR, transactive energy has many advantages and is a promising research field in the near future.

Table 1.2 Comparisons between DR and transactive energy

	DR	Transactive energy
Participants	Flexible loads	Prosumers; smart devices (e.g. TCLs, electric vehicles)
Main regulation methods	Price-driven; Direct Load Control	Transaction driven
Main control/scheduling methods	Centralized	Decentralized
Operation patterns	Leader (the grid) and followers (customers)	P2P and self-interest
Interaction	One way with bad effects	Two-way with good effects
Activeness	Bad	Good
Privacy	Good for price-based DR while bad for direct load control-based DR	Good

References

1. Song, M., & Sun, W. (2021). Applications of thermostatically controlled loads for demand response with the proliferation of variable renewable energy. *Frontiers in Energy*. https://doi.org/10.1007/s11708-021-0732-5
2. Sajjad, I. A., Chicco, G., & Napoli, R. (2016). Definitions of demand flexibility for aggregate residential loads. *IEEE Transactions on Smart Grid, 7*(6), 2633–2643.
3. Radaideh, A., Vaidya, U., & Ajjarapu, V. (2017). Sequential set-point control for heterogeneous thermostatically controlled loads through an extended Markov chain abstraction. *IEEE Transactions on Smart Grid, 10*(1), 116–127.
4. Kohlhepp, P., Harb, H., Wolisz, H., Waczowicz, S., Müller, D., & Hagenmeyer, V. (2019). Large-scale grid integration of residential thermal energy storages as demand-side flexibility resource: A review of international field studies. *Renewable and Sustainable Energy Reviews, 101*, 527–547.
5. Xu, X., Yan, Z., Shahidehpour, M., Li, Z., Yan, M., & Kong, X. (2019). Data-driven risk-averse two-stage optimal stochastic scheduling of energy and reserve with correlated wind power. *IEEE Transactions on Sustainable Energy, 11*(1), 436–447.
6. Erdinc, O., Taşcıkaraoğlu, A., Paterakis, N. G., Eren, Y., & Catalão, J. P. (2016). End-user comfort oriented day-ahead planning for responsive residential HVAC demand aggregation considering weather forecasts. *IEEE Transactions on Smart Grid, 8*(1), 362–372.
7. Kim, Y. J., Norford, L. K., & Kirtley, J. L. (2014). Modeling and analysis of a variable speed heat pump for frequency regulation through direct load control. *IEEE Transactions on Power Systems, 30*(1), 397–408.
8. Song, M., Gao, C., Yan, H., & Yang, J. (2017). Thermal battery modeling of inverter air conditioning for demand response. *IEEE Transactions on Smart Grid, 9*(6), 5522–5534.
9. Hao, H., Wu, D., Lian, J., & Yang, T. (2017). Optimal coordination of building loads and energy storage for power grid and end user services. *IEEE Transactions on Smart Grid, 9*(5), 4335–4345.
10. Pahwa, A., & Brice, C. W. (1985). Modeling and system identification of residential air conditioning load. *IEEE Transactions on Power Apparatus and Systems, 6*, 1418–1425.

11. Song, M., Gao, C., & Su, W. (2016). Modeling and controlling of air-conditioning load for demand response applications. *Automation of Electric Power Systems, 40*(14), 158–167.
12. Molina, A., Gabaldon, A., Fuentes, J. A., & Alvarez, C. (2003). Implementation and assessment of physically based electrical load models: Application to direct load control residential programmes. *IEE Proceedings-Generation, Transmission and Distribution, 150*(1), 61–66.
13. Iacovella, S., Ruelens, F., Vingerhoets, P., Claessens, B., & Deconinck, G. (2015). Cluster control of heterogeneous thermostatically controlled loads using tracer devices. *IEEE Transactions on Smart Grid, 8*(2), 528–536.
14. Hao, H., Corbin, C. D., Kalsi, K., & Pratt, R. G. (2016). Transactive control of commercial buildings for demand response. *IEEE Transactions on Power Systems, 32*(1), 774–783.
15. Wu, X., He, J., Xu, Y., Lu, J., Lu, N., & Wang, X. (2017). Hierarchical control of residential HVAC units for primary frequency regulation. *IEEE Transactions on Smart Grid, 9*(4), 3844–3856.
16. Song, M., Gao, C., Shahidehpour, M., Li, Z., Yang, J., & Yan, H. (2018). Impact of uncertain parameters on TCL power capacity calculation via HDMR for generating power pulses. *IEEE Transactions on Smart Grid, 10*(3), 3112–3124.
17. Wai, C. H., Beaudin, M., Zareipour, H., Schellenberg, A., & Lu, N. (2014). Cooling devices in demand response: A comparison of control methods. *IEEE Transactions on Smart Grid, 6*(1), 249–260.
18. Sinitsyn, N. A., Kundu, S., & Backhaus, S. (2013). Safe protocols for generating power pulses with heterogeneous populations of thermostatically controlled loads. *Energy Conversion and Management, 67*, 297–308.
19. Lu, N., & Zhang, Y. (2012). Design considerations of a centralized load controller using thermostatically controlled appliances for continuous regulation reserves. *IEEE Transactions on Smart Grid, 4*(2), 914–921.
20. Hu, J., Cao, J., Yong, T., Guerrero, J. M., Chen, M. Z., & Li, Y. (2016). Demand response load following of source and load systems. *IEEE Transactions on Control Systems Technology, 25*(5), 1586–1598.
21. Kim, Y. J., Fuentes, E., & Norford, L. K. (2015). Experimental study of grid frequency regulation ancillary service of a variable speed heat pump. *IEEE Transactions on Power Systems, 31*(4), 3090–3099.
22. Song, M., Gao, C., Shahidehpour, M., Li, Z., Lu, S., & Lin, G. (2018). Multi-time-scale modeling and parameter estimation of TCLs for smoothing out wind power generation variability. *IEEE Transactions on Sustainable Energy, 10*(1), 105–118.
23. Malhame, R., & Chong, C. Y. (1985). Electric load model synthesis by diffusion approximation of a high-order hybrid-state stochastic system. *IEEE Transactions on Automatic Control, 30*(9), 854–860.
24. Tindemans, S. H., Trovato, V., & Strbac, G. (2015). Decentralized control of thermostatic loads for flexible demand response. *IEEE Transactions on Control Systems Technology, 23*(5), 1685–1700.
25. Bashash, S., & Fathy, H. K. (2012). Modeling and control of aggregate air conditioning loads for robust renewable power management. *IEEE Transactions on Control Systems Technology, 21*(4), 1318–1327.
26. Kamgarpour, M., Ellen, C., Soudjani, S. E. Z., Gerwinn, S., Mathieu, J. L., Müllner, N., & Lygeros, J. (2013, August). Modeling options for demand side participation of thermostatically controlled loads. In *2013 IREP Symposium Bulk Power System Dynamics and Control-IX Optimization, Security and Control of the Emerging Power Grid* (pp. 1–15).
27. Song, M., Gao, C., Shahidehpour, M., Li, Z., Yang, J., & Yan, H. (2018). State space modeling and control of aggregated TCLs for regulation services in power grids. *IEEE Transactions on Smart Grid, 10*(4), 4095–4106.
28. Song, M., Ciwei, G., Yang, J., Liu, Y., & Cui, G. (2017). Novel aggregate control model of air conditioning loads for fast regulation service. *IET Generation, Transmission & Distribution, 11*(17), 4391–4401.

29. Liu, M., & Shi, Y. (2015). Model predictive control of aggregated heterogeneous second-order thermostatically controlled loads for ancillary services. *IEEE Transactions on Power Systems, 31*(3), 1963–1971.
30. Hu, J., Cao, J., Chen, M. Z., Yu, J., Yao, J., Yang, S., & Yong, T. (2016). Load following of multiple heterogeneous TCL aggregators by centralized control. *IEEE Transactions on Power Systems, 32*(4), 3157–3167.
31. Liu, M., Shi, Y., & Liu, X. (2015). Distributed MPC of aggregated heterogeneous thermostatically controlled loads in smart grid. *IEEE Transactions on Industrial Electronics, 63*(2), 1120–1129.
32. Mathieu, J. L., Koch, S., & Callaway, D. S. (2012). State estimation and control of electric loads to manage real-time energy imbalance. *IEEE Transactions on Power Systems, 28*(1), 430–440.
33. Lu, N. (2012). An evaluation of the HVAC load potential for providing load balancing service. *IEEE Transactions on Smart Grid, 3*(3), 1263–1270.
34. Gilvanejad, M., Abyaneh, H. A., & Mazlumi, K. (2013). Estimation of cold-load pickup occurrence rate in distribution systems. *IEEE Transactions on Power Delivery, 28*(2), 1138–1147.
35. Hao, H., Sanandaji, B. M., Poolla, K., & Vincent, T. L. (2014). Aggregate flexibility of thermostatically controlled loads. *IEEE Transactions on Power Systems, 30*(1), 189–198.
36. Ruelens, F., Claessens, B., Vandael, S., De Schutter, B., Babuska, R., & Belmans, R. (2015). Residential demand response applications using batch reinforcement learning. arXiv preprint arXiv:1504.02125.
37. Vrettos, E., Ziras, C., & Andersson, G. (2016). Fast and reliable primary frequency reserves from refrigerators with decentralized stochastic control. *IEEE Transactions on Power Systems, 32*(4), 2924–2941.
38. Molina-Garcia, A., Bouffard, F., & Kirschen, D. S. (2010). Decentralized demand-side contribution to primary frequency control. *IEEE Transactions on Power Systems, 26*(1), 411–419.
39. Guo, D., Zhang, W., Yan, G., Lin, Z., & Fu, M. (2013, June). Decentralized control of aggregated loads for demand response. In *2013 American Control Conference* (pp. 6601–6606).
40. Meng, K., Wang, D., Dong, Z. Y., Gao, X., Zheng, Y., & Wong, K. P. (2017). Distributed control of thermostatically controlled loads in distribution network with high penetration of solar PV. *CSEE Journal of Power and Energy Systems, 3*(1), 53–62.
41. Saker, N., Petit, M., & Coullon, J. L. (2011, June). Demand side management of electrical water heaters and evaluation of the cold load pick-up characteristics (CLPU). In *2011 IEEE Trondheim PowerTech* (pp. 1–8).
42. Perfumo, C., Braslavsky, J., & Ward, J. K. (2013, September). A sensitivity analysis of the dynamics of a population of thermostatically-controlled loads. In *2013 Australasian Universities Power Engineering Conference (AUPEC)* (pp. 1–6).
43. Wang, D., Jia, H., Wang, C., Lu, N., Fan, M., Miao, W., & Liu, Z. (2014). Performance evaluation of controlling thermostatically controlled appliances as virtual generators using comfort-constrained state-queueing models. *IET Generation, Transmission & Distribution, 8*(4), 591–599.
44. Mahdavi, N., Braslavsky, J. H., & Perfumo, C. (2016). Mapping the effect of ambient temperature on the power demand of populations of air conditioners. *IEEE Transactions on Smart Grid, 9*(3), 1540–1550.
45. Ruiz, N., Cobelo, I., & Oyarzabal, J. (2009). A direct load control model for virtual power plant management. *IEEE Transactions on Power Systems, 24*(2), 959–966.
46. Mathieu, J. L., & Callaway, D. S. (2012, January). State estimation and control of heterogeneous thermostatically controlled loads for load following. In *2012 45th Hawaii International Conference on System Sciences* (pp. 2002–2011).
47. Song, M., Gao, C., Yang, J., & Yan, H. (2018). Energy storage modeling of inverter air conditioning for output optimizing of wind generation in the electricity market. *CSEE Journal of Power and Energy Systems, 4*(3), 305–315.

48. Luo, F., Dong, Z. Y., Meng, K., Wen, J., Wang, H., & Zhao, J. (2016). An operational planning framework for large-scale thermostatically controlled load dispatch. *IEEE Transactions on Industrial Informatics, 13*(1), 217–227.
49. Zhao, L., Zhang, W., Hao, H., & Kalsi, K. (2017). A geometric approach to aggregate flexibility modeling of thermostatically controlled loads. *IEEE Transactions on Power Systems, 32*(6), 4721–4731.
50. Mathieu, J. L., Kamgarpour, M., Lygeros, J., Andersson, G., & Callaway, D. S. (2014). Arbitraging intraday wholesale energy market prices with aggregations of thermostatic loads. *IEEE Transactions on Power Systems, 30*(2), 763–772.
51. Kurucz, C. N., Brandt, D., & Sim, S. (1996). A linear programming model for reducing system peak through customer load control programs. *IEEE Transactions on Power Systems, 11*(4), 1817–1824.
52. Yang, H. T., & Huang, K. Y. (1999). Direct load control using fuzzy dynamic programming. *IEE Proceedings-Generation, Transmission and Distribution, 146*(3), 294–300.
53. Lee, T. F., Cho, M. Y., Hsiao, Y. C., Chao, P. J., & Fang, F. M. (2008). Optimization and implementation of a load control scheduler using relaxed dynamic programming for large air conditioner loads. *IEEE Transactions on Power Systems, 23*(2), 691–702.
54. Trovato, V., Tindemans, S. H., & Strbac, G. (2016). Leaky storage model for optimal multi-service allocation of thermostatic loads. *IET Generation, Transmission & Distribution, 10*(3), 585–593.
55. Chu, C. M., Jong, T. L., & Huang, Y. W. (2007, June). Mitigating DLC constraints of air-conditioning loads using a group-DLC method. In *2007 IEEE Power Engineering Society General Meeting* (pp. 1–6).
56. Wen, Y., Li, W., Huang, G., & Liu, X. (2016). Frequency dynamics constrained unit commitment with battery energy storage. *IEEE Transactions on Power Systems, 31*(6), 5115–5125.
57. Ning, Y., Li, X., Ma, X., Jia, X., & Hui, D. (2016, June). Optimal schedule strategy of battery energy storage systems for peak load shifting based on interior point method. In *2016 12th World Congress on Intelligent Control and Automation (WCICA)* (pp. 2285–2288).
58. Li, N., Uckun, C., Constantinescu, E. M., Birge, J. R., Hedman, K. W., & Botterud, A. (2015). Flexible operation of batteries in power system scheduling with renewable energy. *IEEE Transactions on Sustainable Energy, 7*(2), 685–696.
59. Lujano-Rojas, J. M., Dufo-López, R., Bernal-Agustín, J. L., & Catalão, J. P. (2016). Optimizing daily operation of battery energy storage systems under real-time pricing schemes. *IEEE Transactions on Smart Grid, 8*(1), 316–330.
60. Hutchison, G., Zahawi, B., Harmer, K., Gadoue, S., & Giaouris, D. (2015). Non-invasive identification of turbo-generator parameters from actual transient network data. *IET Generation, Transmission & Distribution, 9*(11), 1129–1136.
61. Bashash, S., & Fathy, H. K. (2011, June). Modeling and control insights into demand-side energy management through setpoint control of thermostatic loads. In *Proceedings of the 2011 American Control Conference* (pp. 4546–4553).
62. Mahdavi, N., Braslavsky, J. H., Seron, M. M., & West, S. R. (2017). Model predictive control of distributed air-conditioning loads to compensate fluctuations in solar power. *IEEE Transactions on Smart Grid, 8*(6), 3055–3065.
63. Vanouni, M., & Lu, N. (2017). A reward allocation mechanism for thermostatically controlled loads participating in intra-hour ancillary services. *IEEE Transactions on Smart Grid, 9*(5), 4209–4219.
64. Renani, Y. K., Ehsan, M., & Shahidehpour, M. (2017). Optimal transactive market operations with distribution system operators. *IEEE Transactions on Smart Grid, 9*(6), 6692–6701.
65. Gregoratti, D., & Matamoros, J. (2014). Distributed energy trading: The multiple-microgrid case. *IEEE Transactions on Industrial Electronics, 62*(4), 2551–2559.

66. Wu, Q., Shahidehpour, M., Li, C., Huang, S., & Wei, W. (2018). Transactive real-time electric vehicle charging management for commercial buildings with PV on-site generation. *IEEE Transactions on Smart Grid, 10*(5), 4939–4950.
67. Trovato, V., Tindemans, S. H., & Strbac, G. (2014, October). Security constrained economic dispatch with flexible thermostatically controlled loads. In *IEEE PES Innovative Smart Grid Technologies, Europe* (pp. 1–6).

Chapter 2
Impact Analysis of Uncertain Parameters on TCL Power Capacity

Abstract TCLs have attracted additional attention in power system operations, mainly due to their flexibility in supplying DR. Determining accurate parameters for the aggregated TCL population is thus crucial. It is often a cumbersome task in practical applications due to inaccurate power metering, end-users' random behavior for consuming energy in real-time, and privacy issues pertaining to the availability and the collection of individuals' data. This chapter investigates the TCL power capacity calculation strategy and the impact of uncertain parameters on TCL power capacity for generating power pulses. First, a TCL power capacity calculation strategy that links on–off switching states with temperature set-point adjustments is proposed to remove TCLs that are deemed unsuitable for DR dispatch as they satisfy end-users' comfort needs. Second, high dimensional model representative (HDMR) is applied to identify important parameters for quantifying the impact of uncertain heterogeneous parameters on TCL power capacities. In addition, a fast framework is provided to reduce the computation burdens in capacity calculation. Simulation results demonstrate that (1) proposed TCL power capacity calculation strategy can maintain end-users in respective comfort ranges; (2) important parameters can be effectively identified by HDMR; (3) TCL power capacity can be calculated quickly via HDMR in various dispatch periods.

Keywords TCL power capacity · Uncertainties · Power pulses · High dimensional model representative (HDMR)

2.1 Introduction

TCLs can store thermal energy in objects and change power consumptions within certain time periods without compromising end-users' comfort [1–3]. The TCL capacity calculation, which is closely tied with TCL control methods, is an essential part of the DR program. TCLs acting as important DR resources have considerable capabilities to provide ancillary services for a reliable and secure power system operation [4–6].

Generally, there are two methods, namely on–off switch control [6, 7] and temperature set-point adjustment [8, 9], that offer direct load control of TCLs in DR

M. Song and C. Gao, *Integration of Distributed Resources in Smart Grids for Demand Response and Transactive Energy*, https://doi.org/10.1007/978-981-16-7170-8_2

programs. Reference [10] points out that on–off switches can instantaneously change TCL adjustments for modifying power consumptions but cannot maintain the adjustment for a long period because of temperature range constraints. On the contrary, thermostat set-point control can regulate power consumption for a longer period, but the impact may not be immediate. Accordingly, on–off switches are mainly utilized to quickly change TCL power consumptions which can shave critical peaks [11] and regulation services [2, 12]. The TCL temperature set-point adjustments are utilized to minimize energy costs by managing energy consumptions [13, 14]. However, on–off switches and temperature set-point adjustments could be considered for some ancillary services in which fast responses and considerable online capacity are required.

The control method that links on–off switches with temperature set-point control will respond quickly with a larger capacity and longer time duration, which helps initiate power pulses in spinning reserve applications. On–off switch control can make TCLs act rapidly to satisfy fast response requirements. Meanwhile, a wider temperature range will enable TCLs to provide power pulses with a longer duration and a larger capacity.

Most studies seldom refer to the combined control methods and mostly focus on either on–off switch control or temperature set-point adjustments in DR programs. An open-loop control algorithm is proposed to generate power pulses in which a subtle temperature set-point adjustment is designed on the basis of on–off switches for significant variations in the aggregated power [11]. Reference [10] presents a hybrid control mechanism that links on–off switches with temperature set-point adjustments to address the maximum capability and period of load reduction. Temperature set-point as an input to the TCL aggregated model is considered in [15] to improve the on–off control of TCLs. Moreover, few references focus on end-users with an out-of-comfort range indoor temperature, which corresponds to heterogeneous TCL parameters when the power pulse is provided. This issue could be troublesome for end-users' comfort.

A large TCL population is required to provide considerable DR capacity in power systems in which case acquiring accurate load parameters is critical to calculate the TCL power capacity effectively. However, erroneous measurements, participants' random behaviors, and customer privacy protection usually make TCL data insufficient and uncertain, which will result in additional difficulties in calculating the DR capacity accurately. Identifying important parameters in the calculation process helps reduce the effort for acquiring less important parameters and pay more attention to pivotal parameters. Meanwhile, it is critical economically, technically, and socially to calculate a fast-online TCL power capacity with fewer TCL parameters in DR programs, which requires less effort and maintains higher accuracy.

However, few references focus on sensitivity analyses of TCL parameters and identifying the equivalent relationship between the capacity of ancillary services and TCL parameters. A multiple linear regression model is employed in [16] where the relationship between the input and output is simply treated as linear to approximate the impact of uncertain TCL physical parameters on the aggregated power. Reference

[17] performs sensitivity analyses and applies simulation methods to identify important relationships between TCL parameters and aggregated dynamics. The detailed expressions between TCL parameters and the aggregated output are not formulated. Despite the analysis of the sensitivity of TCL parameters to power output in [16, 17], two issues remain to be addressed. One is that the impact of uncertain parameters on TCL power capacity as the exact expressions relating uncertain parameters to fast TCL power capacity calculations are often overlooked in the literature. The other is that an effective and powerful tool is required to analyze the independent and cooperative impacts of TCL parameters on its power capacity.

The high dimensional model representative (HDMR) idea was proposed in [18] for building the exact relationships between multi-inputs and the process output. HDMR has offered distinguished capabilities in sensitivity analyses, correlation building, reliability modeling and ease of applications in biology, communication technology, information storage, and electrical engineering [19–23]. In this chapter, we apply HDMR to electric power systems for analyzing the impact of uncertain parameters on calculating TCL power capacity, and focus on power pulse generation by reshaping the TCL power profile. The main work in this chapter is summarized as follows. First, a TCL power capacity calculation strategy that links on–off switches with temperature set-point adjustments is presented for producing power pulses. The end-users who are not dispatched by grid operators will be considered in the TCL power capacity calculation for satisfying end-users' comfort. Second, HDMR is introduced to analyze the impact of uncertain TCL parameters on TCL power capacity. Accordingly, critical parameters are identified and quantified. Third, a fast TCL capacity calculation framework is presented by applying fewer parameters to relieve computation burdens and provide a shortcut for the TCL power capacity calculation.

2.2 TCL Model and Uncertain Parameters

2.2.1 TCL Model

We focus on residential TCLs here and a simplified equivalent thermal parameters (ETP) model [4, 15, 24, 25] is employed to describe TCL thermal dynamics, which is given as:

$$C_a \frac{dT_i}{dt} = \frac{(T_{out} - T_i)}{R} + Q' - S(t) \cdot \eta \cdot P \tag{2.1}$$

TCLs can only work at rated power before being turned off so that indoor temperatures can change periodically within $[T_{\min}, T_{\max}]$ ($T_{\min} = T_{set} - \delta/2$, $T_{\max} = T_{set} + \delta/2$) as depicted in Fig. 2.1 [4]. The switch operating mode is shown as

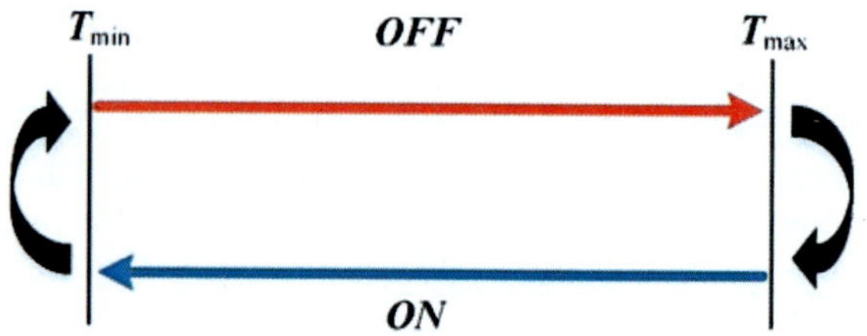

Fig. 2.1 Operation diagram of TCL

$$S(t+\Delta t') = \begin{cases} 1 & T_i(t+\Delta t') > T_{max} \\ 0 & T_i(t+\Delta t') < T_{min} \\ S(t) & T_{min} \le T_i(t+\Delta t') \le T_{max} \end{cases} \quad (2.2)$$

2.2.2 *Uncertain Parameters*

The ETP model (2.1) describes TCL dynamic performances using $\{R, C_a, Q', \eta, P\}$. Since indoor temperature is kept within $[T_{set} - \delta/2, T_{set} + \delta/2]$, both T_{set} and δ contribute to TCL dynamics. Then, a TCL model can be characterized by seven parameters $[R, C_a, Q', \eta, P, T_{set}, \delta]$. However, there is always a varying degree of heterogeneity (e.g., building structures and devices) in TCL population. Even for a homogeneous TCL population with the same building structures and devices, Q' in (2.1) introduces some heterogeneity [17], which causes uncertainty in TCL parameters. Two indices, namely, type of heterogeneity and level of heterogeneity, are employed to address the uncertain parameters of TCLs by simulation.

The type of heterogeneity is the distribution of TCL parameters which governs the dynamic behavior of a TCL population. Generally, normal distribution [26], lognormal distribution [27, 28] and uniform distribution [29] are utilized to describe this kind of heterogeneity, all of which are considered in this chapter. For simplicity, the normal distribution is employed to simulate the dynamics of a TCL population and the other two distributions are discussed in our case studies. Accordingly, the mean and standard deviation of the normal distribution is used to characterize the distribution of each TCL parameter.

Obviously, mean and standard deviation varies with the TCL population, which is characterized by the heterogeneity level. For example, a larger standard deviation means TCL parameters are more heterogeneous and differences among TCL parameters (e.g., R) are larger. In this chapter, we assume mean and standard deviation are uniformly distributed within a certain range. The distributions of $\{R, C_a, Q', P, T_{set}, \delta, \eta\}$ for household air conditioner loads are given in Table 2.1. $N(\mu,\sigma^2)$ represents the normal distribution, in which μ is the mean and σ is the standard deviation. $U(a, b)$ denotes a uniform distribution with lower and upper limits stated as a and b. For generality, we used a multitude of references and took the parameters listed in Table 2.1 from [12, 17, 24, 28, 30]. For example, the values of C_a are set 2 and 3.6 in [12,

Table 2.1 Parameter distribution of TCLs

Parameters	Distribution of parameters	Distribution of μ	Distribution of σ
R	$N(\mu, \sigma^2)$	U(2,4)	U(0.1,0.5)
C_a	$N(\mu, \sigma^2)$	U(2,4)	U(0.1,0.5)
Q'	$N(\mu, \sigma^2)$	U(0.3,0.7)	U(0.01,0.1)
P	$N(\mu, \sigma^2)$	U(6,10)	U(0.5,2)
T_{set}	$N(\mu, \sigma^2)$	U(18,20)	U(0.5,2)
Δ	$N(\mu, \sigma^2)$	U(0.3,1)	U(0.05,0.1)
H	$N(\mu, \sigma^2)$	U(2.5,3)	U(0.1,0.5)

Table 2.2 Distribution of outdoor temperature

Parameter	Distribution of parameter
T_{out}	U(30,40)

28]. Then μ of C_a is assumed to vary within [2, 4] in Table 2.1. The corresponding σ is given within a certain range including $0.1 * (a + b)/2$ (a and b are lower and upper limits of μ, respectively). Similarly, the distributions of μ and σ of other parameters are set in Table 2.1.

In addition to the listed quantities, other parameters representing physical characteristics of buildings and devices can also have a major impact on TCL dynamic behaviors with a direct effect on TCL power consumption. The T_{out} distribution is shown in Table 2.2. Therefore, TCL population uncertainties are described by 15 parameters representing $\{R\_\mu, C_a\_\mu, Q'\_\mu, P\_\mu, T_{set}\_\mu, \delta\_\mu, \eta\_\mu, R\_\sigma, C_a\_\sigma, Q'\_\sigma, P\_\sigma, T_{set}\_\sigma, \delta\_\sigma, \eta\_\sigma, T_{out}\}$, in which, all parameters are assumed to be independent [16].

2.3 TCL Power Capacity Calculation Strategy

Since a small change (ΔT) in temperature set-point cannot aggravate end-users' comfort, on–off switches and temperature set-point adjustments are linked in this chapter to control TCLs for generating power pulses. If indoor temperatures of TCLs are within end-users' comfort range during the control process, TCLs will be suitable for dispatch and the TCL parameters are perfect parameters in the DR program.

As shown in Fig. 2.1, indoor temperatures change periodically within [Tmin, Tmax] when external control signals are not available. Once TCLs are given external control signals, they will act in accordance with the corresponding control signals. In Fig. 2.2a, when down power pluses are served, the turned-on TCLs will be turned off immediately to save power. Accordingly, the TCL switch will be off and TCL indoor temperatures will increase throughout the dispatch period. In Fig. 2.2b, TCL operating behavior is determined by heterogeneous TCL parameters, which may

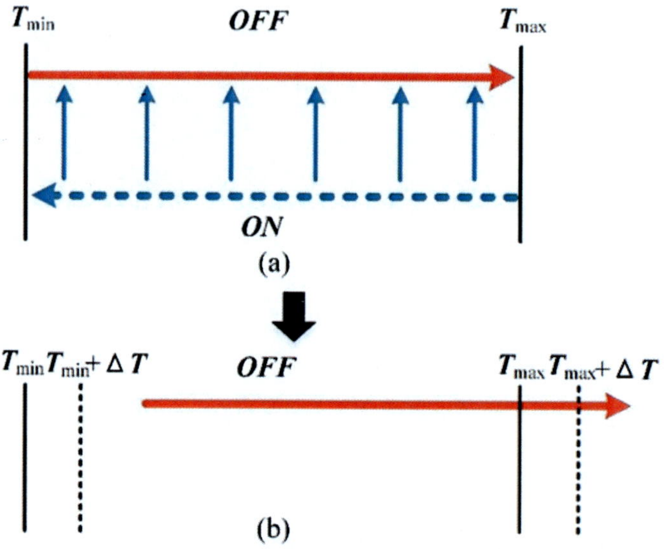

Fig. 2.2 Evolution of TCL with on–off switches and temperature set-point adjustments

allow a few TCL indoor temperatures with imperfect parameters to exceed end-users' comfort range (i.e., indoor temperatures would be higher than $T_{\max} + \Delta T$).

To validate the above TCL behavior, aggregated power and indoor temperature profiles of 10,000 TCLs are simulated, as depicted in Figs. 2.3 and 2.4. Indoor temperatures are normalized according to (2.3). The distribution of TCL parameters is shown in Table 2.3.

$$T_i^s = \frac{T_i - T_{min}}{T_{max} + \Delta T - T_{min}} \tag{2.3}$$

Fig. 2.3 Aggregated power when some TCLs are without perfect parameters

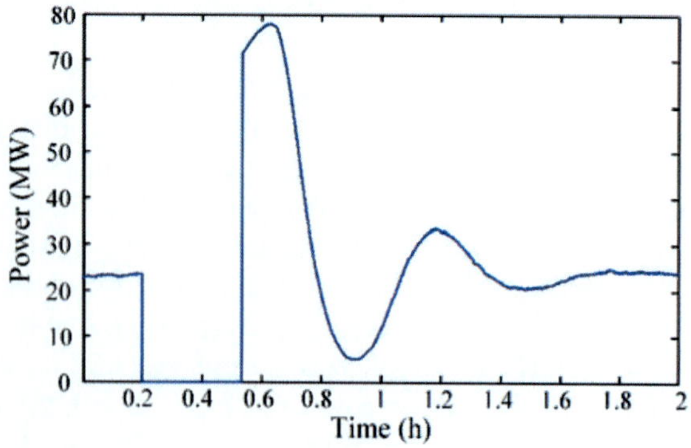

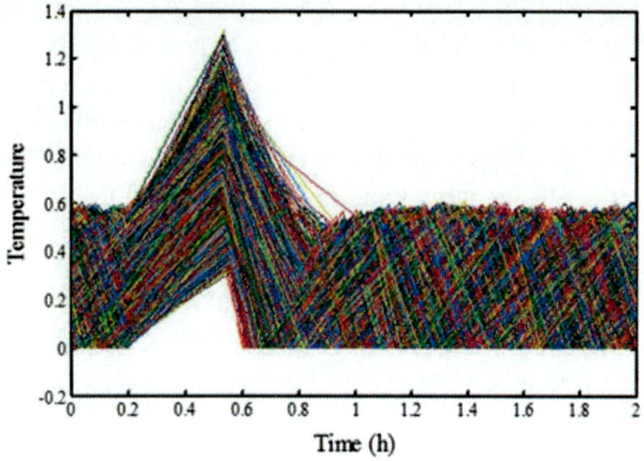

Fig. 2.4 Indoor temperature profiles with a few imperfect TCL parameters

Table 2.3 Distribution of TCL parameters for simulations

Parameters	R	C_a	Q'	P	T_{set}	δ	η
μ	2.5	2.5	0.5	8	20	0.5	3
σ	0.3	0.3	0.05	1	1	0.08	0.3

At $t = 0.2$ h, the turned-on TCLs will be turned off for 20 min to provide power pulse services. The simulation is 2 h, the outdoor temperature is 36 °C, and ΔT is set as 1 °C. In Fig. 2.3, the TCL's aggregated power drops sharply at 0.2 h in response to the TCL control signal. The power is maintained for 20 min until the power rises sharply at about 0.53 h. The corresponding indoor temperatures in Fig. 2.4 begin to increase at 0.2 h and reach the top at the end of the dispatch period. Obviously, some of the normalized TCL indoor temperatures are larger than 1, i.e., temperatures are higher than $T_{max} + \Delta T$, which have an impact on end-users' comfort and could result in more customer complaints. To maintain end-users' comfort, TCLs without perfect parameters should be withdrawn from any control actions during the DR program and the corresponding power capacity be subtracted from the total TCL power capacity.

There is a critical temperature, T_a, which allows a TCL to reach $T_{max} + \Delta T$ after Δt. If the indoor temperature is lower than T_a when the TCL receives the control signal, the corresponding final indoor temperature will be lower than $T_{max} + \Delta T$ after Δt. Otherwise, the corresponding final indoor temperature will be higher than $T_{max} + \Delta T$ after Δt. According to (2.1), T_a is calculated as:

$$T_a = (T_{max} + \Delta T - T_{out} - RQ')e^{\frac{1}{RC_a}\Delta t} + T_{out} + RQ' \quad (2.4)$$

Obviously, only TCLs with indoor temperatures lower than T_a should be controlled to generate power pulses so that end users' comforts are observed. However, we find that, if $T_a < T_{max}$, the aggregated power will increase during Δt when only such TCLs are controlled. This situation will result in power pulses with bad performances and it should be taken into account in the dispatch strategy. Figure 2.5 shows the evolution of a TCL population ($T_a < T_{max}$) when only a fraction of TCLs are regulated. The operating region of TCLs is divided into four parts according to the switch status and indoor temperature, which include A1 (switch status is off and indoor temperature range is $[T_{min}, T_a]$), A2 (switch status is off and indoor temperature range is $[T_a, T_{max}]$), A3 (switch status is on and indoor temperature range is $[T_a, T_{max}]$), and A4 (switch status is on and indoor temperature range is $[T_{min}, T_a]$). When TCLs in A4 are turned off and TCLs in A1 are kept off to generate power pulses, TCLs in A2 and A3 will operate in the original conditions (see Fig. 2.5a). Gradually, more TCLs in A2 will be turned on (see Fig. 2.5b), which will increase the aggregated power during Δt and result in the power pulse service failure. Similarly, if $T_a \le T_{max}$, TCL indoor temperatures will not exceed T_a and TCLs will be dispatched to generate power pulses. Because the number of TCLs is not changed, the aggregated power will be maintained at a steady state, which is required for the power pulse.

Accordingly, TCLs with T_{max} larger than T_a will not be dispatched for generating power pulses in order to keep the aggregated power at a steady level during Δt. Then the total TCLs are divided into two groups according to the relationship between T_a and T_{max}. TCLs in the first group ($T_{max} \le T_a$) have perfect parameters and will be controlled for providing ancillary services. TCLs in the second group ($T_{max} > T_a$) with imperfect parameters will operate as originally planned. The evolution of the two TCL groups is shown in Fig. 2.6. When the control signals are dispatched to the TCL population, the TCLs that are on in the first group will be turned off while the TCLs that are on in the second group will keep their original status (see Fig. 2.6a).

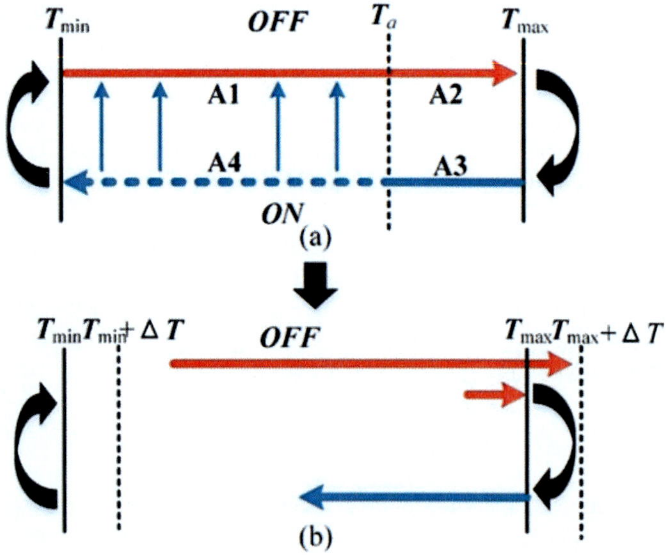

Fig. 2.5 Evolution of a TCL population when some TCLs are controlled ($T_a < T_{max}$)

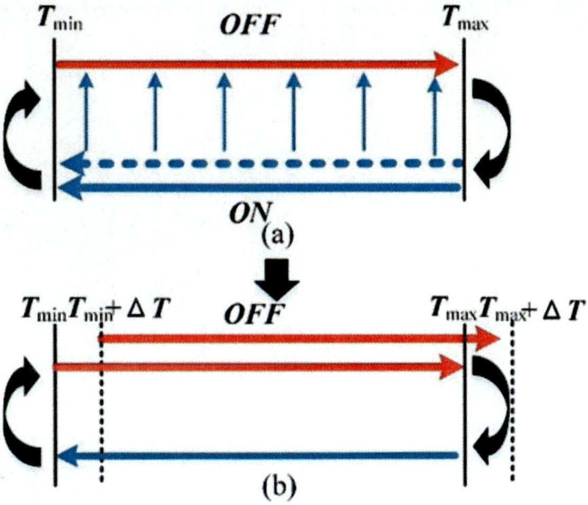

Fig. 2.6 Evolution of two TCL groups to generate power pulses

Then the TCLs in the first group will keep off for Δt and the TCLs in the second group will maintain their original status (see Fig. 2.6b). The TCL's aggregated power and indoor temperatures are given in Figs. 2.7 and 2.8, respectively. Although the total aggregated power is lower as compared with that in Fig. 2.3, the TCL indoor temperatures are all kept within end-users' comfort level. There is a significant cold load pickup in Figs. 2.3 and 2.7 when TCLs are released at the end of a dispatch period. The control strategies for restraining load fluctuations are provided in [11].

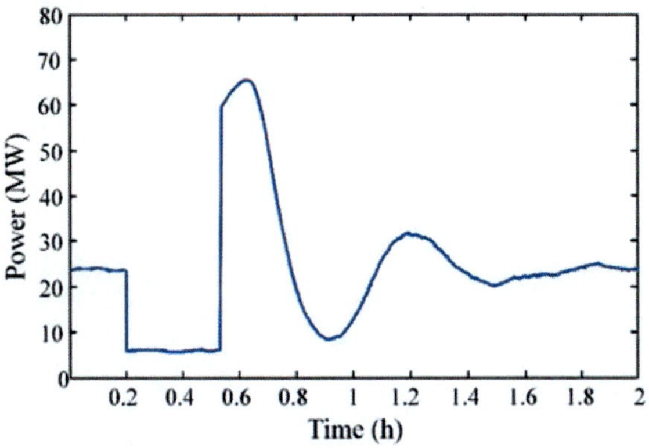

Fig. 2.7 Aggregated power of TCLs with perfect parameters

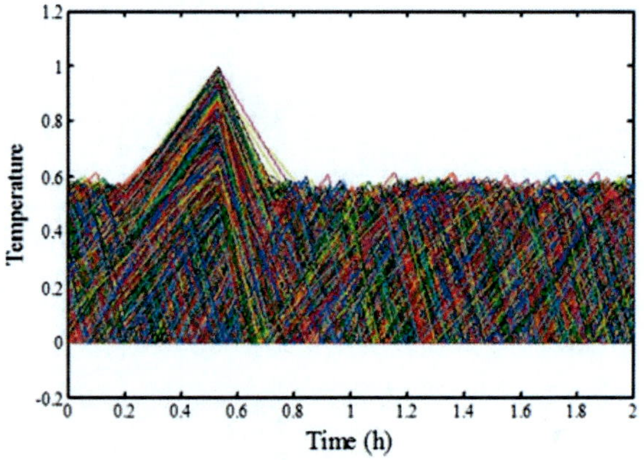

Fig. 2.8 Indoor temperature profiles of TCLs with perfect parameters

2.4 Impact of Uncertain TCL Parameters

2.4.1 HDMR

HDMR is an effective tool to analyze the impact of multi-inputs on system output. Denote inputs as $\boldsymbol{x} = (x_1, x_2, \ldots, x_n)^{\mathrm{T}}$ and the output as $f(\boldsymbol{x})$. Since the impact of inputs on the output is independent and cooperative, $f(\boldsymbol{x})$ is represented as [18, 31]:

$$f(\boldsymbol{x}) \approx f_0 + \sum_{i=1}^{n} f_i(x_i) + \sum_{i<j} f_{ij}(x_i, x_j) + \cdots f_{1,2,\cdots,n}(\boldsymbol{x}) \tag{2.5}$$

For simplicity, inputs are rescaled and $f(\boldsymbol{x})$ is redefined in $X^n = \{(x_1, x_2, \ldots, x_n) | 0 \le x_i \le 1,\ i = 1, 2, \ldots n\}$. Projection operators [18] are employed to describe the component functions in (2.5), which are formulated as:

$$f_0 = \boldsymbol{P}_0 f(\boldsymbol{x}) \tag{2.6a}$$

$$f_i(x_i) = \boldsymbol{P}_i f(\boldsymbol{x}) \tag{2.6b}$$

$$f_{ij}(x_i, x_j) = \boldsymbol{P}_{ij} f(\boldsymbol{x}) \tag{2.6c}$$

Here, $\boldsymbol{P}(\boldsymbol{P}_0, \boldsymbol{P}_i, \boldsymbol{P}_{ij}, \ldots)$ have the following characteristics[18]:

1. Idempotency

$$P^2_{i_1 i_2 \cdots i_l} \equiv P_{i_1 i_2 \cdots i_l} \{i_1, i_2, \ldots, i_l\} \subset \{1, 2, \ldots, n\} \tag{2.7}$$

2. Orthogonality

$$P_{i_1 i_2 \cdots i_l} P_{j_1 j_2 \cdots j_k} = 0 \ \{i_1, i_2, \ldots, i_l\} \not\subset \{j_1, j_2, \ldots, j_k\} \tag{2.8}$$

3. Resolution of the identity

$$P_0 + \sum_i P_i + \sum_{i<j} P_{ij} + \cdots P_{12\cdots n} = 1 \tag{2.9}$$

When inputs are randomly sampled, the component functions in (2.5) are constructed by RS (random sampling)-HDMR [31]. The integral of each component function in (2.6a)–(2.6c) concerning its variables is equal to zero. The Monte Carlo simulation method is applied for sampling and the component functions are evaluated as:

$$f_0 \approx \frac{1}{N_0} \sum_{s=1}^{N_0} f(\boldsymbol{x}^{(s)}) \tag{2.10a}$$

$$f_i(x_i) \approx \frac{1}{N_0} \sum_{s=1}^{N_0} f((x_i, \boldsymbol{x}^i)^{(s)}) - \frac{1}{N_0} \sum_{s=1}^{N_0} f(\boldsymbol{x}^{(s)}) \tag{2.10b}$$

$$f_{ij}(x_i, x_j) \approx \frac{1}{N_0} \sum_{s=1}^{N_0} f((x_i, x_j, \boldsymbol{x}^{ij})^{(s)}) - \frac{1}{N_0} \sum_{s=1}^{N_0} f((x_i, \boldsymbol{x}^i)^{(s)}) - \frac{1}{N_0} \sum_{s=1}^{N_0} f((x_j, \boldsymbol{x}^i)^{(s)}) + \frac{1}{N_0} \sum_{s=1}^{N_0} f(\boldsymbol{x}^{(s)}) \tag{2.10c}$$

However, this direct method for approximating the components is computationally burdensome due to the large number of samples. To address this problem, component functions are expressed by certain functions using HDMR. There are numerous representations for HDMR among which orthonormal polynomials are proved to have the highest accuracy to approximate the component representation [18, 31–33]. Accordingly, $f_i(x_i)$ and $f_{ij}(x_i, x_j)$ in (2.6b) and (2.6c) are expressed as:

$$f_i(x_i) = \sum_{r=1}^{k} \alpha_r^i \phi_r(x_i) \tag{2.11a}$$

$$f_{ij}(x_i, x_j) = \sum_{p=1}^{l} \sum_{q=1}^{m} \beta_{pq}^{ij} \phi_p(x_i) \phi_q(x_j) \tag{2.11b}$$

$\phi(x)$ are orthonormal polynomials which are given as [31]:

$$\phi_1(x) = \sqrt{3}(2x - 1) \tag{2.12a}$$

$$\phi_2(x) = 6\sqrt{5}\left(x^2 - x + \frac{1}{6}\right) \tag{2.12b}$$

$$\phi_3(x) = 20\sqrt{7}\left(x^3 - \frac{3}{2}x^2 + \frac{3}{5}x - \frac{1}{20}\right) \tag{2.12c}$$

Applying the given $\phi(x)$ expressions, the constant coefficients α and β in (2.11a) and (2.11b) are determined for completing the HDMR expansion. To obtain the constant coefficients, we minimize the difference between HDMR and $f(\boldsymbol{x})$ as [18]:

$$\begin{aligned} &\min \int_{X^n} \{f(\boldsymbol{x}) - [f_0 + \sum_i f_i(x_i) + \sum_{i<j} f_{ij}(x_i, x_j) \\ &\quad + \cdots + \sum_{i_1<\cdots i_l} f_{i_1,\ldots,i_l}(x_{i_1}, \ldots, x_{i_l})]\}^2 d\boldsymbol{x} \\ &\text{s.t.} \int_{[x_{i_1},\ldots,x_{i_l}]} f_{i_1,\ldots,i_l}(x_{i_1}, \ldots, x_{i_l}) d\boldsymbol{x}_k, k = i_1, \ldots, i_l \end{aligned} \tag{2.13}$$

Equation (2.13) can be rewritten in a linear form and solved for calculating α and β as:

$$\alpha_r^i = \int_0^1 f_i(x_i)\phi_r(x_i)dx_i \tag{2.14a}$$

$$\beta_{pq}^{ij} = \int_0^1\int_0^1 f_{ij}(x_i, x_j)\phi_p(x_i)\phi_q(x_i)dx_i dx_j \tag{2.14b}$$

Samples are obtained using the Monte Carlo simulation method, α and β are approximated as:

$$\alpha_r^i \approx \frac{1}{N_0}\sum_{s=0}^{N_0} f(\boldsymbol{x}^{(s)})\phi_r(x_i^{(s)}) \tag{2.15a}$$

$$\beta_{pq}^{ij} \approx \frac{1}{N_0}\sum_{s=1}^{N_0} f(\boldsymbol{x}^{(s)})\phi_p(x_i^{(s)})\phi_q(x_j^{(s)}) \tag{2.15b}$$

For most physical systems, the output dependents primarily on low-order functions. Hence, the relationship between multiple inputs and the output of a system is approximated as [31, 34]:

$$f(x) = f_0 + \sum_{i=1}^{n}\sum_{r=1}^{k}\alpha_r^i\phi_r(x_i) + \sum_{i<j}\sum_{p=1}^{l}\sum_{q=1}^{m}\beta_{pq}^{ij}\phi_p(x_i)\phi_q(x_j) \tag{2.16}$$

where D, D_i, and D_{ij} are defined as [35]:

$$D = \sum_i D_i + \sum_{i<j} D_{ij} \tag{2.17a}$$

$$D_i \approx \sum_{r=1}^{k}(\alpha_r^i)^2 \tag{2.17b}$$

$$D_{ij} \approx \sum_{p=1}^{l}\sum_{q=1}^{m}(\beta_{pq}^{ij})^2 \tag{2.17c}$$

Also, S_i and S_{ij} are defined as [35]:

$$S_i = \frac{D_i}{D} \tag{2.18a}$$

$$S_{ij} = \frac{D_{ij}}{D} \tag{2.18b}$$

2.4.2 HDMR Sensitivity Analyses

According to the HDMR modeling process shown in Sect. 2.4.1, a large number of samples are required to build the equivalent relationship between multiple inputs and the output of a system in (2.16) and evaluate the corresponding sensitivity indices of inputs in (2.17). In this chapter, a TCL population is presented as an unknown set. The inputs are 15 TCL parameters $\{R\_\mu, C_a\_\mu, Q^{'}\_\mu, P\_\mu, T_{set}\_\mu, \delta\_\mu, \eta\_\mu, R\_\sigma, C_a\_\sigma, Q^{'}\_\sigma, P\_\sigma, T_{set}\_\sigma, \delta\_\sigma, \eta\_\sigma, T_{out}\}$ and the output is the power capacity. When an input set is given, parameters of the TCL population will be generated according to the predefined distribution described in Sect. 2.2.2. Then the power capacity of each TCL will be calculated using the proposed strategy in Sect. 2.3. The total power capacity of the TCL population is the system output. When N_0 pairs of input–output samples (N_0 inputs are uniformly distributed in Table 2.1) are obtained, they will calculate the HDMR of the TCL system as shown in Fig. 2.9. The corresponding coefficients α and β of HDMR will be calculated to identify the sensitive TCL parameters and build the relationship between TCL parameters and the power capacity. Accordingly, the TCL system can be approximately replaced by the corresponding HDMR, which provides a better insight into the TCL system.

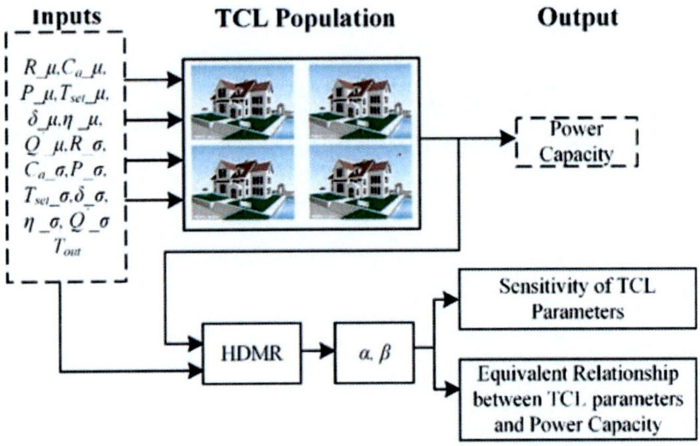

Fig. 2.9 HDMR modeling for sensitivity analysis and the equivalent relationship between TCL parameters and power capacity

We apply a Quasi Monte Carlo instead of Monte Carlo sampling method to generate input–output pairs for faster convergence [34]. The steps for evaluating the impact of TCL parameters on the power capacity are described as follows.

1. Quasi Monte Carlo sampling methods are employed to generate N_0 input samples according to ranges shown in Tables 2.1 and 2.2; then the inputs of the TCL population system are described as:

$$\begin{aligned} \boldsymbol{x}^{(s)} = [&R\_\mu^{(s)}, C_a\_\mu^{(s)}, Q'\_\mu^{(s)}, P\_\mu^{(s)}, T_{set}\_\mu^{(s)}, \delta\_\mu^{(s)}, \eta\_\mu^{(s)}, \\ &R\_\sigma^{(s)}, C_a\_\sigma^{(s)}, Q'\_\sigma^{(s)}, P\_\sigma^{(s)}, T_{set}\_\sigma^{(s)}, \delta\_\sigma^{(s)}, \eta\_\sigma^{(s)}, T_{out}^{(s)}]^T \\ &(s = 1, 2, \ldots, N_0) \end{aligned} \tag{2.19}$$

2. For each input sample, the parameters of 10,000 TCLs are generated according to the actual parameter distributions (i.e., normal, lognormal or uniform distribution). The total capacity of the TCL population for generating power pulses is calculated by individual simulations using the strategy given in Sect. 2.3. There will be N_0 outputs which are given as;

$$f(\boldsymbol{x}^{(s)}) = C^{(s)}(s = 1, 2, \ldots, N_0) \tag{2.20}$$

3. x in (2.13) is required to vary within [0,1]. Then N_0 input samples are normalized as:

$$x_i^* = \frac{x_i - x_{\min}}{x_{\max} - x_{\min}} \tag{2.21}$$

4. N_0 pairs of normalized input–output samples are utilized to build HDMR using GUI-HDMR [35], in which the constant-coefficient α, β and the associated

sensitivity indices are obtained; For example, the first-order coefficient of the independent impact of inputs, α_1, is calculated as:

$$\begin{bmatrix} \alpha_1^1 \\ \alpha_1^2 \\ \vdots \\ \alpha_1^{15} \end{bmatrix} = \frac{1}{N_0} \begin{bmatrix} \sum_{s=0}^{N_0} C^{(s)} \phi_1 (R\_\mu^{(s)}) \\ \sum_{s=0}^{N_0} C^{(s)} \phi_1 (C_a\_\mu^{(s)}) \\ \vdots \\ \sum_{s=0}^{N_0} C^{(s)} \phi_1 (T_{out}^{(s)}) \end{bmatrix} \tag{2.22a}$$

The second-order coefficient of the independent impact of inputs, α_2, is expressed as:

$$\begin{bmatrix} \alpha_2^1 \\ \alpha_2^2 \\ \vdots \\ \alpha_2^{15} \end{bmatrix} = \frac{1}{N_0} \begin{bmatrix} \sum_{s=0}^{N_0} C^{(s)} \phi_2 (R\_\mu^{(s)}) \\ \sum_{s=0}^{N_0} C^{(s)} \phi_2 (C_a\_\mu^{(s)}) \\ \vdots \\ \sum_{s=0}^{N_0} C^{(s)} \phi_2 (T_{out}^{(s)}) \end{bmatrix} \tag{2.22b}$$

Accordingly, the higher-order coefficients of the independent impact of inputs will also be calculated. The cooperative impact of inputs on the output is described by β. When the first order coefficient of the cooperative impact of inputs, β_{11}, is calculated, there are 105 combinations, which are given as:

$$\begin{bmatrix} \beta_{1,1}^{1,2} \\ \beta_{1,1}^{1,3} \\ \vdots \\ \beta_{1,1}^{14,15} \end{bmatrix} = \frac{1}{N_0} \begin{bmatrix} \sum_{s=0}^{N_0} C^{(s)} \phi_1 (R\_\mu^{(s)}) \phi_1 (C_a\_\mu^{(s)}) \\ \sum_{s=0}^{N_0} C^{(s)} \phi_1 (R\_\mu^{(s)}) \phi_1 (Q'\_\mu^{(s)}) \\ \vdots \\ \sum_{s=0}^{N_0} C^{(s)} \phi_1 (\eta\_\sigma^{(s)}) \phi_1 (T_{out}^{(s)}) \end{bmatrix} \tag{2.22c}$$

Similarly, the other order coefficients of the cooperative impact of inputs on the output will be calculated. When α and β are obtained, sensitivity indices of independent and cooperative inputs will be obtained according to (2.17) and (2.18). TCL parameters with larger sensitivity indices have greater impacts on the power capacity of the TCL population.

5. The HDMR modeling results are validated. Sensitive parameters, sensitivity indices and the associated coefficients are stored for practical applications.

2.4.3 *Framework for Fast TCL Power Capacity Calculation*

When TCLs are controlled to partake in DR programs, their power capacity under different dispatch periods will be evaluated according to power system requirements. The number of TCLs is varying because of end-users' random behavior and TCL temperature set-points are adjusted in response to varying outdoor temperature, resulting in a fluctuating TCL power capacity. However, the computation burden will be huge if the TCL power capacity is calculated through individual simulations as certain parameters are adjusted. Moreover, in case of insufficient data, it is difficult to accurately obtain the TCL power capacity. Therefore, a quick and effective real-time TCL power capacity calculation method is needed.

HDMR applies orthonormal polynomials to provide parameter sensitivity indices and build analytical relationships between multi inputs and outputs. The proposed approach will reduce computation efforts for TCL power calculation in random conditions. The proposed fast TCL capacity calculation framework which uses HDMR is shown in Fig. 2.10.

2.5 Case Studies

Table 2.1 shows the parameter distribution of 10,000 TCLs. Table 2.2 shows the distribution of outdoor temperature. ΔT is 1 °C and the dispatch period is 10 min. When HDMR is applied, k in (2.8) is 3, $l(m)$ is 2, and N_0 is 10,000. Matlab R2014a is utilized to simulate the individual power capacity of TCLs. The CPU model of the desktop for these studies is Intel(R) Core (TM) i7-6700 and the frequency is 3.40 GHz.

2.5.1 *HDMR Modeling Results*

The sensitivity indices of TCL parameters via HDMR are given in Table 2.4, in which the parameters with sensitivity indices less than 1% are ignored and the modeling results show that the cooperative effect of parameters can be ignored. In this case, only the effects of R_μ, T_{out} and η_μ acting independently upon TCL power capacity are significant. The analytical relationship between the capacity and multi-inputs is given in (2.23), considering coefficients larger than 0.5.

$$\begin{aligned} C_t = 22.5756 - 4.1195\phi_1(R\_\mu) + 0.5698\phi_2(R\_\mu) \\ + 3.6635\phi_1(T_{out}) - 1.2143\phi_1(\eta\_\mu) \end{aligned} \tag{2.23}$$

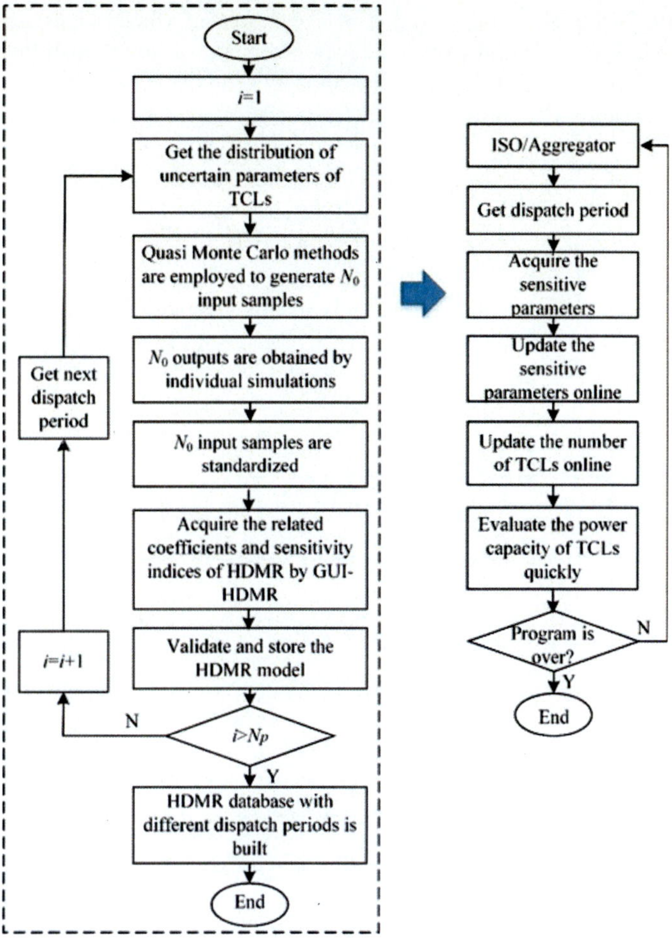

Fig. 2.10 Fast capacity calculation framework

Table 2.4 Sensitivity indices of TCL parameters via HDMR

Parameter	R_μ	T_{out}	η_μ
Sensitivity index	0.5164	0.3902	0.0445

To validate the sensitivity of TCL parameters and the analytical expression of the TCL power capacity, five scenarios are designed and 100 samples are considered in each scenario.

Scenario 1 (S1): All parameters which are the same as those in Table 2.3 are kept constant. 100 samples are chosen to evaluate the power capacity of 10,000 TCLs. This is the base case that is used to show the power capacity of the TCL population when the 15 parameters are fixed.

Scenario 2 (S2): R_μ, T_{out} and η_μ are kept constant (their values are the same as those in Table 2.3) and the other 12 parameters are randomly distributed within their ranges. This case is to identify the impact of the other 12 TCL parameters on the power capacity. The comparison of TCL power capacities for S1 and S2 is shown in Fig. 2.11.

Scenario 3: Only R_μ changes randomly within its range while other parameters are kept constant. This case is utilized to describe the independent impact of R_μ on the power capacity. The relationship between TCL power capacity and R_μ is shown in Fig. 2.12.

Scenario 4: Only T_{out} changes randomly within its range while the other parameters are kept constant. This scenario is to describe how the power capacity is changed when Tout varies. The relationship between TCL power capacity and Tout is shown in Fig. 2.13.

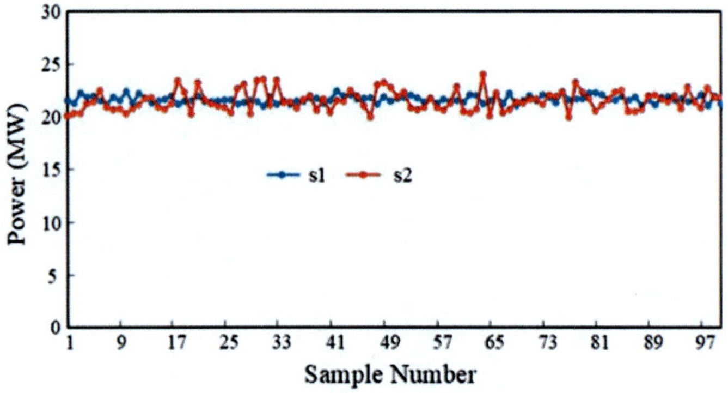

Fig. 2.11 TCL power capacity comparisons between S1 and S2

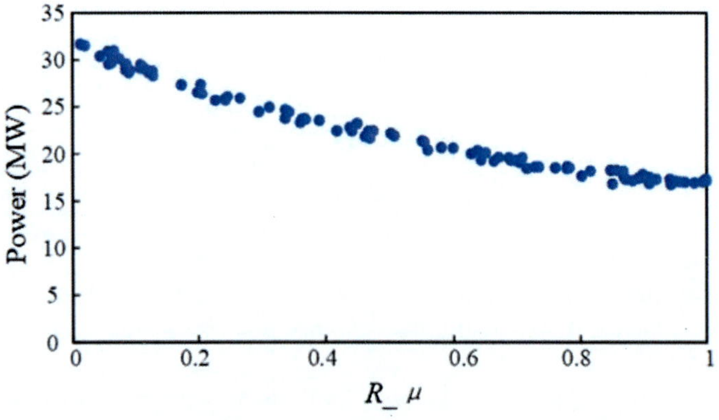

Fig. 2.12 Relationship between TCL power capacity and R_μ

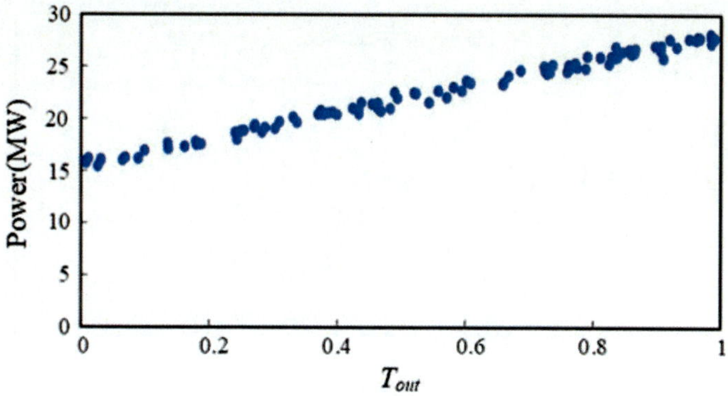

Fig. 2.13 Relationship between TCL power capacity and T_{out}

Scenario 5: Only η_μ changes randomly within its range while the other parameters are kept constant. This case is to show the impact of η_μ on the power capacity. The relationship between TCL power capacity and η_μ is shown in Fig. 2.14.

In Fig. 2.11, the power capacity of 10,000 TCLs in S1 is almost fixed when 100 samples are simulated. The small fluctuations are caused by normal distribution errors in each sample. Although 12 parameters in S2 are variable, the capacity differences between S1 and S2 are very small, which demonstrates that the 12 parameters play minute impacts on TCL power capacity.

Figure 2.12 shows that R_μ plays a significant role in the TCL power capacity calculation. The relationship between capacity and R_μ is approximately represented by a quadratic function. The TCL power capacity is lowered by about 15 MW when R_μ increases from 0 to 1, which concurs with (2.22c). The relationship between TCL power capacity and T_{out} in Fig. 2.13 appears linear. When T_{out} changes from 0 to 1, the TCL power capacity will increase by about 12 MW, which concurs with the modeling result in (2.23). η_μ has a smaller impact on the TCL power capacity according to Fig. 2.14. Their relationship is also linear and the capacity will be

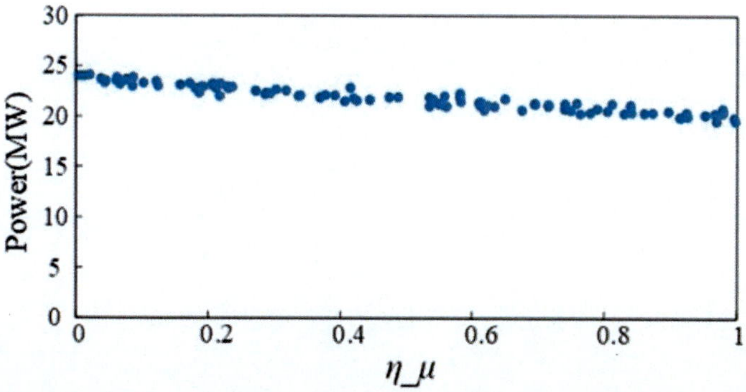

Fig. 2.14 Relationship between TCL power capacity and η_μ

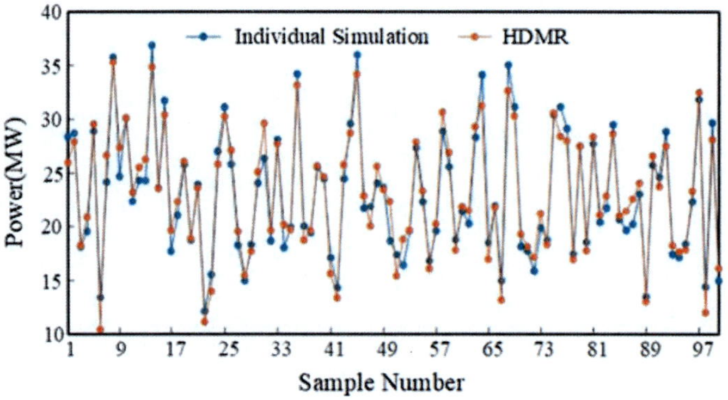

Fig. 2.15 TCL capacity for individual simulations and HDMR

reduced by 4 MW when η_μ changes from 0 to 1, which again concurs with (2.23). The simulation results demonstrate that only R_μ, T_{out} and η_μ play an important role in the TCL capacity calculation and (2.23) can describe the efforts of R_μ, T_{out} and η_μ acting independently on the TCL power capacity calculation.

To further validate the analytical expression in (2.23), 100 samples, are individually simulated with 15 parameters randomly distributed within their ranges. Figure 2.15 shows the comparison of TCL power capacities using individual simulations and HDMR in (2.23). HDMR can accurately evaluate the TCL power capacity with an average relative error of 4.38% in (2.24). As R_μ, T_{out} and η_μ are altered, (2.23) is employed to calculate the TCL power capacity directly according to the latest values of parameters. Thus, more efforts should be paid to measure R_μ, T_{out} and η_μ. However, there will be no need to update the TCL power capacity, when other parameters with minute impacts (i.e., T_{set_μ}) are varying, which lowers the computation efforts.

$$r = \frac{\sum_{i=1}^{N_s} |x_{ind} - x_{\text{HDMR}}|/x_{ind}}{N_s} \tag{2.24}$$

When the number of TCLs is changed, the TCL power capacity is calculated according to (2.25).

$$C = \frac{N_a}{N_t} C_t \tag{2.25}$$

2.5.2 *Effect of Population Heterogeneity and Dispatch Period*

We consider the following cases in this section:

(1) Different types of TCL heterogeneity: The heterogeneity of TCL population is characterized by a normal distribution in Sect. 2.2. Two other distributions in the literature govern TCL behavior, including lognormal distribution [27, 28] and uniform distribution [29]. The effects of two heterogeneity types are considered for the 15 parameters. The modeling results for lognormal distribution are shown in Tables 2.5 and (2.26). The sensitivity indices and analytical expression for uniform distribution are given in Table 2.6 and (2.27). 100 samples are chosen to calculate the TCL power capacity using (2.23), (2.26) and (2.27). The power capacity errors for the three types of population heterogeneity are compared in Fig. 2.16. The average errors of lognormal and uniform distributions are 4.65% and 4.55%, respectively. It is obvious that R_μ, Tout and η_μ have significant impacts on the TCL power capacity and sensitivity indices in the three distributions. The results depicted in Fig. 2.16 are almost identical which demonstrates that the heterogeneity types do not have substantial impacts on the TCL power capacity.

$$C_l = 22.8595 - 4.1821\phi_1(R\_\mu) + 0.5698\phi_2(R\_\mu)$$

Table 2.5 Sensitivity indices of TCL parameters with lognormal distribution

Parameter	R_μ	T_{out}	η_μ
Sensitivity indices	0.5140	0.3765	0.0467

Table 2.6 Sensitivity indices of TCL parameters with uniform distribution

Parameter	R_μ	T_{out}	η_μ
Sensitivity indices	0.5031	0.3797	0.0432

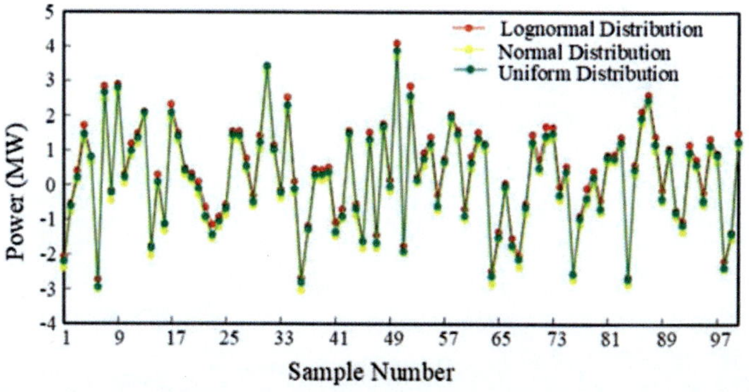

Fig. 2.16 Power capacity errors comparisons using various distributions

$$+ 3.6097\phi_1(T_{out}) - 1.2723\phi_1(\eta\_\mu) \tag{2.26}$$

$$C_u = 22.7136 - 4.1611\phi_1(R\_\mu) + 0.6948\phi_2(R\_\mu) + 3.6648\phi_1(T_{out}) - 1.2357\phi_1(\eta\_\mu) \tag{2.27}$$

(2) Different dispatch periods: The relationships between different parameters and dispatch periods are shown in Figs. 2.17, 2.18 and 2.19 when the dispatch period is changed from 1 to 30 min. Obviously, TCL power capacity is affected by several parameters in each dispatch period. Dispatch periods have a similar tendency with respect to R_μ and T_{out}. When the dispatch period is less than 10 min, R_μ and T_{out} have significant impacts on TCL power capacity. Therefore, additional attention must be devoted to the two important parameters (especially T_{out}) for a more accurate calculation of TCL capacity. If the TCL population is maintained, the capacity should be updated according to changes in T_{out}. As the dispatch period increases, the impact

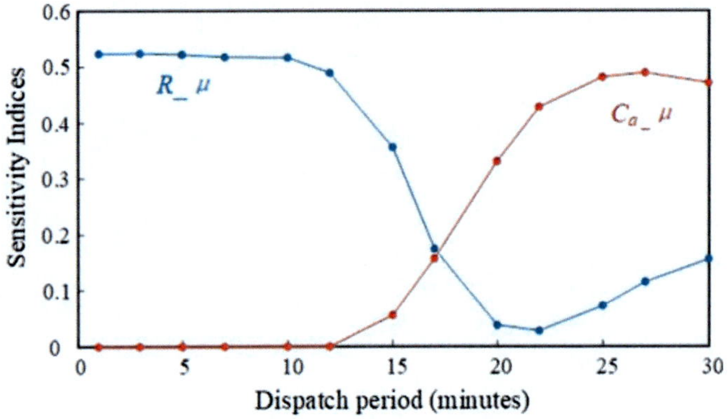

Fig. 2.17 Relationship between the parameters (R_μ, C_a_μ) and dispatch period

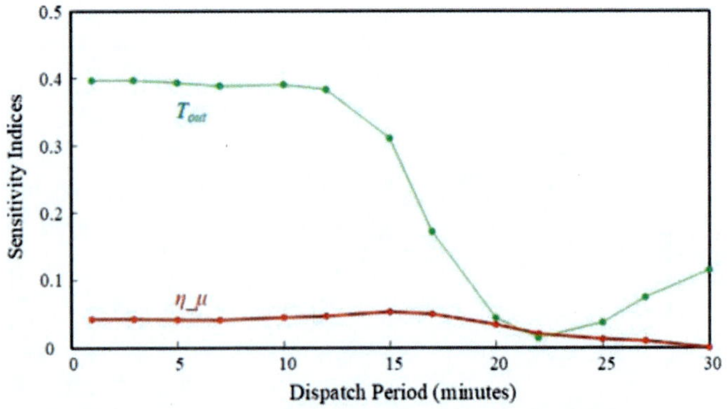

Fig. 2.18 Relationship between the parameters (T_{out}, η_μ) and dispatch period

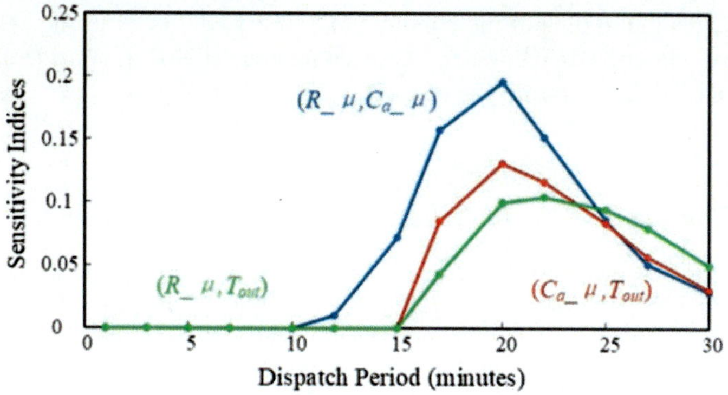

Fig. 2.19 Relationship between cooperative parameters and dispatch period

of R_μ and T_{out} will decrease until the dispatch period reaches about 22 min when their impacts start to increase again.

The TCL power capacity is almost unaffected by C_{a}_μ when the dispatch period is shorter than 12 min. As the dispatch period increases, the impact on TCL power capacity heightens, which is opposite to those of R_μ and T_{out}. Therefore, more attention should be devoted to C_a_μ in longer dispatch periods. The impact of η_μ on TCL power capacity remains small when the dispatch period is less than 17 min. As the dispatch period increases, the sensitivity index of η_μ becomes smaller, which will reduce to 0 when the dispatch period is 30 min. Although η is an important TCL parameter, it does not affect the TCL power capacity significantly.

Figure 2.19 shows the cooperative impacts of parameters (R_μ, C_{a}_μ, T_{out}) on the TCL power capacity. Here, cooperative impacts of parameters begin to work when the dispatch period is longer than 10 min. Also, sensitivity indices of cooperative parameters are not less than 0.1 when the dispatch period is 20 min, which means that the cooperative impacts play an important role that cannot be ignored in the TCL capacity calculation.

The simulation results show that the heterogeneity types of TCL has no substantial effect on the TCL power capacity while the dispatch period significantly affects the TCL power capacity. Thus, a real-time update of the TCL heterogeneity types is not necessary for HDMR. The HDMR of the TCL power capacity in various dispatch periods can be built offline with the relative coefficients stored in a database for real-time applications. When the dispatch period is stated by the dispatch center, the corresponding HDMR coefficients will be obtained quickly from the database and the power capacity of the TCL population will be calculated according to the latest critical TCL parameters. Accordingly, TCL power capacity can be obtained in real-time via HDMR.

Here, the equivalent relationship between the power capacity and multi TCL parameters built by HDMR is a simple polynomial with several critical TCL parameters, which is succinct enough to achieve the fast calculation of TCL power capacity. The simulation time of 10,000 TCLs for the power capacity calculation is about 6.59 s while the calculation time of the polynomial built by HDMR is so short that it can be practically ignored. In particular, the saved time via HDMR is considerable when a larger TCL population is dispatched.

2.5.3 Comparisons with Other Methods

Few studies have focused on the equivalent relationship between the capacity of ancillary services and TCL parameters. A similar problem can only be solved using multiple linear regression to bridge the relationship between the aggregate power and TCL parameters [16]. Here, multiple linear regression is employed to compare to HDMR for calculating TCL power capacity. 10,000 pairs of input–output samples are generated to build the equivalent expression of TCL power capacity via the two methods when the dispatch period is 20 min. The modeling result of multiple linear regression is stated in (2.28), in which the TCL parameters with coefficients smaller than 1 are ignored. The HDMR modeling result is described in (2.29).

$$C_t = 15.2148 - 2.8005R\_\mu + 10.5304C_a\_\mu - 3.5961\eta\_\mu + 3.4624T_{out} \tag{2.28}$$

$$\begin{aligned} C_t = {} & 19.0983 - 0.8046\phi_1(R\_\mu) - 0.7198\phi_2(R\_\mu) + 3.0534\phi_1(C_a\_\mu) \\ & - 1.1210\phi_2(C_a\_\mu) - 1.0429\phi_1(\eta\_\mu) + 1.0904\phi_3(\eta\_\mu) \\ & + 0.9917\phi_1(T_{out}) - 0.6399\phi_2(T_{out}) - 2.3807\phi_1(R\_\mu)\phi_1(C_a\_\mu) \\ & + 1.5957\phi_1(R\_\mu)\phi_1(T_{out}) + 1.9782\phi_1(C_a\_\mu)\phi_1(T_{out}) \end{aligned} \tag{2.29}$$

In this study, 100 samples are randomly chosen to validate the modeling results of the two methods. The average relative error of multiple linear regression and HDMR is 15.03% and 5.52%, respectively. The comparison of relative error for the two methods is shown in Fig. 2.20, in which HDMR has a higher modeling accuracy.

Multiple linear regression can only build the independent impact of inputs on the system output, with a linear relationship between the inputs and output. Thus, the cooperative effects of inputs and the nonlinear relationship between the inputs and output cannot be accurately modeled via multiple linear regression. In contrast, HDMR can model not only the cooperative effect of inputs acting collectively but also the nonlinear effect of single input acting independently upon the system output, which enables a broader applications for HDMR. Therefore, the modeling accuracy of HDMR is higher than that of multiple linear regression when analyzing the relationship between TCL parameters and power capacity.

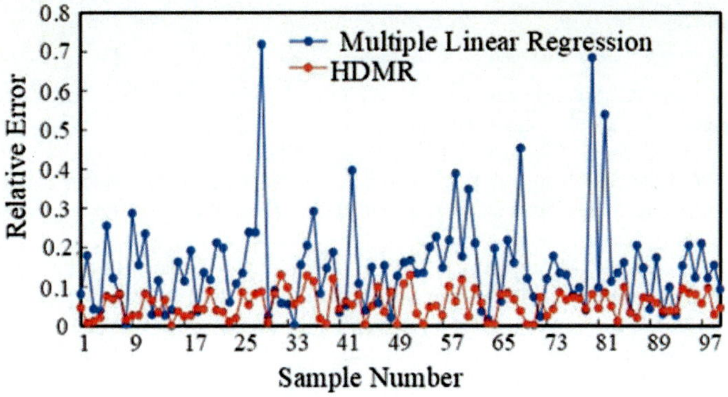

Fig. 2.20 Comparison of relative error of multiple linear regression and HDMR

2.6 Conclusions

A calculation strategy is proposed to generate power pulses, in which, TCLs unsuitable for dispatch are considered to satisfy end users' comfort requirements. HDMR is employed for sensitivity analyses and taken as the key step in evaluating the TCL power capacity to identify important factors when TCL parameter uncertainties are considered. A fast TCL power capacity calculation framework is also presented using HDMR, and significant results are obtained. According to the simulations, we draw the following conclusions: (1) The proposed TCL capacity calculation strategy, in which TCLs without perfect parameters are removed, can maintain end-users' comfort levels; (2) The equivalent expression relating TCL power capacity to multi-parameters using HDMR can calculate fast TCL capacity as certain parameters are altered, which helps reduce computation burdens; (3) The power capacity of a TCL population is affected differently by TCL parameters when the dispatch period is altered. In periods shorter than 15 min, the capacity is mainly affected by R_μ and T_{out}. The most significant effect of Ca_μ was observed when the scheduling period exceeded 22 min. The cooperative impacts of R_μ, Ca_μ and T_{out} also play substantial roles when the dispatch period is between 15 and 22 min. (4) Irrespective of the duration of the dispatch period, TCL power capacity is mainly affected by R_μ, Ca_μ, T_{out} and η_μ. This implies that more accurate R_μ, T_{out}, and η_μ can benefit the accuracy of capacity calculation for a TCL population; (5) HDMR offers great advantages in modeling the cooperative effect of inputs acting collectively and the nonlinear effect of single input acting independently on the system output. HDMR can build the equivalent relationship between TCL parameters and the power capacity with higher accuracy as compared with multiple linear regression.

References

1. Song, M., Gao, C., Shahidehpour, M., Li, Z., Yang, J., & Yan, H. (2018). Impact of uncertain parameters on TCL power capacity calculation via HDMR for generating power pulses. *IEEE Transactions on Smart Grid, 10*(3), 3112–3124.
2. Zhang, W., Lian, J., Chang, C. Y., & Kalsi, K. (2013). Aggregated modeling and control of air conditioning loads for demand response. *IEEE Transactions on Power Systems, 28*(4), 4655–4664.
3. Kara, E. C., Bergés, M., & Hug, G. (2015). Impact of disturbances on modeling of thermostatically controlled loads for demand response. *IEEE Transactions on Smart Grid, 6*(5), 2560–2568.
4. Lu, N. (2012). An evaluation of the HVAC load potential for providing load balancing service. *IEEE Transactions on Smart Grid, 3*(3), 1263–1270.
5. Trovato, V., Teng, F., & Strbac, G. (2017). Role and benefits of flexible thermostatically controlled loads in future low-carbon systems. *IEEE Transactions on Smart Grid, 9*(5), 5067–5079.
6. Sanandaji, B. M., Vincent, T. L., & Poolla, K. (2015). Ramping rate flexibility of residential HVAC loads. *IEEE Transactions on Sustainable Energy, 7*(2), 865–874.
7. Lu, N., & Zhang, Y. (2012). Design considerations of a centralized load controller using thermostatically controlled appliances for continuous regulation reserves. *IEEE Transactions on Smart Grid, 4*(2), 914–921.
8. Bashash, S., & Fathy, H. K. (2012). Modeling and control of aggregate air conditioning loads for robust renewable power management. *IEEE Transactions on Control Systems Technology, 21*(4), 1318–1327.
9. Wang, D., Jia, H., Wang, C., Lu, N., Fan, M., Miao, W., & Liu, Z. (2014). Performance evaluation of controlling thermostatically controlled appliances as virtual generators using comfort-constrained state-queueing models. *IET Generation, Transmission & Distribution, 8*(4), 591–599.
10. Wai, C. H., Beaudin, M., Zareipour, H., Schellenberg, A., & Lu, N. (2014). Cooling devices in demand response: A comparison of control methods. *IEEE Transactions on Smart Grid, 6*(1), 249–260.
11. Sinitsyn, N. A., Kundu, S., & Backhaus, S. (2013). Safe protocols for generating power pulses with heterogeneous populations of thermostatically controlled loads. *Energy Conversion and Management, 67*, 297–308.
12. Hao, H., Sanandaji, B. M., Poolla, K., & Vincent, T. L. (2014). Aggregate flexibility of thermostatically controlled loads. *IEEE Transactions on Power Systems, 30*(1), 189–198.
13. Hong, Y. Y., Lin, J. K., Wu, C. P., & Chuang, C. C. (2012). Multi-objective air-conditioning control considering fuzzy parameters using immune clonal selection programming. *IEEE Transactions on Smart Grid, 3*(4), 1603–1610.
14. Erdinc, O., Taşcıkaraoğlu, A., Paterakis, N. G., Eren, Y., & Catalão, J. P. (2016). End-user comfort oriented day-ahead planning for responsive residential HVAC demand aggregation considering weather forecasts. *IEEE Transactions on Smart Grid, 8*(1), 362–372.
15. Hu, J., Cao, J., Chen, M. Z., Yu, J., Yao, J., Yang, S., & Yong, T. (2016). Load following of multiple heterogeneous TCL aggregators by centralized control. *IEEE Transactions on Power Systems, 32*(4), 3157–3167.
16. Sun, Y., Elizondo, M., Lu, S., & Fuller, J. C. (2014). The impact of uncertain physical parameters on HVAC demand response. *IEEE Transactions on Smart Grid, 5*(2), 916–923.
17. Perfumo, C., Braslavsky, J., & Ward, J. K. (2013, September). A sensitivity analysis of the dynamics of a population of thermostatically-controlled loads. 2013 Australasian Universities Power Engineering Conference (AUPEC) (pp. 1–6).
18. Rabitz, H., & Aliş, Ö. F. (1999). General foundations of high-dimensional model representations. *Journal of Mathematical Chemistry, 25*(2), 197–233.

19. Miller, M., Feng, X., Li, G., & Rabitz, H. (2009, September). Nonlinear bionetwork structure inference using the random sampling-high dimensional model representation (RS-HDMR) algorithm. In *2009 Annual International Conference of the IEEE Engineering in Medicine and Biology Society* (pp. 6412–6415).
20. Yücel, A. C., Bağcı, H., & Michielssen, E. (2015). An ME-PC enhanced HDMR method for efficient statistical analysis of multiconductor transmission line networks. *IEEE Transactions on Components, Packaging and Manufacturing Technology, 5*(5), 685–696.
21. Khalid, F., & Speidel, J. (2014). Robust hybrid precoding for multiuser MIMO wireless communication systems. *IEEE Transactions on Wireless Communications, 13*(6), 3353–3363.
22. Tipcharoen, W., Warisarn, C., & Kovintavewat, P. (2016). Effects of island volume and hotspot position fluctuation for heated-dot magnetic recording. *IEEE Magnetics Letters, 7*, 1–4.
23. Sridharan, J., & Chen, T. (2006, March). Modeling multiple input switching of CMOS gates in DSM technology using HDMR. In *Proceedings of the Design Automation & Test in Europe Conference* (pp. 1–6).
24. Mathieu, J. L., Koch, S., & Callaway, D. S. (2012). State estimation and control of electric loads to manage real-time energy imbalance. *IEEE Transactions on Power Systems, 28*(1), 430–440.
25. Sanandaji, B. M., Hao, H., & Poolla, K. (2014, January). Fast regulation service provision via aggregation of thermostatically controlled loads. In *2014 47th Hawaii International Conference on System Sciences* (pp. 2388–2397).
26. Bashash, S., & Fathy, H. K. (2011, June). Modeling and control insights into demand-side energy management through setpoint control of thermostatic loads. In *Proceedings of the 2011 American Control Conference* (pp. 4546–4553).
27. Callaway, D. S. (2009). Tapping the energy storage potential in electric loads to deliver load following and regulation, with application to wind energy. *Energy Conversion and Management, 50*(5), 1389–1400.
28. Mahdavi, N., Braslavsky, J. H., Seron, M. M., & West, S. R. (2017). Model predictive control of distributed air-conditioning loads to compensate fluctuations in solar power. *IEEE Transactions on Smart Grid, 8*(6), 3055–3065.
29. Mathieu, J. L., & Callaway, D. S. (2012, January). State estimation and control of heterogeneous thermostatically controlled loads for load following. In *2012 45th Hawaii International Conference on System Sciences* (pp. 2002–2011).
30. Kamgarpour, M., Ellen, C., Soudjani, S. E. Z., Gerwinn, S., Mathieu, J. L., Müllner, N., & Lygeros, J. (2013, August). Modeling options for demand side participation of thermostatically controlled loads. In *2013 IREP Symposium Bulk Power System Dynamics and Control-IX Optimization, Security and Control of the Emerging Power Grid* (pp. 1–15).
31. Li, G., Wang, S. W., & Rabitz, H. (2002). Practical approaches to construct RS-HDMR component functions. *The Journal of Physical Chemistry A, 106*(37), 8721–8733.
32. Alış, Ö. F., & Rabitz, H. (2001). Efficient implementation of high dimensional model representations. *Journal of Mathematical Chemistry, 29*(2), 127–142.
33. Rabitz, H., Aliş, Ö. F., Shorter, J., & Shim, K. (1999). Efficient input—output model representations. *Computer Physics Communications, 117*(1–2), 11–20.
34. Feil, B., Kucherenko, S., & Shah, N. (2009, September). Comparison of Monte Carlo and quasi Monte Carlo sampling methods in high dimensional model representation. In *2009 First International Conference on Advances in System Simulation* (pp. 12–17).
35. Ziehn, T., & Tomlin, A. S. (2009). GUI–HDMR–A software tool for global sensitivity analysis of complex models. *Environmental Modelling & Software, 24*(7), 775–785.

Chapter 3
Time-Dependent Cold Load Pickup of TCLs and Its Application in Distribution System Load Restoration

Abstract Service restoration is one of the critical functions to achieve a future self-healing distribution system. An accurate characterization of realistic system operating conditions is required for timely and reliable restoration of the distribution system. In this chapter, two main factors having a significant impact on distribution system restoration (DSR) in practice are investigated. First, cold load pickup (CLPU), generally caused by TCLs, is a common phenomenon after an outage and is shaped by the outage duration. The time-dependent behaviors of CLPU however are rarely considered in the literature. In this chapter, the operating state evolution of TCLs after an outage is analyzed to characterize time-dependent CLPU; the time-dependent CLPU is analytically embedded in DSR to accurately represent the actual behaviors of the restored loads. Second, The lack of real-time measurement data makes it difficult to predict fluctuating loads during DSR. Accordingly, a robust DSR based on the information gap decision theory (IGDT) is proposed to address this challenge by sufficiently considering the uncertainty of CLPU. The proposed models are tested in IEEE 13-node and 123-node test feeders. Simulation results demonstrate that the time-dependent CLPU model and the uncertainty modeling of CLPU can accurately capture the actual behaviors of loads with TCLs after an outage, which greatly improves DSR decisions in practice.

Keywords Cold load pickup · Distribution system restoration · Robust optimization

3.1 Introduction

The more frequent natural disasters and cyber-attackssss have posed great challenges to enhancing the resilience of power systems. Statistics show that the duration of outages caused by distribution system failures takes the overwhelming majority of the total outage time every year [1]. After a failure happens to distribution systems, the advanced distribution management system (ADMS) can rapidly detect and isolate the fault, and enable service restoration to recover the outage area. Faced with the evolving active distribution systems, there are tremendous challenges

M. Song and C. Gao, *Integration of Distributed Resources in Smart Grids for Demand Response and Transactive Energy*, https://doi.org/10.1007/978-981-16-7170-8_3

and opportunities in improving distribution system restoration (DSR) to achieve self-healing.

In DSR, one important characteristic of loads after an outage is the cold load pickup (CLPU) phenomenon. CLPU generally happens in the area with a high penetration of TCLs, and it is usually caused by the diversity loss of TCL operating states [2]. When TCLs work under normal conditions, their operating states are uniformly distributed with steady power consumption. After a blackout happens, the diversity of TCL operating states is destroyed, and more TCLs will be turned on once energization starts [3]. This will drive the total power consumption of TCLs much larger than their normal value [4, 5].

The excessive loads (i.e. CLPU) after an outage have been analytically modeled in the literature. One of the earliest methods is the RC circuit model [6] that represents the decay of CLPU power consumption. However, this model is inaccurate due to the oversimplified assumption, which makes it inapplicable in the real distribution system. Physical models [2, 7] are utilized to represent CLPU effects by the first or second-order differential expressions of TCLs. Although these physical models can provide a great insight into CLPU, they only track the temperatures of TCLs and the conversion to the power consumption is still required. The most common model of CLPU is the delayed exponential model (DEM) [8–12]. For example, two blocks are used in [11] to represent the exponential decay delay of CLPU for less computational burden. Authors in [12] apply a delayed exponentially decaying function of CLPU to characterize the actual operating behaviors of loads after an outage. In all the existing CLPU models, the CLPU profile is described as a fixed curve invariant with the outage time.

However, these models can only apply to the extended outage, during which the diversity of TCL operating states is completely lost and all TCLs have the same operating states. For distribution systems equipped with ADMS, the outage area can be restored soon after a failure occurs. With the increasing penetration of distributed energy resources (DERs) to enable blackstart or improve self-healing in the future, the outage duration may be as short as several minutes. Under these circumstances, it has been shown that the diversity of TCL operating states is not completely lost and the CLPU curve depends on the outage time. Hence, the previous models cannot accurately characterize the real behaviors of CLPU. And a more general and realistic model that can represent the actual CLPU performances under any outage duration is required. Therefore, instead of the common DEM, the time dependent model of CLPU should be employed in DSR, as a first try in this chapter.

Besides CLPU, another critical issue is the uncertainty of loads during restoration. It is difficult to accurately predict loads due to the random behaviors of customers [13]. This becomes one of the uncertainty sources for restoration decision-making [14]. The stochastic optimization with uncertain load parameters can be modeled to address load uncertainty. In [15], a probabilistic method is utilized to characterize the uncertain parameters of loads. Authors in [16] propose a probabilistic approach based on the two-point estimation method to describe the load uncertainty. In [17], the demand uncertainty is captured by the fuzzy set theory and maximum membership functions. However, it is difficult to obtain the required probabilistic distribution of uncertain load parameters in practice. Information gap decision theory (IGDT) only requires the uncertainty set without probability distribution for decision-making under the worst case, which is a good alternative to address this issue [1, 18]. The IGDT-based robust method seeks the maximal uncertainty horizon while the target behaviors are guaranteed [18, 19]. It can provide robust solutions to be immune against uncertainty by sacrificing some performances [1]. Compared to other robust methods [20, 21] that need an explicit-boundary uncertainty set, the IGDT-based robust method requires less information on the uncertainty structure. In fact, it is difficult to characterize the CLPU fluctuations caused by the load uncertainty using an explicit boundary. Thus, the IGDT-based robust method is applied in this chapter to formulate DSR with the uncertainty associated with CLPU.

The main work of this chapter includes, (1) the timed-dependent CLPU model is developed by analyzing the operating state evolution of TCLs after an outage to characterize the time-dependent behavior of CLPU; (2) the DSR with time-dependent CLPU is proposed by accurately modeling the actual performances of loads with TCLs; (3) the robust DSR is developed to characterize the load demand uncertainties. Different from the IGDT-based robust model of load restoration in [22] which only considers the uncertainty of non-TCLs, this chapter model the uncertainties associated with CLPU caused by TCLs in detail.

3.2 Time-Dependent CLPU Modeling

Simulation results of the CLPU for TCLs with parameters from [23] under different outage durations are shown in Fig. 3.1. It shows that the behaviors of CLPU depend on the outage duration in DSR. The evolution process of CLPU is shown in Fig. 3.2. It is noted that the operating states of TCLs are determined by the switch state and

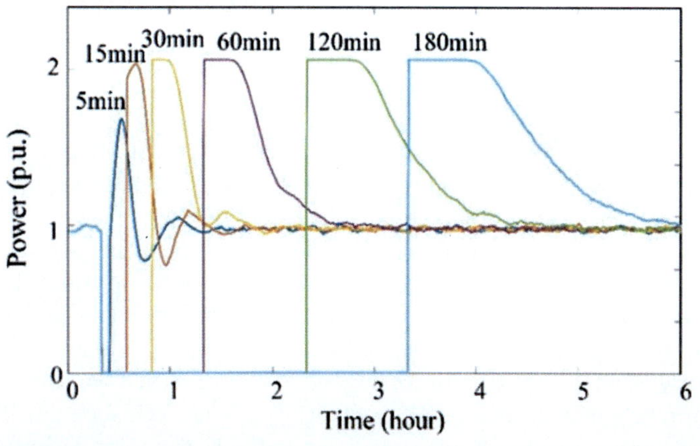

Fig. 3.1 CLPU profiles under different outage durations

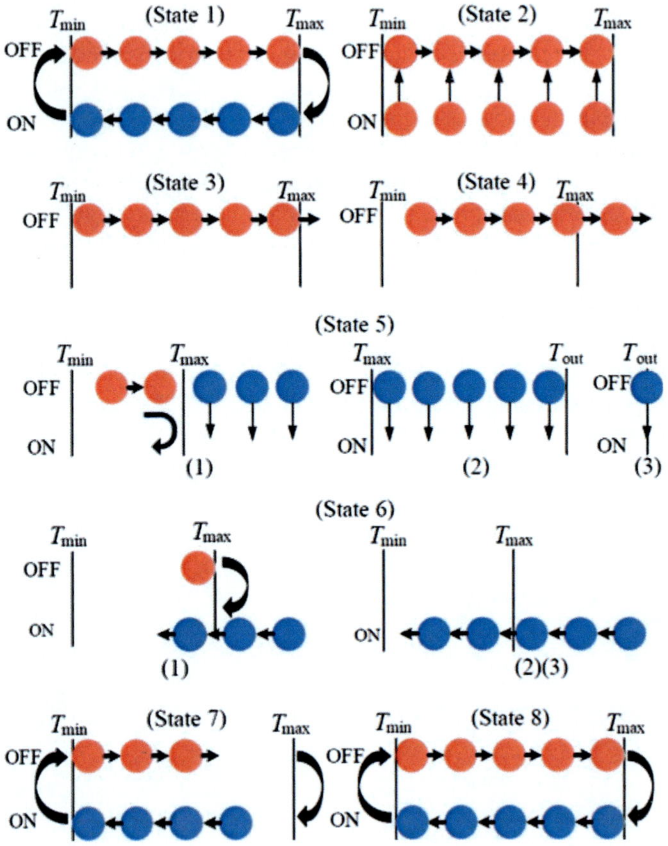

Fig. 3.2 The evolution of TCL operating states after an outage

the internal temperature. Thus, the colors in Fig. 3.2 are utilized to represent the switch states of TCLs. The on and off switch states of TCLs are described by the red and blue solid circles, respectively. The horizontal location in Fig. 3.2 is employed to characterize the internal temperature of TCLs. For example, different red solid circles in State 1 of Fig. 3.2 mean that TCLs have different internal temperatures. Additionally, TCLs will have higher internal temperatures when the red circles are closer to T_{max}. The detailed analysis is given as follows.

Stage 1—Initialization and Outage: When TCLs work under normal conditions (State 1), their operating states are diversified and uniformly distributed; and their aggregated power remains constant. Specifically, there are only two power states for TCLs, either rated power (on) or 0 (off) [24–26]. When the internal temperature is higher than T_{max}, TCLs in cooling mode will be turned on, and the internal temperature begins to decrease until reaching T_{min}. Then TCLs are turned off, and the internal temperature begins to increase. Accordingly, the internal temperatures of

TCLs periodically change within $[T_{\min}, T_{\max}]$, and the power consumption of TCLs is stable.

When TCLs lose the power supply due to an outage, the periodic operating states of TCLs will be destroyed. Once an outage happens, TCLs with the original switch state of on will be forced to change to off (State 2). Then all TCLs are off (State 3), and the total power consumption is 0 with the diversity of operating states lost. Then the internal temperatures of TCLs begin to increase as a result of the heat transfer between internal and external surroundings (State 4).

Stage 2—Re-energization: Under different outage durations, TCL internal temperatures have three different scenarios:

(1) *Short outage duration*: Due to the short outage duration, only some TCLs' internal temperatures are higher than $T_{\max}$, and the other TCLs' internal temperature is kept within $[T_{\min}, T_{\max}]$ (State 5-(1)). Once re-energized, the TCLs with an internal temperature higher than $T_{\max}$ will be immediately turned on, and the TCLs with an internal temperature lower than $T_{\max}$ will be regulated by their nature operating laws (State 6-(1)). The collective power of TCLs immediately increases until reaching the peak, shown as the CLPU profiles with the outage duration of 5 and 15 min in Fig. 3.1.

(2) *Medium outage duration*: As the outage duration becomes longer, there are more TCLs with an internal temperature higher than $T_{\max}$. Finally, the internal temperatures of all TCLs are higher than $T_{\max}$ but lower than T_{out} (State 5-(2)). When TCLs are re-energized, all of them will be turned on immediately, which makes the initial total power equal to the rated power of all TCLs. Then, the internal temperatures of TCLs begin to gradually decrease (State 6-(2)). The total power will remain constant until the TCLs begin to be turned off, which the duration of constant power increases with longer outage time, shown as the CLPU profiles with the outage duration of 30, 60 and 120 min in Fig. 3.1.

(3) *Long outage duration*: When the outage duration is long enough, internal temperatures of all TCLs are the same as *Tout* (State 5-(3)). All TCLs will be turned on once re-energized, with the aggregated power equal to the total rated TCL power. Therefore, the internal temperatures of TCLs will gradually decrease (State 6-(2)). As the difference between T_{out} and TCLs' internal temperature preferences is fixed, the duration of the total power remaining constant is a fixed time, which is independent of the outage duration. In this scenario, the power profile of CLPU stays unchanged, which can be modeled by the DEM.

Stage 3—Diversity Restoration: After re-energization, the internal temperatures of TCLs begin to decrease until $T_{\min}$; then, they are turned off under their natural laws (State 7). The diversity of TCL operating states begins to restore until fully restored with a stable total power (State 8). During this process, the CLPU may show fluctuations due to that not all operating states of TCLs are completely restored.

The generalized time-dependent model of CLPU is shown in Fig. 3.3. Before t_0, TCLs operate under normal conditions with the baseline power of Z^{D0}. When an outage happens at t_0, the diversity of their operating states is gradually lost. After a

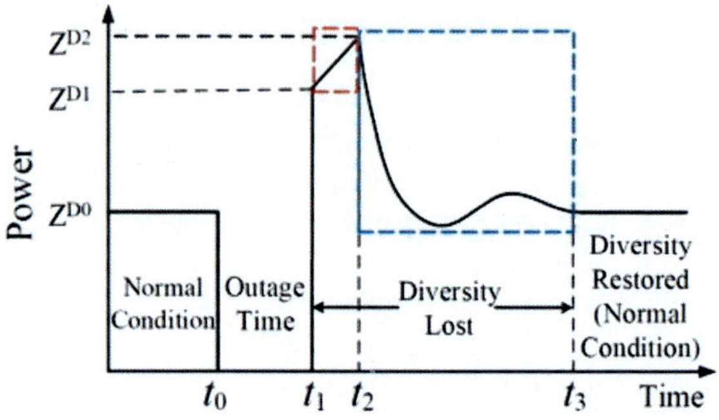

Fig. 3.3 The generalized time-dependent CLPU

while, TCLs are re-energized at t_1 with the initial power of Z^{D1}, which goes up to Z^{D2} at t_2. The diversity of TCL operating states begins to restore until completely restored at t_3. Our simulation results show that polynomial functions can accurately capture the behaviors of the time-dependent CLPU curve shown in Fig. 3.3 can be formulated as follows:

$$Z = \begin{cases} Z_1, \ t_1 \le t \le t_2 \\ Z_2, \ t_2 \le t \le t_3 \\ Z_3, \ t_3 \le t \end{cases} \tag{3.1a}$$

$$Z_1 = \frac{Z^{D2} - Z^{D1}}{t_2 - t_1} t + Z^{D1} \tag{3.1b}$$

$$Z_2 = a_n t^n + a_{n-1} t^{n-1} + \cdots + a_1 t + a_0 \tag{3.1c}$$

$$Z_3 = Z^{D0} \tag{3.1d}$$

where $a_n, a_{n-1}, \ldots, a_1, a_0$ are coefficients of polynomial functions, and n is the number of polynomial functions. t_1, t_2, t_3, Z^{D1} and Z^{D2} depend on the outage duration.

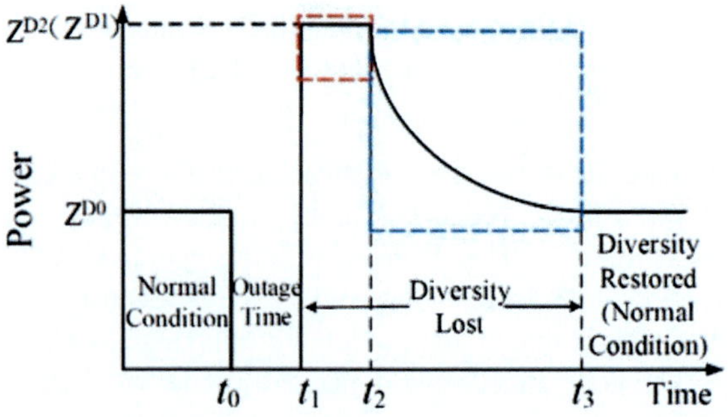

Fig. 3.4 The DEM of CLPU

Although DEM (shown in Fig. 3.4) as the common model of CLPU has been widely utilized in literature, it cannot accurately represent CLPU behaviors, especially under short outage duration. In contrast, the time-dependent CLPU model is more general and can accurately capture the real operating behaviors of CLPU under any outage duration. As stated in Stage 2, DEM is one special case of the time-dependent CLPU model under the extended outage duration. Only when the outage duration is long enough, the two models are equivalent, in which $Z^{D2} = Z^{D1}$ is satisfied and the difference between t_2 and t_1 is fixed (see the red box in Fig. 3.4). The CLPU power gradually decays with time and then reaches at a steady status from t_2 to t_3 (see the blue box in Fig. 3.4). As the distribution system can rapidly recover the outage area after an outage with the help of ADMS, the time-dependent CLPU model rather than DEM should be applied in DSR.

3.3 Problem Formulation

Generally, a distribution system restoration can be categorized into emergency service restoration and blackout service restoration [27]. The emergency service restoration requires isolation of faulted areas and energization of un-faulted out-of-service areas. However, in the blackout service restoration, the whole network is without power, and the restoration procedure depends on the start-up of generators, energization of lines and substations, and performing load restoration in distribution networks. This chapter focuses on the load restoration phase, which aims to minimize the outage impacts by finding optimal sets of loads to be restored. In this procedure, all sectionalizing switches are closed and all tie-switches are open, while the network topology is similar to the one under normal conditions. However, the restoration consists of several time steps as dictated from the transmission side.

3.3.1 DSR with Time-Dependent CLPU

DSR is modeled as a mixed-integer convex optimization problem. All variables with an upper arrow bar denote the variables of three-phase unbalanced distribution networks. The objective function is to maximize total restored base loads as below:

$$\max \sum_{t \in T} \sum_{i \in L} \omega_i s_{i,t}^D \vec{P}_{i,t}^D \tag{3.2}$$

Loads consist of non-TCLs and TCLs. As stated in Sect. 3.2, the actual power consumption of TCLs is either larger or smaller than the normal value upon re-energization after an outage. Accordingly, the actual power consumption of TCLs with CLPU phenomenon is equal to their baseline power plus the power fluctuations.

Therefore, the restored base loads of the objective function refer to the sum of non-TCLs and the baseline power of TCLs under normal operation.

The problem is subjects to the following constraints.

(1) *CLPU modeling*: The sampling method [12] is applied to calculate the load demand of a given CLPU curve. The CLPU curve is assumed to be equally sampled with the sampling interval Δt, which is also the time step of DSR. We first derive the equations with $\Delta Z(m)$, which is then replaced by $\Delta\ P^{TD}$ and $\Delta\ Q^{TD}$ to calculate P^{TD} and Q^{TD}.

$$\Delta Z(m) = \begin{cases} Z(m) - Z(m-1), & 1 < m \leq N \\ 0, & m = 1 \end{cases} \tag{3.3}$$

For a given CLPU curve at node i, shown in Fig. 3.3 and modeled in (3.1), $Z_i(m)$ can be modeled as:

$$Z_i(m) = \left[\frac{Z_i^{D2} - Z_i^{D1}}{t_{i,2} - t_{i,1}}(m-1)\Delta t + Z_i^{D1}\right]\left(1 - u\left(C_i^1(m)\right) + f\left(C_i^1(m)\right)\right.$$
$$\left.\left(C_i^1(m)\right)\left(1 - u\left(C_i^2(m)\right)\right) + Z_i^{D0} u\left(C_i^2(m)\right)\right) \tag{3.4a}$$

$$f(x) = a_n x^n + a_{n-1} x^{n-1} + \cdots + a_1 x + a_0 \tag{3.4b}$$

$$C_i^1(m) = (m-1)\Delta t - \left(t_{i,2} - t_{i,1}\right) \tag{3.4c}$$

$$C_i^2(m) = (m-1)\Delta t - \left(t_{i,3} - t_{i,1}\right) \tag{3.4d}$$

$$u(x) = \begin{cases} 1, & x > 0 \\ 0, & x \leq 0 \end{cases} \tag{3.4e}$$

where $C_i^1(m)$ and $C_i^2(m)$ are the duration between m^{th} sampling step and t_2 and t_3, respectively.

$A_{i,t,r}$ and $B_{i,t,r}$, defined as the active and reactive power of CLPU curve at time t when the load is restored at time step r ($1 < r < N$), respectively, can be formulated as:

$$A_{i,t,r} = \begin{cases} Z_{i,r}^{D1} - \sum_{m=r}^{t} \Delta P_{i,r}^{TD}(m), & r < t \\ 0, & r \geq t \end{cases} \tag{3.5a}$$

$$B_{i,t,r} = \begin{cases} Z_{i,r}^{D1} - \sum_{m=r}^{t} \Delta Q_{i,r}^{TD}(m), & r < t \\ 0, & r \geq t \end{cases} \tag{3.5b}$$

where $Z_{i,r}^{D1}$ is the value of Z_i^{D1} when the load is restored at time step r. $\Delta P_r^{TD}(m)$ and $\Delta Q_r^{TD}(m)$ are the value of $\Delta P^{TD}(m)$ and $\Delta Q^{TD}(m)$ when the load is restored at time step r, which can be calculated by replacing $\Delta Z(m)$ in (3.3) and prior to DSR.

As CLPU dependents on the restoration duration, there are N different CLPU curves for a given load when this load is restored at different time steps. $A_{i,t}$ and $B_{i,t}$ are the active and reactive power of CLPU at time step t. Thus, there are t different values in each of $A_{i,t}$ and $B_{i,t}$ based on (3.5a) and (3.5b) ($A_{i,t,r}$ and $B_{i,t,r}$ are 0 when $r > t$), which are stated as

$$A_{i,t} = \left[A_{i,t,1}, A_{i,t,2}, \ldots, A_{i,t,r}, \ldots, A_{i,t,t}\right]^T \tag{3.6a}$$

$$B_{i,t} = \left[B_{i,t,1}, B_{i,t,2}, \ldots, B_{i,t,r}, \ldots, B_{i,t,t}\right]^T \tag{3.6b}$$

Accordingly, CLPU is determined by the load restoration time, which is shown as:

$$P_{i,t}^{TD} = \left[s_{i,1}^D, s_{i,2}^D - s_{i,1}^D, \ldots, s_{i,t}^D - s_{i,t-1}^D\right] \cdot A_{i,t} \tag{3.7a}$$

$$Q_{i,t}^{TD} = \left[s_{i,1}^D, s_{i,2}^D - s_{i,1}^D, \ldots, s_{i,t}^D - s_{i,t-1}^D\right] \cdot B_{i,t} \tag{3.7b}$$

Note that CLPU modeling is developed in the scalar form, which will be changed to the vector form for the 3-phase model.

(2) *Sequencing constraints*: Once a load is energized, it should not be shed during the restoration procedure.

$$s_{i,t+1}^D \geq s_{i,t}^D \tag{3.8}$$

(3) *Voltage limit*: Nodal voltage should be kept within the minimum and maximum voltage limit:

$$\left(V^{\min}\right)^2 \leq \vec{U}_{i,t}(ph) \leq (V^{\max})^2, \quad ph \in \phi \tag{3.9}$$

where $\vec{U}_{i,t} = \left|\vec{V}_{i,t}\right| \odot 2$ represents the square of voltage and ph is the phase. The element-wise product is denoted by $\odot$.

(4) *Branch capacity*: The power flow of distribution lines should be limited based on their thermal capacities.

$$-\vec{P}_k^{L\_\max} \leq \vec{P}_{k,t}^L \leq \vec{P}_k^{L\_\max} \tag{3.10a}$$

$$-\vec{Q}_k^{L\_\max} \leq \vec{Q}_{k,t}^L \leq \vec{Q}_k^{L\_\max} \tag{3.10b}$$

(5) *Substation capacity limit*: The restorative capacity of the substation is limited as below:

$$0 \le \tilde{P}_{1,t}^{L} \le P_t^{sub}, \quad 0 \le \tilde{Q}_{1,t}^{L} \le Q_t^{sub} \tag{3.11}$$

(6) *DG operational constraints*: The operational constraints of the distributed generator (DG) including capacity limits and ramping-up limits are provided by (3.12) and (3.13), respectively.

$$-s_{i,t}^{G}\vec{P}_{i,t}^{G\_\min} \le \vec{P}_{i,t}^{G} \le s_{i,t}^{G}\vec{P}_{i,t}^{G\_\max}, \quad i \in G \tag{3.12a}$$

$$-s_{i,t}^{G}\vec{Q}_{i,t}^{G\ \min} \le \vec{Q}_{i,t}^{G} \le s_{i,t}^{G}\vec{Q}_{i,t}^{G}\max, i \in G \tag{3.12b}$$

$$-\vec{P}_i^{ramp} \le \vec{P}_{i,t+1}^{DG} - \vec{P}_{i,t}^{DG} \le \vec{P}_i^{ramp}, i \in G \tag{3.13}$$

(7) *Capacitor bank limit*: It is assumed that capacitor banks have continuous operations within their maximum limits.

$$0 \le \vec{Q}_{i,t}^{cap} \le \vec{Q}_i^{cap\_\max}, \quad i \in C \tag{3.14}$$

(8) *Voltage regulator modeling*: In this chapter, type-B voltage regulators are chosen, and the relationship between the primary voltage at node i and the secondary voltage at node j can be expressed as below:

$$\vec{U}_{i,t} = \vec{c}^{\odot 2} \odot \vec{U}_{j,t} \tag{3.15}$$

where $\vec{c} \in \mathbb{R}^{3\times 1}$ stands for the vector of voltage ratio, which can be calculated using the method described in [14].

(9) *Power flow constraints*: The branch flow model is applied to provide the closed form formulation for three-phase unbalanced power flow equations in distribution networks [27]. This method was originally proposed in [28], and then the angle relaxation is proposed in [29] to provide a closed form formulation for balanced networks. Afterwards, this model is further developed for unbalanced networks as a second order cone programming model [27, 30]. Moreover, the model can be simplified by neglecting the much smaller quadratic branch loss term to attain a linear model, which has been extensively used in the literature [31, 32].

$$\vec{U}_{i,t} = \vec{U}_{j,t} - 2\left(\tilde{\boldsymbol{r}}_{ji}\vec{P}_{ji,t}^{L} + \tilde{\boldsymbol{x}}_{ji}\vec{Q}_{ji,t}^{L}\right) \tag{3.16a}$$

$$\vec{P}_{i,t}^{TD\_baseline} = \sigma_{i,t}\vec{P}_{i,t}^{D} \tag{3.16b}$$

$$\vec{Q}_{i,t}^{TD\_baseline} = \sigma_{i,t}\vec{Q}_{i,t}^{D} \tag{3.16c}$$

$$\vec{P}_{ji,t}^{L} = \sum_{k\in K(i,\cdot)} \vec{P}_{k,t}^{L} + \vec{P}_{i,t}^{TD} + s_{i,t}^{D}\left(\vec{P}_{i,t}^{D} - \vec{P}_{i,t}^{TD\_baseline}\right) - \vec{P}_{i,t}^{G} \tag{3.16d}$$

$$\vec{Q}_{ji,t}^{L} = \sum_{k\in K(i,\cdot)} \vec{Q}_{k,t}^{L} + \vec{Q}_{i,t}^{TD} + s_{i,t}^{D}\left(\vec{Q}_{i,t}^{D} - \vec{Q}_{i,t}^{TD\_baseline}\right) - \vec{Q}_{i,t}^{G} - \vec{Q}_{i,t}^{cap} \tag{3.16e}$$

$\widetilde{\mathbf{r}}_{ji} + j\widetilde{\mathbf{x}}_{ji}$ is the unbalanced line impedance matrix and can be calculated as (3.17) [27]. Voltage drop is modeled as (3.16a), where node j is the parent node of node i. (3.16b) and (3.16c) are employed to calculate the baseline power of TCL loads, which is Z^{D0} in Fig. 3.3. Active and reactive power balances are modeled by (3.16d) and (3.16e), where $\vec{P}^{D} - \vec{P}^{TD\_baseline}$ and $\vec{Q}^{D} - \vec{Q}^{TD\_baseline}$ are the active and reactive power of non-TCL loads, respectively.

$$\widetilde{\mathbf{r}_{ji}} = Re\{\boldsymbol{\alpha} \odot \mathbf{r}_{ji}\} - \mathrm{Im}\{\boldsymbol{\alpha} \odot \mathbf{x}_{ji}\} \tag{3.17a}$$

$$\widetilde{\mathbf{x}_{ji}} = \mathrm{Re}\{\boldsymbol{\alpha} \odot \mathbf{x}_{ji}\} + \mathrm{Im}\{\boldsymbol{\alpha} \odot \mathbf{r}_{ji}\} \tag{3.17b}$$

$$\boldsymbol{\alpha} = \begin{bmatrix} 1 & e^{-j2\pi/3} & e^{j2\pi/3} \\ e^{j2\pi/3} & 1 & e^{-j2\pi/3} \\ e^{-j2\pi/3} & e^{j2\pi/3} & 1 \end{bmatrix} \tag{3.17c}$$

3.3.2 IGDT-Based Robust DSR Model

This chapter focused on the impacts of loads with time-dependent CLPU on DSR. Thus, only the uncertainty of loads is considered, while DG uncertainty can be included in future work. Envelop-bound models, energy-bound models and Minkowski-norm models are usually employed to describe the uncertain parameters [33], which are the key parts for IGDT modeling. As the absolute value of load demand uncertainty is our focus, the envelop-bound model is utilized to characterize this uncertainty, which can be modeled as follows.

$$B(\mu, b_0) = \{b : b = (1+\mu)b_0\} \tag{3.18}$$

where b is the uncertain parameter, b_0 is the nominal value, and μ is the extent of the fluctuation. The corresponding load uncertainty characterized by (3.18) is given in (3.28a) and (3.28b).

The IGDT-based robust DSR can be formulated as follows. The objective function is to maximize the allowed fluctuations of load demand, while the restored load is guaranteed to be equal to or larger than δR_0. δ is the discount coefficient, within (0, 1] to represent a discounted performance. R_0 is the objective value of the original deterministic restoration model.

$$\max .\mu \tag{3.19}$$

$$\text{s.t. } \mu > 0 \tag{3.20}$$

$$\sum_{t\in T}\sum_{i\in L}\omega_i s_{i,t}^D \vec{P}_{i,t}^D \geq \delta R_0 \tag{3.21}$$

$$\text{Over } (3.8) - (3.16); (3.25) - (3.27); (3.29) - (3.31); \tag{3.22}$$

There are two main differences between the deterministic model described in Sect. III.A and the IGDT-based robust model. First, the objective function of the IGDT-based robust model in (3.19) is to maximize the extent of load uncertainties (i.e.μ). Second, a new constraint (3.21) is introduced in the IGDT-based robust model, which is to ensure that the restored loads are not worse than the discounted value of the original restored loads in the deterministic model. Therefore, the IGDT-based robust model makes a trade-off between the amount of restored loads and the immunity against load uncertainties.

(1) *CLPU uncertainty modeling*: Customers' random behaviors have great impacts on TCL operating states, which are the main causes of TCL uncertainty. For example, some customers keep their TCLs off and choose to turn on their TCLs in the DSR process after re-connecting to the grid, which will cause the CLPU power to increase. Thus, this chapter focuses on the uncertainty modeling of TCLs caused by customers' random behaviors in the DSR process. The IGDT-based robust method is utilized to characterize the uncertainty of loads. Thus, it is the worst case when the loads become larger than their original value. The load uncertainty here refers to the upward fluctuations of loads. Specifically, the internal temperatures of the increased TCLs are kept equal to the ambient temperature before they are turned on. When the increased TCLs are energized as a result of customers' random behaviors, they will show an extended CLPU phenomenon, which can be characterized by DEM [12]. The sampling method is used to calculate the load demand of the new CLPU curve. $\Delta P_i^{TD\_et}$ and $\Delta Q_i^{TD\_et}$ can be calculated as follows, where $C_i^{et}(m)$ is the duration between the sampling step t_2^{et} and mth.

$$\Delta Z^{et}(m) = \begin{cases} Z^{et}(m) - Z^{et}(m-1), & 1 < m \leq N \\ 0, & m = 1 \end{cases} \tag{3.23}$$

And Z_i^{et} is given as:

$$Z_i^{et}(m) = Z_i^{D1\_et}\left(1 - C_i^{et}(m)\right) + h\left(C_i^{et}(m)\right)u\left(C_i^{et}(m)\right) \tag{3.24a}$$

$$h(x) = Z_i^{D0\_et} + \left(Z_i^{D1\_et} - Z_i^{D0\_et}\right)e^{-\alpha_i \cdot x} \tag{3.24b}$$

$$C_i^{et}(m) = (m-1)\Delta t - \left(t_{i,2}^{et} - t_{i,1}^{et}\right) \tag{3.24c}$$

$E_{i,t,r}$ and $F_{i,t,r}$, defined as the active and reactive power of the extended CLPU curve at time step t when the load is restored at time step r, are stated as:

$$E_{i,t,r} = \begin{cases} Z_i^{D1\_et} - \sum\limits_{m=r}^{t} \Delta P_i^{TD\_et}(m), & r < t \\ 0, & r \geq t \end{cases} \tag{3.25a}$$

$$F_{i,t,r} = \begin{cases} Z_i^{D1\_et} - \sum\limits_{m=r}^{t} \Delta Q_i^{TD\_et}(m), & r < t \\ 0, & r \geq t \end{cases} \tag{3.25b}$$

In fact, the shapes of $E_{i,t,r}$ and $F_{i,t,r}$ do not change with r. $E_{i,t}$ and $F_{i,t}$ are the active and reactive power of extended CLPU at time step t. When a load is restored at different time steps $E_{i,t}$ and $F_{i,t}$ will have t values, shown as follows:

$$E_{i,t} = \left[E_{i,t,1}, E_{i,t,2}, \ldots, E_{i,t,r}, \ldots, E_{i,t,t}\right]^T \tag{3.26a}$$

$$F_{i,t} = \left[F_{i,t,1}, F_{i,t,2}, \ldots, F_{i,t,r}, \ldots, F_{i,t,t}\right]^T \tag{3.26b}$$

Once a load is restored, it will remain energized in the following time steps. And there will be uncertainties of loads in the following each time step. Then the accumulative CLPU caused by the increased load is formulated as:

$$P_{i,t}^{TD\_et} = \left[s_{i,1}^D, s_{i,2}^D, \ldots, s_{i,t}^D\right] \cdot E_{i,t} \tag{3.27a}$$

$$Q_{i,t}^{TD\_et} = \left[s_{i,1}^D, s_{i,2}^D, \ldots, s_{i,t}^D\right] \cdot F_{i,t} \tag{3.27b}$$

Note that CLPU modeling is developed in the scalar form, which will be changed to the vector form for the 3-phase model.

(2) *Power balance constraints*: Voltage drop is formulated as (3.16a). And the baseline power of TCLs is the same as those in (3.16b) and (3.16c). The active and reactive power balance is given as follows:

$$\vec{P}_{ji,t}^{L} = \sum_{k\in K(i,\cdot)} \vec{P}_{k,t}^{L} + \vec{P}_{i,t}^{TD} + \mu\vec{P}_{i,t}^{TD\_et} + s_{i,t}^{D}(1+\mu)\left(\vec{P}_{i,t}^{D} - \vec{P}_{i,t}^{TD\_baseline}\right) - \vec{P}_{i,t}^{DG} \quad (3.28a)$$

$$\vec{Q}_{ji,t}^{L} = \sum_{k\in K(i,\cdot)} \vec{Q}_{k,t}^{L} + \vec{Q}_{i,t}^{TD} + \mu\vec{Q}_{i,t}^{TD\_et} + s_{i,t}^{D}(1+\mu)\left(\vec{Q}_{i,t}^{D} - \vec{Q}_{i,t}^{TD\_baseline}\right) - \vec{Q}_{i,t}^{DG} - \vec{Q}_{i,t}^{cap} \quad (3.28b)$$

where $\vec{P}_{i,t}^{TD} + \mu\vec{P}_{i,t}^{TD\_et}$ and $\vec{Q}_{i,t}^{TD} + \mu\vec{Q}_{i,t}^{TD\_et}$ are the active and reactive power of TCLs with uncertainty; $\mu\vec{P}_{i,t}^{TD\_et}$ and $\mu\vec{Q}_{i,t}^{TD\_et}$ are the active and reactive power of the increased TCLs, which are characterized by DEM; $s_{i,t}^{D}(1+\mu)\left(\vec{P}_{i,t}^{D} - \vec{P}_{i,t}^{TD\_baseline}\right)$ and $s_{i,t}^{D}(1+\mu)\left(\vec{Q}_{i,t}^{D} - \vec{Q}_{i,t}^{TD\_baseline}\right)$ are the active and reactive power of the non-TCLs with uncertainty.

It is obvious that constraints (3.28a) and (3.28b) are non-convex, and the big M method is employed to convexify the problem. Additional variable y is utilized to replace $s\,\mu$, which is described as follows:

$$y_{i,t}^{D} = s_{i,t}^{D}\mu \quad (3.29)$$

Then constraints (3.28a) and (3.28b) can be rewritten as:

$$\vec{P}_{ji,t}^{L} = \sum_{k\in K(i,\cdot)} \vec{P}_{k,t}^{L} + \vec{P}_{i,t}^{TD} + \sum_{r=1}^{t} y_{i,r}^{D}\vec{E}_{i,t,r} + s_{i,t}^{D}\vec{P}_{i,t}^{D} + y_{i,t}^{D}\vec{P}_{i,t}^{D} - \left(s_{i,t}^{D} + y_{i,t}^{D}\right)\vec{P}_{i,t}^{TD\_baseline} - \vec{P}_{i,t}^{DG} \quad (3.30a)$$

$$\vec{Q}_{ji,t}^{L} = \sum_{k\in K(i,\cdot)} \vec{Q}_{k,t}^{L} + \vec{Q}_{i,t}^{TD} + \sum_{r=1}^{t} y_{i,r}^{D}\vec{F}_{i,t,r} + s_{i,t}^{D}\vec{Q}_{i,t}^{D} + y_{i,t}^{D}\vec{Q}_{i,t}^{D} - \left(s_{i,t}^{D} + y_{i,t}^{D}\right)\vec{Q}_{i,t}^{TD\_baseline} - \vec{Q}_{i,t}^{DG} - \vec{Q}_{i,t}^{cap} \quad (3.30b)$$

Additional constraints on y are stated as:

$$0 \le y_{i,t}^{D} \le s_{i,t}^{D}M \quad (3.31a)$$

$$0 \le \mu - y_{i,t}^{D} \le \left(1 - s_{i,t}^{D}\right)M \quad (3.31b)$$

The other constraints are the same as those in Sect. 3.3.1. Finally, the IGDT-based robust DSR is modeled as a mixed-integer linear programming problem.

3.4 Numerical Results

The developed robust DSR with time-dependent CLPU is validated by testing on the modified IEEE 13-node and IEEE-123 node test feeders. The CLPU of TCLs is simulated on Matlab R2018b, and DSR simulations are implemented on Python with Gurobi solver. Each time step is 5 min and the number of time steps is set as 10 for DSR. All power values have been normalized based on the base power. To obtain the CLPU models for DSR, the operating behaviors of TCLs under different outage durations are simulated in MATLAB. TCL parameters are from [23] and the ambient temperature is set as 36 °C. Accordingly, CLPU performances at different time steps can be characterized in terms of (3.1a)–(3.1d), whose parameters are extracted from simulation results. For IGDT modeling, δ is 0.8, and R_0 is obtained from the optimization result of the deterministic model (3.2).

3.4.1 The Impact of Time-Dependent CLPU on IEEE 13-Node Test Feeder

(1) *Comparisons with other models*: It is assumed that each load has the same penetration level (40%) of TCLs under normal operating conditions. The base power is 80,000 kVA, and the outage duration is 10 min. Three cases are designed and formulated as deterministic optimization models to analyze the impact of time-dependent CLPU on DSR including, (1) CLPU is not considered in DSR and all loads are assumed to be non-TCLs (as a benchmark); (2) CLPU is considered and characterized by DEM [12], and (3) CLPU is modeled by the proposed time-dependent model. Figure 3.5 shows the modified IEEE 13-node test network, with one 400 kW DG connected at node 680. Furthermore, Table

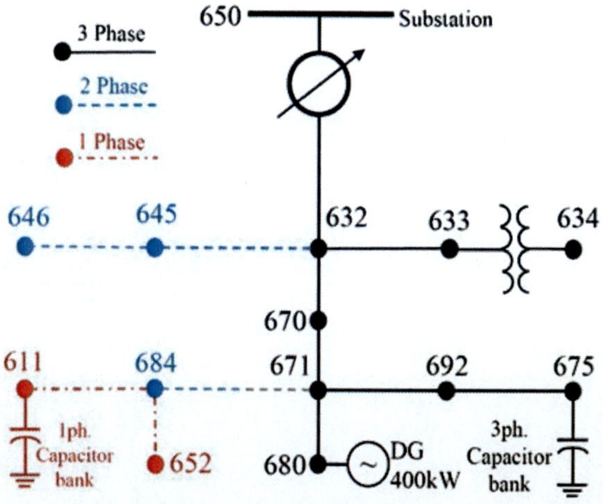

Fig. 3.5 Modified IEEE 13-node test feeders

3.1 summarizes the optimal results of load restoration sequence in different cases.

It is noted that the total generation is the same for all three cases. However, the generation that has been utilized to restore the loads is different in three cases, as different CLPU models are used in each case. The utilized generation (including both from transmission system (TS) and DG) and restored baseload in Cases 1–3 are provided in Fig. 3.6. The differences between generation and the restored base loads are provided in Fig. 3.7, which shows the power fluctuations of TCLs.

The actual power of loads with TCLs is generally larger than their normal value as a result of CLPU during the restoration process. Thus, more loads without TCLs

Table 3.1 Load restoration sequences in different cases

Load	Location (node)	Time Step (Case 1)	Time Step (Case 2)	Time Step (Case 3)
634a	634	1	1	1
634b	634	7	4	7
634c	634	9	2	7
645	645	2	6	9
646	646	1	1	1
670a	670	1	7	3
670b	670	6	5	4
670c	670	2	9	9
671	671	5	10	6
692	692	6	1	3
675a	675	8	6	10
675b	675	6	3	2
675c	675	1	7	10
611	611	7	5	1
652	652	1	8	2

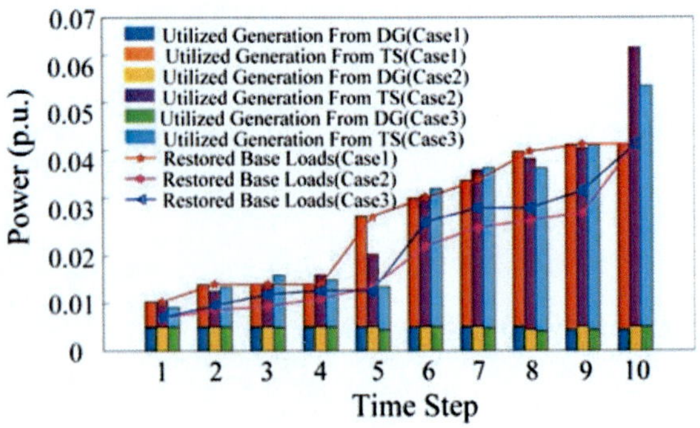

Fig. 3.6 Utilized generation and restored baseload in Cases 1–3

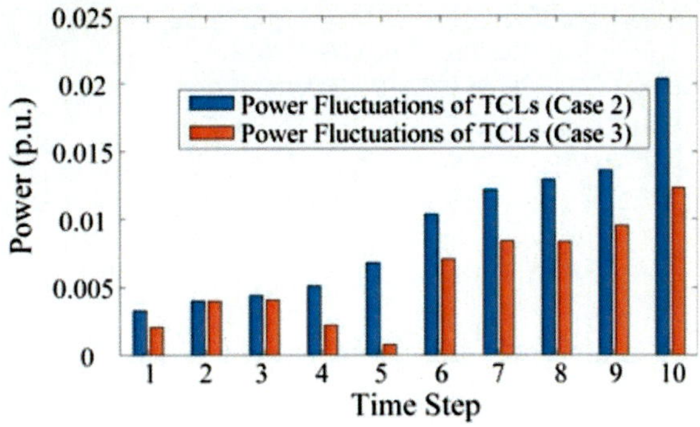

Fig. 3.7 Power fluctuations of TCLs in Cases 2 and 3

will be picked up compared to those with TCLs when the available generation in the distribution system is limited. As shown in Fig. 3.6, the restored base loads in Case 1 (0.277) are more than those in Case 3 (0.218). Even though, the required generation in Case 3 is occasionally equal to (at time step 9) or more than (at time steps 3, 6 and 7) that in Case 1 as a result of CLPU, as shown in Fig. 3.6. Moreover, it can be seen that the restored loads in Case 3 remain unchanged from time steps 4 to 5 (see Fig. 3.6) while the required generation becomes less (see Fig. 3.7). This is because the actual power of loads with TCLs begins to decrease after it reaches the peak upon re-energization, as shown in Fig. 3.1. It is shown that as the critical characteristics of loads with TCLs, CLPU plays an important role in distribution system restoration. If the loads with TCLs are picked up without considering CLPU, the restored loads will be much more than the total dispatchable generation. This may result in system security violations or load shedding. Compared to the result of Case 3, fewer loads in Case 2 (0.197) are restored. This is because CLPU is characterized by DEM in Case 2, in which the load power keeps at a high level for a longer period and more generation is required. For example, at time step 1 of Fig. 3.6, the restored loads in Case 2 are almost equal to those in Case 3, while the generation in Case 2 is more than that in Case 3, as shown in Fig. 3.7. Accordingly, there will be fewer loads to be restored using the DEM of CLPU. This will result in the generation in distribution systems not being fully employed and slowing down the restoration process.

Moreover, the power profiles of loads at nodes 634a, 645 and 671 during the DSR process are presented in Fig. 3.8 to show the time-dependent behaviors of CLPU. The load at node 634a is restored at the first time step. Its power profile goes up and then goes down, which is consistent with the CLPU profile under the short outage duration (e.g. 15 min) in Fig. 3.1. Similarly, the loads at nodes 671 and 645 are picked up at the sixth and ninth-time steps, respectively. Their profiles tend to keep at their peak and then go down, which have similar shapes to those in Fig. 3.1 when the outage duration is 60 or more minutes. These simulation results validate the time-dependent behaviors of CLPU during the DSR process.

In summary, the restoration plans greatly depend on the actual load profiles. CLPU is inevitable for loads, especially for those with high penetration of TCLs. When CLPU is not properly modeled, it will result in a series of problems, e.g.

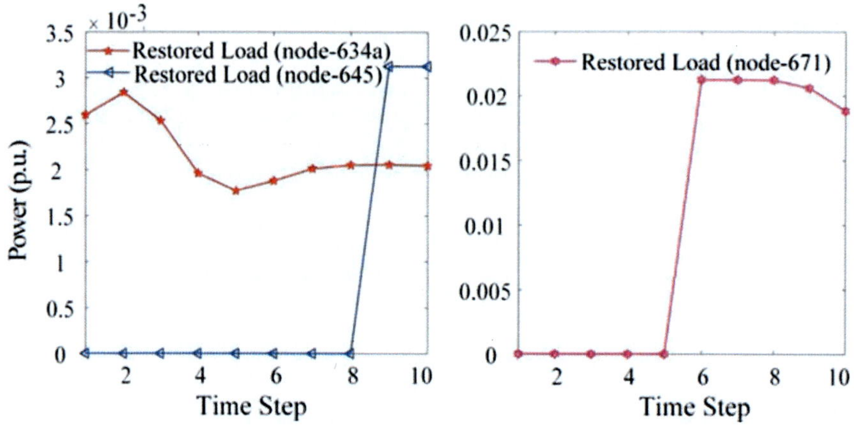

Fig. 3.8 Power profiles of the loads at different nodes (634a is phase A of the node 634)

load shedding, voltage and power flow violations, and prolonged restoration process. Thus, time-dependent CLPU should be accurately embedded in the restoration model.

(2) *The impact of outage duration*: The total restored loads (i.e. the objective value in (3.2)) under different outage durations are presented in Fig. 3.9. It shows that the restored loads become less with a longer outage duration. As the outage duration becomes longer, the CLPU profile tends to gradually approach its peak and keeps at the highest level for a longer time, as shown in Fig. 3.1. Thus, when the outage duration is long enough, the CLPU profile will not drastically change to cause any significant impacts on restoration decisions. The proposed time-dependent CLPU model will greatly affect DSR when the outage duration is short, especially considering different efforts on resilience enhancement and self-healing. This is the main application scenario of the proposed model.

(3) *The impact of TCL penetration level*: The total restored loads (i.e. the objective value in (3.2)) with different penetration levels of TCLs are provided in Fig. 3.10. When the TCL penetration level is higher, the actual power of loads

Fig. 3.9 Restored load under different outage duration

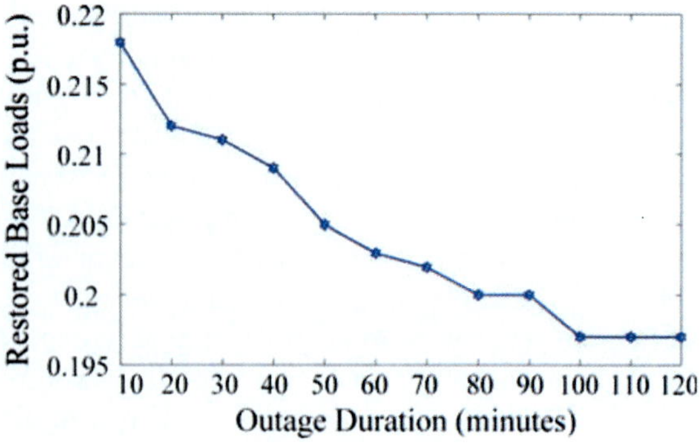

Fig. 3.10 Restored base loads with different TCL

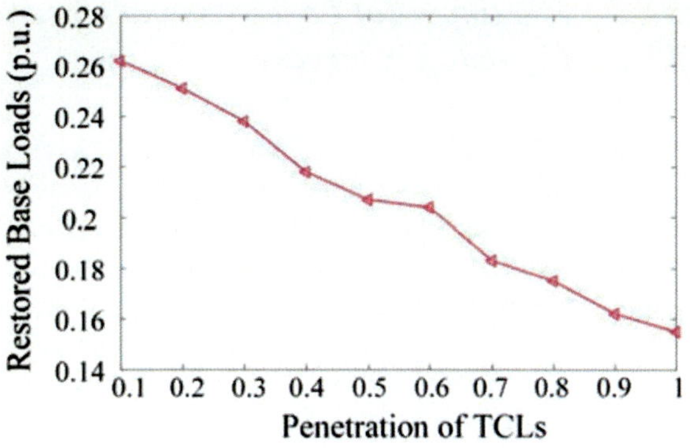

will increase, which requires more generation to restore. Thus, the restored base loads become less as the penetration level of TCLs is higher. It means that the restoration process will be slow when there are too many TCLs connected to distribution systems. To reduce the impacts of CLPU on the restoration process, some control strategies can be applied to regulate TCLs for re-shaping CLPU profiles. For example, authors in [4] propose TCL control methods to alleviate the CLPU phenomenon after an outage or TCL temperature deviations, which could be helpful for restoration decision-making. Regulating TCLs to improve the resilience of distribution systems will be considered in future work.

(4) *The impact of ambient temperature*: To show the impacts of ambient temperature on the restoration process, the restored loads at each time step are compared in Fig. 3.11, when the ambient temperature is 30 °C, 34 °C and 38 °C, respectively. The corresponding total restored loads are 0.258, 0.231 and 0.213, respectively. It can be seen that there are fewer loads to be restored when the ambient temperature becomes higher. This is because TCLs with a higher ambient temperature are easier to deviate further from their temperature-set points after an outage. Then, the CLPU power tends to be larger, which will result in fewer loads being picked up in DSR.

Fig. 3.11 Comparisons of restored base loads under different ambient temperatures

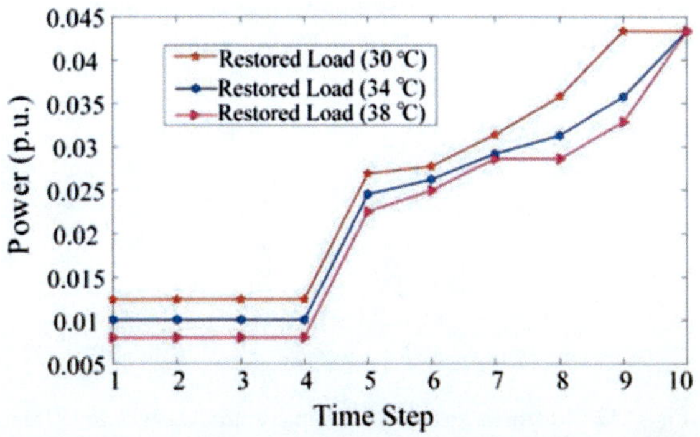

3.4.2 The Impact of Load Demand Uncertainty on IEEE 13-Node Test Feeder

As the time-dependent characteristics of CLPU have been analyzed, this section will validate the un-certainty modeling of CLPU. Loads are composed of non-TCLs and TCLs, and the uncertainty modeling of non-TCL loads has been extensively studied in the literature (e.g. [1]). Thus, this section focuses on validating the uncertainty modeling of TCLs which is rarely studied yet. It is noted that the uncertainty caused by the increased TCLs should be characterized by DEM, which has been explained in Sect. 3.3.2. Two cases are designed for comparisons to analyze the load demand uncertainty including, (A) only the uncertainty of non-TCLs is considered, but not for TCLs; (B) the uncertainties of both non-TCLs and TCLs are considered, and CLPU uncertainty caused by newly increased TCLs is described by DEM.

The objective value μ in cases A and B is 31.3% and 14.1%, respectively. It is the maximum allowed fluctuation extent of loads in the two cases, while the restored loads are not less than 80% of those in the deterministic restoration model. The corresponding restored loads and the objective value μ in cases A and B are provided in Fig. 3.12. It means that when the loads fluctuate within $[b_0, (1 + \mu)\, b_0]$, the restored loads will be equal to or larger than 80% of 0.218. b_0 is the nominal value of the loads, and 0.218 is the total restored loads of the deterministic restoration model. In Case A, as the uncertainty of TCLs is not considered, no extra generation is required to balance the CLPU caused by the newly increased TCLs. Thus, the maximum allowed load fluctuation in Case A is larger than that in Case B.

To check the system performance under load uncertainties, all loads are increased to be $(1 + \mu)b_0$. Loads are restored following the optimization results of Case A and Case B, in which the corresponding μ is 31.3% and 14.1%, respectively. Simulation results show that the voltage of some nodes in Case A tends to violate their lower bounds. For example, the voltage profile of node 692 (in Case A) is given in Fig. 3.13,

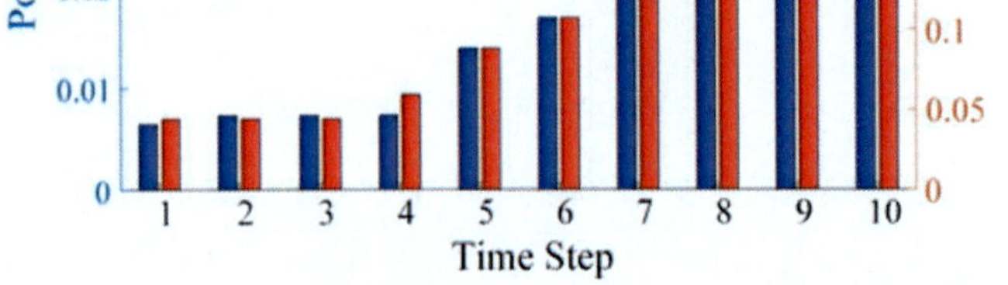

Fig. 3.12 Restored base load and maximum allowed load fluctuation extent in Cases A–B

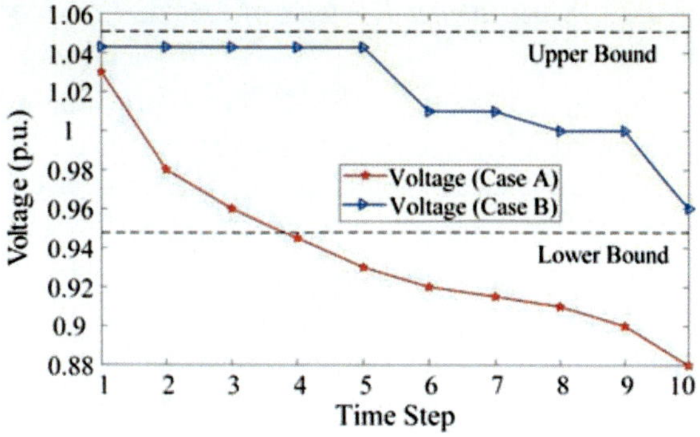

Fig. 3.13 Voltage profiles of node 692 under different scenarios

which shows that the voltage is violating its lower bound from the time step 4. This is because that the maximum allowed fluctuation extent is improperly estimated when the uncertainty of CLPU is ignored in Case A. The actual maximum allowed fluctuation extent is much lower than 31.3%. In contrast, there are no voltage violations when the CLPU uncertainty is properly modeled in Case B. As shown in Fig. 3.13, the voltage profile of node 692 (in Case B) is maintained within its upper and lower bounds. It means that no less than 80% of the restored loads in the deterministic DRO model can be picked up when all loads fluctuate within $[b_0, 1.141b_0]$. Thus, the CLPU uncertainty should be properly embedded in DSR model; otherwise, it may result in security problems, e.g. voltage violation or even load shedding.

The relationship between δ and μ is given in Fig. 3.14. It shows that μ becomes smaller with larger δ, which means that the maximum allowed load fluctuation is smaller when the minimum value of restored loads becomes bigger. Compared to the range of [0.1, 0.4] for δ, μ is less sensitive to δ when δ is within the range of [0.5, 0.9]. It demonstrates that the restored loads will significantly decrease even though the loads slightly fluctuate around the estimated value. Thus, it is important to accurately forecast loads for reliable and efficient restoration.

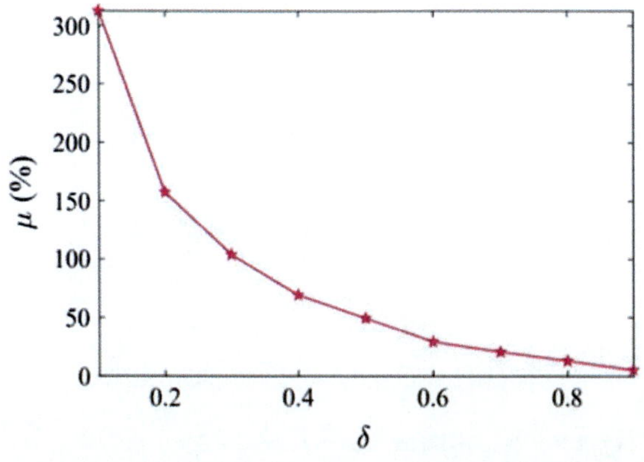

Fig. 3.14 Relationship between δ and μ

3.4.3 Simulation Results of IEEE 123-Node Test Feeder

To further validate the scalability of the proposed models, the three cases defined in Sect. 3.4.1, and the two cases defined in Sect. 3.4.2 are also applied to the IEEE 123-node test feeder. Simulation results are provided in Figs. 3.15 and 3.16. The base power is 5000 kVA.

(1) *The impact of time-dependent CLPU*: In Fig. 3.15, it shows that DSR with the time-dependent CLPU model in Case 3 picks up fewer base loads compared to that without CLPU in Case 1, but restores more base loads compared to that with DEM in Case 2. The value of the objective function in Cases 1–3 is 5.43, 4.11 and 4.77, respectively. Simulation results are consistent with those in IEEE 13-node test feeder. It validates the effectiveness of the proposed DSR with time-dependent CLPU.

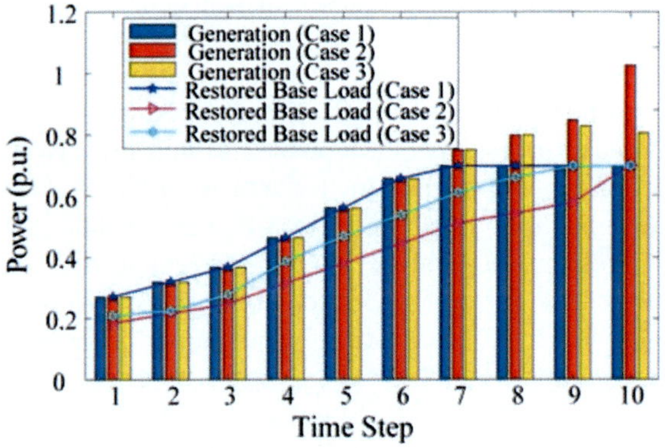

Fig. 3.15 Utilized generation and restored baseload in Cases 1–3

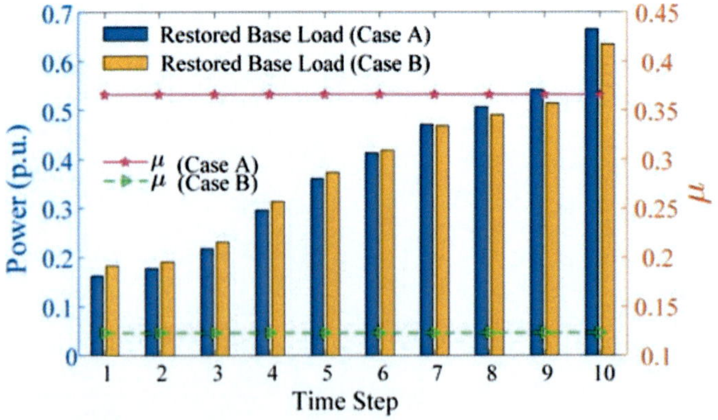

Fig. 3.16 Restored base load in Cases A–B

Table 3.2 DSR optimization model complexity

	Number of constraints	Number of continuous variables	Number of binary variables	Number of non-zero elements	Number of simplex iterations	Execution time (s)	Optimality gap
No CLPU (case 1)	3017	1880	847	11,266	2230	0.3	0.0059
DEM (case 2)	3202	2143	858	12,152	12,485	1.4	0.0048
Time-dependent model (case 3)	4141	3335	910	17,541	9661	2.0	0.0059

(2) *The impact of load demand uncertainty*: In Fig. 3.16, μ in Cases A–B is 36.6% and 12.3%, respectively. It means that the allowed fluctuations of the proposed method in Case B are less than those in Cases A, when a certain amount of loads are guaranteed to be restored. This is also consistent with the simulation results in Sect. 3.4.2. It shows that the proposed robust DSR with time-dependent CLPU is scalable in large-scale test feeders.

(3) *DSR optimization problem complexity*: The optimization models for Cases 1–3 have been solved using Gurobi solver, in which the branch-and-bound method has been used. Table 3.2 provides a complexity assessment of the mathematical modeling and solution procedure for various case studies in IEEE 123-node test network. It is clear that adding the CLPU phenomenon into the DSR problem formulation with both DEM and time-dependent models causes the expansion of the optimization model, including the number of constraints, variables, and nonzero elements. However, compared to the network scale, this evolution is insignificant as a negligible increase in the number of binary variables. Furthermore, considering the expansion of the optimization model, the solution procedure requires more simplex iterations and exploring further branch-and-bound nodes, while both CLPU models reach an acceptable optimality gap in less than 2 s. This execution time could be acceptable considering the scalability of the test feeder and the detailed modeling of the CLPU effect on the DSR problem. Moreover, owing to the higher modeling accuracy, the DSR problem with the time-dependent CLPU model has more constraints, continuous variables, binary variables and non-zero elements compared to that with DEM, so the corresponding execution time is longer.

3.5 Conclusions

As an inevitable phenomenon of loads after an outage, the CLPU of TCLs depends on outage duration. This chapter analyzes the evolution process of CLPU under different outage durations and develops the DSR model with time-dependent CLPU. Moreover, the uncertainty of loads is fully considered in IGDT-based robust DSR.

Therefore, loads can be restored in a fast and secure way by accurately modeling the actual operating behaviors of loads. Simulation results on IEEE 13-node and 123-node test feeders show that the modeling approach of CLPU has a great impact on DSR, and the actual performances of CLPU (including time-dependent behaviors and uncertainty) are required for the accurate characterization. Also, the impacts of outage duration and TCL penetration level on DSR are analyzed. In future work, the control strategies of TCLs under emergency or restorative conditions will be developed to alleviate the impact of CLPU on system recovery and improve the resilience of distribution systems.

References

1. Meng, S., & Sun, W. (2021). Robust distribution system load restoration with time-dependent cold load pickup. *IEEE Transactions on Power Systems, 36*(4), 3204–3215.
2. Schneider, K. P., Sortomme, E., Venkata, S., Miller, M. T., & Ponder, L. (2015). Evaluating the magnitude and duration of cold load pick-up on resi-dential distribution using multi-state load models. *IEEE Transactions on Power Systems, 31*(5), 3765–3774.
3. Song, M., Gao, C., Shahidehpour, M., Li, Z., Yang, J., & Yan, H. (2018). Impact of uncertain parameters on tcl power capacity calculation via hdmr for generating power pulses. *IEEE Transactions on Smart Grid, 10*(3), 3112–3124.
4. Sinitsyn, N. A., Kundu, S., & Backhaus, S. (2013). Safe protocols for generat-ing power pulses with heterogeneous populations of thermostatically controlled loads. *Energy Conversion and Management, 67*, 297–308.
5. Bu, F., Dehghanpour, K., Wang, Z., & Yuan, Y. (2019). A data-driven frame-work for assessing cold load pick-up demand in service restoration. *IEEE Transactions on Power Systems, 34*(6), 4739–4750.
6. Mirza, O. (1997). Usage of clpu curve to deal with the cold load pickup problem. *IEEE transactions on power delivery, 12*(2), 660–667.
7. Ihara, S., & Schweppe, F. C. (1981). Physically based modeling of cold load pickup. *IEEE Transactions on Power Apparatus and Systems, 9*, 4142–4150.
8. Uc¸ak, C. U., & Pahwa, A. (1995). Optimal step-by-step restoration of distribution systems during excessive loads due to cold load pickup. *Electric Power Systems Research, 32*(2), 121–128.
9. Wang, Y., Yao, M., Chen, L., & Li, X. (2018). Islanding strategy for restoring electric power supply by means of electric vehicle idle power against cold load pickup. *IET Generation, Transmission & Distribution, 13*(5), 613–625.
10. Kumar, V., HC, R. K., Gupta, I., & Gupta, H.O. (2009). Dg integrated approach for service restoration under cold load pickup. *IEEE Transactions on Power Delivery, 25*(1), 398–406.
11. Arif, A., Ma, S., Wang, Z., Wang, J., Ryan, S. M., & Chen, C. (2018). Optimizing service restoration in distribution systems with uncertain repair time and demand. *IEEE Transactions on Power Systems, 33*(6), 6828–6838.
12. Chen, B., Chen, C., Wang, J., & Butler-Purry, K. L. (2017). Multi-time step service restoration for advanced distribution systems and microgrids. *IEEE Transactions on Smart Grid, 9*(6), 6793–6805.
13. Tsai, M.-S. (2008). Development of an object-oriented service restoration expert system with load variations. *IEEE Transactions on Power Systems, 23*(1), 219–225.
14. Kersting, W. H. (2006). *Distribution system modeling and analysis*. CRC Press.
15. Xu, X., Yan, Z., Shahidehpour, M., Wang, H., & Chen, S. (2017). Power system voltage stability evaluation considering renewable energy with correlated variabilities. *IEEE Transactions on Power Systems, 33*(3), 3236–3245.

16. Kavousi-Fard, A., & Niknam, T. (2013). Optimal distribution feeder recon-figuration for reliability improvement considering uncertainty. *IEEE Transactions on Power Delivery, 29*(3), 1344–1353.
17. Popovic, D. S., & Popovic, Z. N. (2004). A risk management procedure for supply restoration in distribution networks. *IEEE Transactions on Power Systems, 19*(1), 221–228.
18. Cao, X., Wang, J., & Zeng, B. (2017). A chance constrained information-gap decision model for multi-period microgrid planning. *IEEE Transactions on Power Systems, 33*(3), 2684–2695.
19. Aghaei, J., Agelidis, V. G., Charwand, M., Raeisi, F., Ahmadi, A., Nezhad, A. E., & Heidari, A. (2016). Optimal robust unit commitment of chp plants in electricity markets using information gap decision theory. *IEEE Transactions on Smart Grid, 8*(5), 2296–2304.
20. Zhao, J., Wang, H., Hou, Y., Wu, Q., Hatziargyriou, N. D., Zhang, W., & Liu, Y. (2020). Robust distributed coordination of parallel restored subsystems in wind power penetrated transmission system. *IEEE Transactions on Power Systems, 35*(4), 3213–3223.
21. Xu, X., Yan, Z., Shahidehpour, M., Li, Z., Yan, M., & Kong, X. (2019). Data-driven risk-averse two-stage optimal stochastic scheduling of energy and reserve with correlated wind power. *IEEE Transactions on Sustainable Energy, 11*(1), 436–447.
22. Xie, Y., Chen, X., Wu, Q., & Zhou, Q. (2019). Second-order conic programming model for load restoration considering uncertainty of load increment based on information gap decision theory. *International Journal of Electrical Power & Energy Systems, 105*, 151–158.
23. Mathieu, J., Kamgarpour, M., Lygeros, J., Andersson, G., & Callaway, D. (2015). Arbitraging intraday wholesale energy market prices with aggregations of thermostatic loads. *IEEE Transactions on Power Systems, 30*(2), 763–772.
24. Lu, N. (2012). An evaluation of the hvac load potential for providing load balancing service. *IEEE Transactions on Smart Grid, 3*(3), 1263–1270.
25. Song, M., Gao, C., Shahidehpour, M., Li, Z., Yang, J., & Yan, H. (2018). State space modeling and control of aggregated tcls for regulation services in power grids. *IEEE Transactions on Smart Grid, 10*(4), 4095–4106.
26. Song, M., Sun, W., Wang, Y., Shahidehpour, M., Li, Z., & Gao, C. (2020). Hierarchical scheduling of aggregated tcl flexibility for transactive energy in power systems. *IEEE Transactions on Smart Grid, 11*(3), 2452–2463.
27. Nejad, R. R., & Sun, W. (2019). Distributed load restoration in unbal-anced active distribution systems. *IEEE Transactions on Smart Grid, 10*(5), 5759–5769.
28. Baran, M. E., & Wu, F. F. (1989). Optimal capacitor placement on radial distribution systems. *IEEE Transactions on Power Delivery, 4*(1), 725–734.
29. Farivar, M., & Low, S. H. (2013). Branch flow model: Relaxations and convexification—Part i. *IEEE Transactions on Power Systems, 28*(3), 2554–2564.
30. Low, S. H. (2014). Convex relaxation of optimal power flow—Part i: For-mulations and equivalence. *IEEE Transactions on Control of Network Systems, 1*(1), 15–27.
31. Robbins, B. A., & Domínguez-García, A. D. (2016). Optimal reactive power dispatch for voltage regulation in unbalanced distribution systems. IEEE Transactions on Power Systems, 31(4):2903–2913.
32. Ding, T., Lin, Y., Li, G., & Bie, Z. (2017). A new model for resilient distribution systems by microgrids formation. *IEEE Transactions on Power Systems, 32*(5), 4145–4147.
33. Ben-Haim, Y. (2006). *Info-gap decision theory: Decisions under severe uncer-tainty*. Elsevier.

Chapter 4
Aggregated Control of TCLs Based on Modified State Space Model

Abstract TCLs which possess thermal inertia have attracted additional attention in providing ancillary services to power systems. In this regard, the aggregated model for TCLs is critical for evaluating the performance and optimizing e the control of a population of TCLs. We learn by theoretical analyses and simulations that in most cases, the accuracy of the state-space model is deteriorated by applying longer TCL dispatch periods. Thus, a modified aggregated TCL model is proposed and a new state transition matrix is derived in this paper to improve the TCL modeling accuracy for the provision of grid regulation services. The proposed control model for the aggregated TCL takes into account compressor time delays, which will enhance the collective behavior of TCL. A probabilistic control strategy is presented to reduce the adverse impacts on the durability of TCL compressors and communication requirements for implementing the TCL control signal. Simulation results demonstrate that the proposed aggregated TCL model with control signals can closely follow the actual TCL behavior and the proposed control strategy can reduce the number of TCL switching for providing ancillary services to power systems.

Keywords Aggregated model · State space model · Regulation services

4.1 Introduction

Regulation is an indispensable ancillary service that is deployed to balance short-term power fluctuations (up to one minute) [1–7], which can be achieved by TCL controls. To effectively manage TCLs for regulation services, an aggregated control model is developed in [8] to track the dynamic behavior of the TCL in 15 s intervals. In [9], a 50 s interval is chosen to simulate and compare the accuracy of non-uniform and uniform bin models for fast regulation services. Model predictive control strategies of TCLs, which provides regulation and balance services, are formulated to track the 1-minute fluctuations in solar power [10]. Additionally, a distributed model predictive control strategy of TCLs is proposed to provide regulation services with a 40 s prediction horizon [11]. TCLs are capable of changing their power consumption through certain control strategies to follow regulation signals. Thus, the accurate characterization of

M. Song and C. Gao, *Integration of Distributed Resources in Smart Grids for Demand Response and Transactive Energy*, https://doi.org/10.1007/978-981-16-7170-8_4

actual operating behaviors of TCLs in seconds to one-minute intervals is the key control issue for TCLs in regulation services.

The large number of TCLs and the small capacity of each TCL means that certain aggregation techniques are needed to characterize the collective performance of TCLs to implement control strategies. One category considerable research has been focused on aggregated TCL models, which are generally classified into two categories according to their applications.is to model TCLs as batteries for the grid integration [7, 12–18]. These battery models of TCLs can describe the relationship between the energy storage level which corresponds to indoor temperatures and power adjustments based on the steady-state power baseline of TCLs. Thus, they are mainly applied to dispatch optimizations in power systems and are not suitable to provide rich information on TCL operating states for control. The second category of aggregated TCL models describes the temperature and switching evolution of the TCL population, which is characterized by TCL transition probabilities in different state bins [8, 9, 19–22]. Detailed derivations of transition matrices are presented based on state-space models and then aggregated models are developed to analyze on/off TCL controls for regulation services [8, 9, 21].

A 2-D state bin is presented in [19] to model the TCL dynamic performance. Reference [20] considers uncertainties caused by building structures, ambient temperatures and customer behaviors in the state queuing model. A modification method of the state transition matrix is also provided and Kalman filtering is utilized to estimate both states and joint parameters of the state-space model of TCLs [22]. State-space models of aggregated TCLs perform well at capturing detailed operating performances of TCLs with control signals. Accordingly, state-space models are mainly applied in the real-time control of TCLs for providing ancillary services in short intervals, which is the focus of this chapter. Although state-space models have been massively studied, the corresponding modeling accuracy with different dispatch periods is rarely studied and the methods to improve the modeling accuracy are seldom documented in the literature. Meanwhile, the time delay of TCL compressors is often ignored in the state-space model of TCLs [19, 20, 22], which may result in large discrepancies between the aggregated model and the actual operating behavior of TCLs.

Based on the aggregated TCL model, control signals are generated to guide the collective TCL power to track regulation signals. The number of TCL switching should be examined as a control strategy to reduce adverse impacts on the durability of TCL compressors [6, 14]. References [6, 7, 14] proposed a priority-stack control method that takes indoor TCL temperatures into account and arranges TCLs with different priorities to reduce the number of TCL switching. However, the approach introduces heavy communication requirements between the central controller and TCLs to distribute signals among individual heterogeneous TCLs [9] which could result in time delays to respond to TCL control signals. To reduce communication burdens for control signals, Ref. [8, 9, 23] proposed a probability control method in which the central controller sends a global control signal to all TCLs to decide whether to change switching states using the same probability. The method is helpful to relieve communication burdens for the provision of control signals. However, the number of TCL switching is yet to be reduced. Thus, the existing problem is to reduce the number of TCL switching and communication requirements for implementing the control strategy.

In this chapter, the aggregated TCL model and its control strategy are studied for regulation services in a power grid. Reference [24] analyzes the accuracy of the state-space model of aggregated TCLs using different dispatch periods and proposes a modification method to improve the modeling accuracy. The critical TCL issue in power system operations is to control TCLs to follow the desired power profile, while the aggregated model in [24] cannot be utilized directly to regulate TCLs for this purpose. Thus, we extend our earlier work and propose an aggregated model for control that would follow the dynamic behavior of TCLs. The corresponding control strategy is also presented.

The main work of this chapter includes (1) A modified state-space model of aggregated TCLs is proposed to improve the TCL modeling accuracy for grid regulation services. The analytical derivations of the state transition matrix are also presented; (2) A control strategy, in which TCLs receive the same control signal and respond based on their local indoor temperatures and switch states, is presented to reduce the adverse impacts of control strategy on the durability of TCL compressors and lower communication requirements between the central controller and TCLs; (3) The control strategy is embedded in the aggregated TCL model. The corresponding aggregated control model of TCLs is proposed to track the collective TCL performance considering compressor time delays when control signals are applied.

4.2 Control Frameworks for TCLs

In general, two main control methods are available, including central control and decentralized control [25, 26]. Central control allows for easy implementation with good predictability and control, while having drawbacks in terms of communication requirements and protection of customer privacy [26, 27]. By contrast, decentral control is adept at protecting the customer's privacy and maintaining a fast response while it may not obtain a globally optimal solution because of its limited information on system-level resources [28, 29]. Thus, the two control methods fulfill a set of common objectives but share conflicting features [25]. A combination of central and decentralized control has been used in some studies for better performance [30, 31], which can also be used to regulate TCL populations.

Many residential TCLs, each with a small capacity, are unable to participate in the electricity market for ancillary services. To address this problem, TCLs in a community that usually shares similar parameters are managed by an aggregator [32]. The corresponding aggregated model will represent TCLs to reap the benefits of DR programs. Accordingly, the central control with its controllability and predictability features [26] is applied at the aggregator level to manage TCLs. The central control allows TCLs to follow the desired power profile closely as compared to decentral control [25]. When the TCL population is controlled at the power system level, participating aggregators are managed by decentral control for better performances.

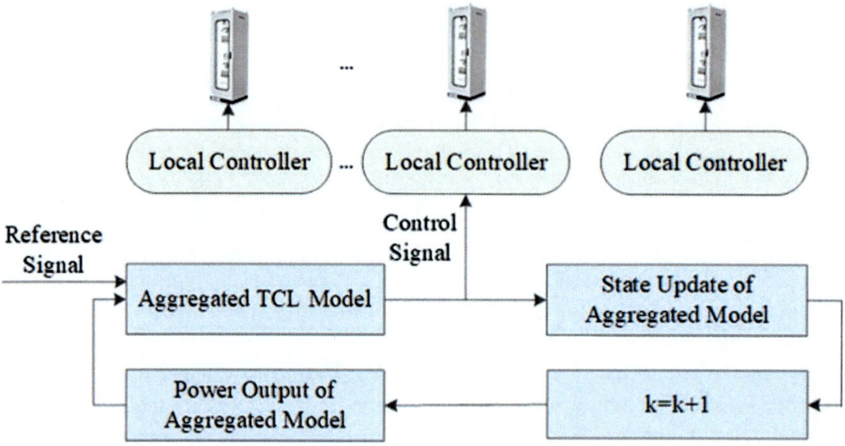

Fig. 4.1 Control framework of TCLs

Considering the control of multiple aggregators is not our focus, the TCL control at the level of the aggregator is discussed in this chapter. Meanwhile, the control proposed strategy is practical, which can reduce the communication requirements of the central control.

The control framework of TCLs managed by an aggregator is given in Fig. 4.1. When regulation signals are followed by an aggregator, control signals will be generated according to the aggregated TCL model and then distributed to individual TCLs. The corresponding control signals will update the TCL operating states in the aggregated model. Then the power output of aggregated TCL model will be compared with the reference signal to generate the next control signal. The aggregated TCL model will be updated periodically with actual TCL measurements for developing an accurate TCL behavior [6, 9].

4.3 Aggregated TCL Model Without Control Signals

4.3.1 Thermodynamic Model of TCL System

The two-state hybrid model of a TCL is described as [33]:

$$C_a \frac{dT_i}{dt} = \frac{(T_o - T_i)}{R} - S(t) \cdot \eta \cdot P \tag{4.1}$$

There are only two TCL operating states, i.e., rated power P and 0, which place indoor temperature changes within $[T_{\min}, T_{\max}]$ ($T_{\min} = T_{\text{set}} - \delta$, $T_{\max} = T_{\text{set}} + \delta$).

From (4.1), t_{on} and t_{off} are calculated as:

$$t_{on} = -RC_a \ln \frac{T_o - \eta PR - T_{\min}}{T_o - \eta PR - T_{\max}} \tag{4.2a}$$

$$t_{off} = -RC_a \ln \frac{T_o - T_{\max}}{T_o - T_{\min}} \tag{4.2b}$$

Accordingly, the P_{baseline} of the TCL population is given as [34]:

$$P_{baseline} = \sum_{i=1}^{N} P_i \frac{t_{on}^i}{t_{on}^i + t_{off}^i} \tag{4.3}$$

4.3.2 Aggregation and Control Problem Formulation

Using the aggregated TCL model, the TCL operating states are divided into several state bins according to their switch state and indoor temperature [8, 21, 35]. Each TCL state bin is represented by indoor temperature and switch status. The temperature interval is the same in each state bin. The number of state bins where the switch state is off is equal to that of on. The state bins are denoted by 1, 2… N, $N + 1$,…, $2N$, as shown in Fig. 4.2. When there are no TCL control signals, the TCL state transition is described as [36]:

$$\frac{d\boldsymbol{x}(t)}{dt} = \boldsymbol{G} \cdot \boldsymbol{x}(t) \tag{4.4}$$

where $\boldsymbol{x}(t)$ is expressed by $[n_1(\text{t})\ n_2(\text{t})\ldots n_N\ (\text{t})\ n_{N+1}(\text{t})\ldots n_{2\text{N}}\ (t)]^T$. $n_i(\text{t})$ is the number of TCLs in the state bin i at time t. $\boldsymbol{G}$ is a $2N \times 2N$ matrix.

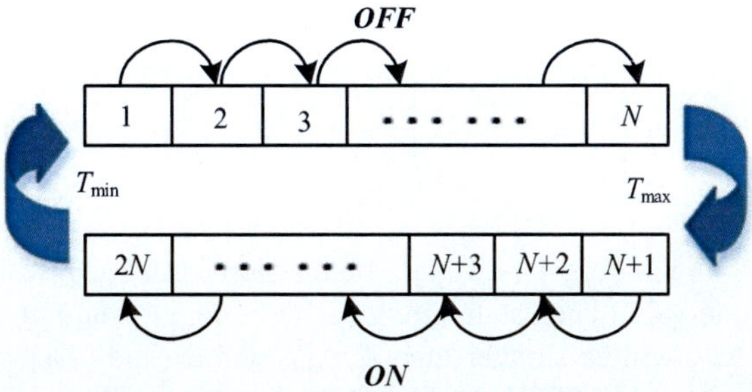

Fig. 4.2 State bins of the aggregated TCLs

TCLs in a state bin will be transferred to the next state bin or stay in the same state bin. We denote (d, q) in matrix G, by g_{dq}, which is the speed at which TCLs in the state bin q are transferred to the state bin d [9]. Thus, G is a large sparse matrix and g_{dq} is approximately given as [8, 9]:

$$g_{dq} = \begin{cases} -\frac{1}{\Delta t_{dq}}, & if\ d = q \\ \frac{1}{\Delta t_{dq}}, & if\ (d = q + 1\&1 \le q \le 2N - 1)\ or\ (d = 1\&q = 2N) \\ 0, & otherwise \end{cases} \tag{4.5}$$

Here, Δt_{dq} depends on state bins d and q, which is stated as:

$$\Delta t_{dq} = \begin{cases} -RC_a ln\frac{T_o - \eta PR - T_d^m}{T_o - \eta PR - T_q^m}, & if\ S(t) = 1 \\ -RC_a ln\frac{T_o - T_d^m}{T_o - T_q^m}, & if\ S(t) = 0 \end{cases} \tag{4.6}$$

The TCL transition speed within a temperature range $[T_{\min}, T_{\max}]$ is closely related to the indoor temperature. When (4.4) is utilized to describe the aggregated TCL performance, the TCL transition speed between state bins p and d is replaced by an average speed g_{dq} in (4.5), which introduces additional modeling errors. In such a case, more state bins (i.e. shorter Δt_{dq}) are helpful to allow the average speed to follow the transient speed more closely [8, 22].

Generally, regulation services are deployed in seconds (up to 1 min) to compensate for short-term power fluctuations [7]. Accordingly, we consider discrete dispatch period Δt [10, 11], in which (4.4) is stated as

$$\boldsymbol{x}(k+1) = \boldsymbol{A} \cdot \boldsymbol{x}(k) \tag{4.7a}$$

$$\boldsymbol{A} = 1 + \boldsymbol{G} \cdot \Delta t \tag{4.7b}$$

where the state vector $x(k)$ includes the number of TCLs in a state bin in kth dispatch period. $\boldsymbol{A}$ is a $2N \times 2N$ matrix whose elements represent the TCL transition probability from a state bin [9]. Therefore, each A element, denoted by a_{dq}, is the TCL transition probability from the state bin q to d and $0 \le a_{dq} \le 1$, which according to (4.5)–(4.7) is stated as

$$a_{dq} = \begin{cases} 1 - \frac{\Delta t}{\Delta t_{dq}}, & if\ d = q\&1 \le q \le 2N \\ \frac{\Delta t}{\Delta t_{dq}}, & if\ (d = q + 1\&1 \le q \le 2N - 1)\ or\ (d = 1\&q = 2N) \\ 0, & others \end{cases} \tag{4.8}$$

According to (4.8), Δt_{dq} will be longer than Δt, which means that the temperature interval of each state bin needs to be long enough to accommodate Δt. Otherwise, Δt_{dq} will be smaller than Δt, a_{dq} will exceed [0,1], and (4.7) will fail to track the aggregated TCL behavior. Hence, more state bins can improve the accuracy of the

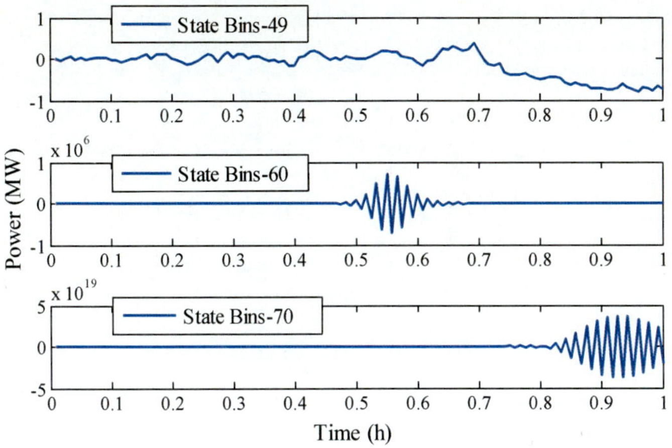

Fig. 4.3 Aggregated power errors of TCLs with different number of state bins

aggregated model in (4.4)–(4.6) while they will increase the risk of modeling failure in (4.7)–(4.8). Thus, the selection of N is critical to enable (4.4)–(4.8) to track the TCL aggregated behavior closely.

When Δt is 30 s, the aggregated power of 10,000 TCLs using (4.8) is compared to that of individual simulations. N is 49, 60 and 70, respectively. The corresponding power errors are shown in Fig. 4.3. R is 2 °C/kW. C_a is 2 kJ/°C. η is 2.5. P is 6 kW. T_{set} is 25 °C. δ is 2 °C. The outdoor temperature is 36 °C. The initial TCL indoor temperatures are randomly distributed within $[T_{min}, T_{max}]$. In Fig. 4.3, power errors are much smaller when N is 49, which means that the collective TCL power can be followed by (4.7). However, the power errors grow when N is 60 and 70, which means that (4.7) cannot describe the collective TCL behavior.

Accordingly, extensive simulations have shown that the maximum value of N is 100, 49 and 24 when Δt is 15 s, 30 s and 60 s, respectively. The aggregated model (4.7) will fail to follow the TCL behavior if N is larger than its maximum. The corresponding error between the aggregated power using (4.7) and individual simulations are given in Fig. 4.4 when Δt is 15 s, 30 s and 60 s, respectively. The cumulative effect of modeling errors allows power errors to enlarge when the simulation time is about 0.7 h. Obviously, TCL power errors grow with a longer Δt. This is because the maximum value of N is reduced when Δt is longer, which aggravates the tracking behavior in (4.4)–(4.6).

In summary, two issues need to be considered when using the state-space model to describe the collective TCL performance for regulation services. One is that N should be chosen properly. Otherwise, the state-space model cannot follow the aggregated TCL behavior. The other is that the accuracy of the state space model deteriorates as Δt increases, which increases the power error between the aggregated model and actual TCL. In this case, the regulation signals cannot be tracked closely by TCLs. To address these two issues, $\boldsymbol{A}$ is recalculated as described in Sect. 4.3.3.

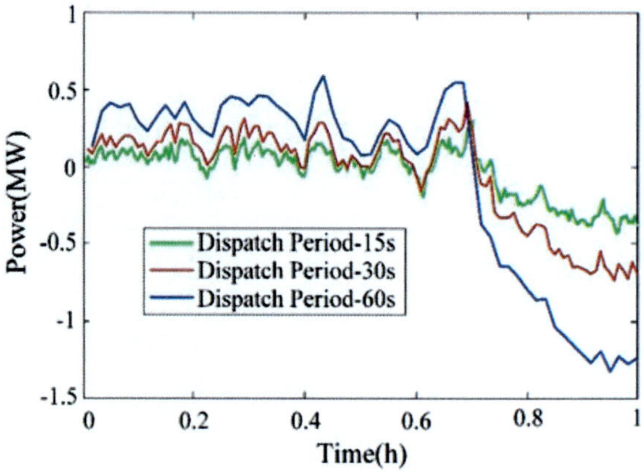

Fig. 4.4 Power error profiles of state-space model in various dispatch periods

4.3.3 Derivation of A

In this section, we propose a method to derive **A**, which will release the limit on N and increase the accuracy of the state-space model (4.7). When TCLs in the state bin q are transferred to other state bins during Δt, there are 6 different cases to be considered according to the original and final TCL switch states. (*i*) $1 \leq q \leq N$ (i.e., original TCL switch states are off) and TCLs are still off after Δt; (*ii*) $1 \leq q \leq N$ and only a fraction of TCLs are still off after Δt; (*iii*) $1 \leq q \leq N$ and all TCLs are switched on after Δt; (*iv*) $N < q \leq 2N$ (i.e., original TCL switch states are on) and TCLs are still on after Δt; (*v*) $N < q \leq 2N$ and only a fraction of TCLs are still on after Δt; (*vi*) $N < q \leq 2N$ and all TCLs are switched off after Δt. For simplicity, only **A** in cases (*i*)–(*iii*) will be discussed in the following, as those in cases (*iv*)–(*vi*) can be derived similarly.

(i) In this case, TCL switch states in the state bin q are not changed during Δt. The TCL temperature range changes from $[T_q, T_{q+1}]$ to $[T'_q, T'_{q+1}]$, which is shown in Fig. 4.5. Using (4.1), T'_q and T'_{q+1} are calculated as:

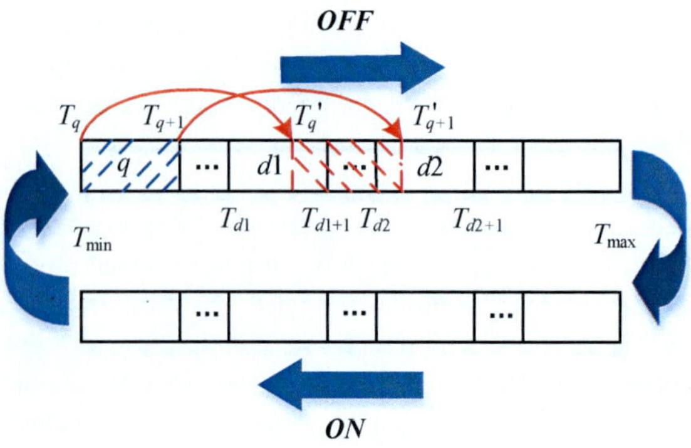

Fig. 4.5 Transition diagram of TCLs in the case (*i*)

$$\begin{cases} T_q^{'} = (T_q - T_o)e^{-\frac{1}{RC_a}\Delta t} + T_o \\ T_{q+1}^{'} = (T_{q+1} - T_o)e^{-\frac{1}{RC_a}\Delta t} + T_o \end{cases} \tag{4.9}$$

Accordingly, T'_q is in the state bin $d1$ and T'_{q+1} is in the state bin $d2$. $1 \leq d1 \leq N$ and $1 \leq d2 \leq N$ are both satisfied.

Assuming that TCLs are uniformly distributed with temperature, the $\boldsymbol{A}$ elements are calculated as follows.

(1) When $d1 = d2$, all TCLs in state bin q are transferred to state bin $d1$ ($d2$). Thus, the TCL transition probability from the state bin q to $d1$ ($d2$), a_{d1q} (a_{d2q},), is 1, which is described as:

$$a_{d1q} = a_{d2q} = 1 \tag{4.10}$$

(2) When $d2 - d1 = 1$, TCLs in the state bin q are transferred to the state bins $d2$ and $d1$ after Δt. The corresponding transition probabilities are related to the positions of T'_q in the state bin $d1$ and T'_{q+1} in the state bin $d2$, which are given as:

$$\begin{cases} a_{d1q} = \frac{T_{d1+1} - T_q^{'}}{T_{q+1}^{'} - T_q^{'}} \\ a_{d2q} = \frac{T_{q+1}^{'} - T_{d2}}{T_{q+1}^{'} - T_q^{'}} \end{cases} \tag{4.11}$$

(3) When $d2 - d1 \geq 2$, TCLs in the state bin q are transferred to more than 2 state bins, i.e., state bins $d1$, da, ($d1 + 1 \leq da \leq d2 + 1$) and $d2$. a_{d1q} and a_{d2q} are calculated as in (4.11) and a_{daq} is stated as:

$$a_{daq} = \frac{\Delta T}{T_{q+1}^{'} - T_q^{'}} \tag{4.12}$$

(ii) Using (4.9), $T_{d1} \leq T'_q \leq T_{d1+1}$ and $T'_{q+1} > T_{\max}$, which means that some TCLs in the state bin q are switched on while the others are off during Δt. T'_{q+1} in (4.9) is recalculated as:

$$T_{q+1}^{'} = (T_{\max} - T_o + \eta RP)e^{-\frac{\Delta t - \Delta t_1}{RC_a}} + T_o \tag{4.13a}$$

$$\Delta t_1 = -RC_a \ln \frac{T_o - T_{\max}}{T_o - T_{q+1}} \tag{4.13b}$$

The TCL transition diagram in (*ii*) is shown in Fig. 4.6, in which $1 \leq d1 \leq N$ and $N + 1 \leq d2 \leq 2N$ are satisfied. The $\boldsymbol{A}$ elements are calculated as follows.

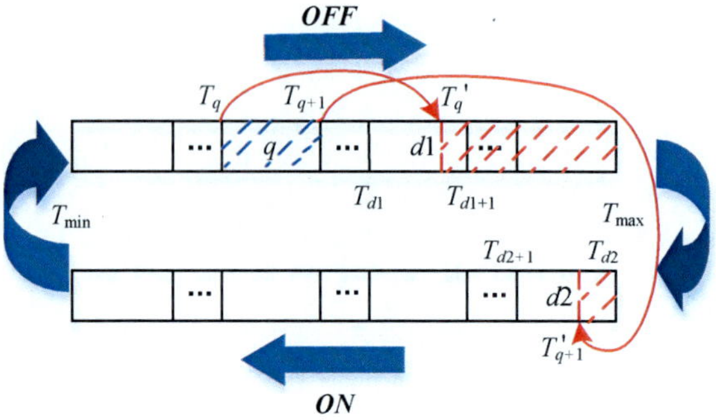

Fig. 4.6 Transition diagram of TCLs in the case (*ii*)

(1) When $d2 - d1 = 1$, TCLs in the state bin q are transferred to the state bins $d1$ and $d2$. a_{d1_q} and a_{d2_q} are given as:

$$\begin{cases} a_{d1q} = \frac{T_{d1+1} - T_q'}{2T_{\max} - T_{q+1}' - T_q'} \\ a_{d2q} = \frac{T_{d_2} - T_{q+1}'}{2T_{\max} - T_{q+1}' - T_q'} \end{cases} \tag{4.14}$$

(2) When $d2 - d1 \geq 2$, TCLs in state bin q are transferred to state bins $d1$, da, $(d1 + 1 \leq da \leq d2 + 1)$ and $d2$. a_{d1_q} and a_{d2_q} are calculated as in (4.14) and a_{daq} is calculated as:

$$a_{daq} = \frac{\Delta T}{2T_{\max} - T_{q+1}' - T_q'} \tag{4.15}$$

(iii) In this case, the TCL switch states in state bin q are changed from off to on during Δt. Using (4.9), $T'_q > T_{\max}$ and T'_q is recalculated as:

$$T_q' = (T_{\max} - T_o + \eta RP)e^{-\frac{1}{RC_a}(\Delta t - \Delta t_2)} + T_o \tag{4.16a}$$

$$\Delta t_2 = -RC_a \ln \frac{T_o - T_{\max}}{T_o - T_q} \tag{4.16b}$$

The TCL state transition diagram is shown in Fig. 4.7. $N + 1 \leq d2 \leq 2N$ and $N + 1 \leq d2 \leq 2N$ are both satisfied. The entries in $\boldsymbol{A}$ are calculated as follows.

(1) When $d2 = d1$, TCLs in state bin q are transferred to state bin $d1$ ($d2$) with a transition probability of 1, which is formulated as:

$$a_{d1q} = a_{d2q} = 1 \tag{4.17}$$

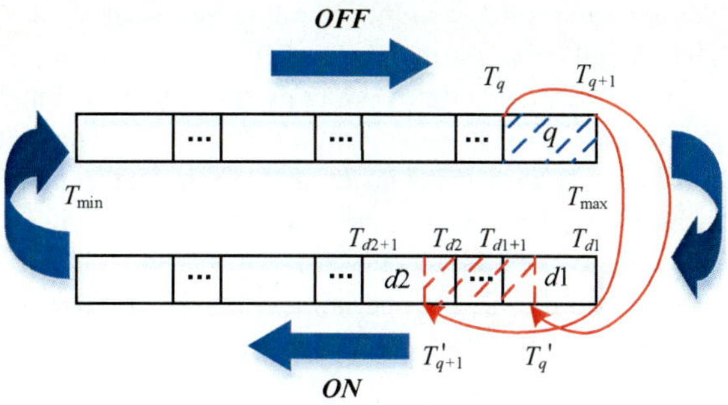

Fig. 4.7 Transition diagram of TCLs (*iii*)

(2) When $d2 - d1 = 1$, TCLs in state bin q are transferred to two-state bins $d1$ and $d2$ during Δt and the corresponding transition probabilities are given as:

$$\begin{cases} a_{d1q} = \frac{T'_q - T_{d1+1}}{T'_q - T'_{q+1}} \\ a_{d2q} = \frac{T_{d2} - T'_{q+1}}{T'_q - T'_{q+1}} \end{cases} \tag{4.18}$$

(3) When $d2 - d1 \geq 2$, TCLs in state bin q are transferred to state bins $d1$, da, $(d1 + 1 \leq da \leq d2 + 1)$ and $d2$. a_{d1q} and a_{d2q} are calculated as in (4.18) and a_{daq} is stated as:

$$a_{daq} = \frac{\Delta T}{T'_q - T'_{q+1}} \tag{4.19}$$

According to the analysis provided in Sect. 4.3.2, it is clear that a long Δt (up to 1 min) restricts N and lowers the modeling accuracy during the calculation of $\boldsymbol{A}$ in (4.5)–(4.8). To solve this problem, $\boldsymbol{A}$ is re-calculated in terms of actual TCL operating behavior, in which N has no relation with Δt. Thus, N can be freely chosen to improve the modeling accuracy.

4.4 Aggregated TCL Model with Control Signals

4.4.1 Problem Formulation

When aggregators receive power signals, the dynamics of the collective TCL behavior will be modeled to generate control signals to regulate the aggregated TCL power. This section focuses on the aggregated control model of TCLs to show the collective TCL behavior with control signals. The time delay of TCL compressors [7, 37] will also be taken into account, which is helpful for the aggregated model to track the TCL actual behavior more closely.

A probabilistic control strategy that sends a global control signal to all TCLs is proposed to reduce the communication requirements between the central controller and TCLs. Here, it is not necessary to generate and distribute different control signals to heterogeneous TCLs individually, which reduces the communication burdens. Meanwhile, the global control signal is treated differently by individual TCLs which is based on the TCL's operating state (i.e., indoor temperature, switch status, and compressor time-delay state), which helps to decrease the number of TCL switching and enhance the durability of compressors.

4.4.2 Aggregated TCL Model with Control Signals

Excessive cycling of TCL compressors can result in low efficiency and additional compressor failures [37]. Generally, a compressor time delay is considered to keep the compressor off for a minimum time (i.e., locked time, t_{lock}) when the TCL is switched off. This phenomenon is usually referred to as the lockout effect [8]. The lockout effect does not affect the normal TCL operations when there are no external control signals. However, frequent switching controls can interrupt the normal behavior of TCLs and affect the collective TCL performance significantly. Thus, it is essential in practical applications to take the compressor lockout effect into account in the aggregated TCL model.

When the TCL off duration is less than t_{lock}, TCLs will not respond to any control signals and will be locked until it is unlocked. Thus, total TCLs, $\boldsymbol{x}(k)$, are divided into two groups: unlocked TCLs, $x^a(k)$, which can respond to control signals immediately, and locked TCLs, $\boldsymbol{X}^L(k)$, which cannot respond to control signals. $x^a(k)$ is a $2N \times 1$ matrix which represents the number of unlocked TCLs in each state bin. $\boldsymbol{X}^L(k)$ is a $2N \times nL$ ($nL = t_{lock}/\Delta t$) matrix which is given as:

$$\boldsymbol{X}^L(k) = [\boldsymbol{x}_1^L(k)\ \boldsymbol{x}_2^L(k) \cdots \boldsymbol{x}_p^L(k) \cdots \boldsymbol{x}_{nL}^L(k)] \tag{4.20a}$$

$$\boldsymbol{x}_p^L(k) = [n_{p,1}^L(k)\ n_{p,2}^L(k) \cdots n_{p,N}^L(k)\ 0 \cdots 0]^T \tag{4.20b}$$

where $\boldsymbol{X}^L(k)$ is a $2N \times 1$ matrix to represent locked TCLs that will be unlocked after $nL - p + 1$ dispatch periods ($1 \leq p \leq N$).

Here, $u_0(k) \in [-1,1]$ as the global control signal will be broadcasted to all TCLs. When the TCL receives $u_0(k)$, it will decide whether to change the TCL switch state according to its current operating state. Different from the response probability given in [8], local controllers here will act based on local indoor temperature, switch states and the locked time to lessen the impact on the compressor durability. When $u_0(k) >$ 0, some unlocked TCLs in off state will be turned on. The TCLs with higher indoor temperatures will be turned on with a larger probability. When $u_0(k) \leq 0$, TCLs in on will be turned off and TCLs with lower indoor temperatures will more likely be turned off.

In the following, the proposed probabilistic control strategy is discussed specifically. At the level of end-users, each TCL has its response function $f(T_i(k), S(k), u_0(k))$ which is the probability of changing the switch state. For locked TCLs, $f(T_i(k), S(k), u_0(k))$ is 0 while $f(T_i(k), S(k), u_0(k))$ varies with $T_i(k)$, $S(k)$ and $u_0(k)$ when TCLs are not locked. $f(T_i(k), S(k), u_0(k))$ has the following properties.

(1) When $u_0(k)$ is 0, $f(T_i(k), S(k), u_0(k))$ is 0, which means that the TCL will not change its switch state;
(2) When $u_0(k)$ is 1, $f(T_i(k), S(k), u_0(k))$ is 1 if s is 0 and the TCL is unlocked. It means that the unlocked TCL in off state will be turned on; When $u_0(k)$ is -1, $f(T_i(k), S(k), u_0(k))$ is 1 if s is 1. It means that the TCL in on state will be turned off;
(3) When $0 < u_0(k) < 1$, $f(T_i(k), S(k), u_0(k))$ with the higher T_i is larger if s is 0 and the TCL is unlocked; otherwise, $f(T_i(k), S(k), u_0(k))$ is 0;
(4) When $-1 < u_0(k) < 0$, $f(T_i(k), S(k), u_0(k))$ with the lower T_i is larger if s is 1; otherwise, $f(T_i(k), S(k), u_0(k))$ is 0.

Therefore, $f(T_i(k), S(k), u_0(k))$ is given as:

$$f(T_i(k), S(k), u_0(k)) = \begin{cases} u_0(k)\left(\frac{T_{\max}-T_i(k)}{T_{\max}-T_{\min}}(u_0(k))^w + \frac{T_i(k)-T_{\min}}{T_{\max}-T_{\min}}\right), & if\ u_0(k) \geq 0\ \&\ S(k)=0\ \&\ \text{unlocked} \\ |u_0(k)|\left(\frac{T_i(k)-T_{\min}}{T_{\max}-T_{\min}}|u_0(k)|^w + \frac{T_{\max}-T_i(k)}{T_{\max}-T_{\min}}\right), & if\ u_0(k) < 0\ \&\ S(k)=1 \\ 0, & otherwise \end{cases} \tag{4.21}$$

When unlocked TCLs respond to $u_0(k)$, the switch states of some TCLs will be changed. The impacts of $u_0(k)$ on the TCL population are embedded in the aggregated model, which is described by $\boldsymbol{u}(k)$. $\boldsymbol{u}(k)$ is a $2N \times 1$ matrix that represents the number of TCLs that are controlled to be turned on/off at the kth dispatch period. The indoor TCL temperatures in a state bin are assumed to be identical and equal to the intermediate temperature of the state bin. According to the response function of TCLs in (4.21), the relationship between $\boldsymbol{u}(k)$ and $u_0(k)$ is stated as:

$$\boldsymbol{u}(k) = diag(\boldsymbol{x}^a(k) \cdot \boldsymbol{F}(k)^T) \tag{4.22a}$$

$$\boldsymbol{F}(k) = \begin{cases} \boldsymbol{I}_1 \cdot u_0(k)^{w+1} + \boldsymbol{I}_2 \cdot u_0(k), & u_0(k) \geq 0 \\ \boldsymbol{I}_3 \cdot |u_0(k)^{w+1}| + \boldsymbol{I}_4 \cdot |u_0(k)|, & u_0(k) < 0 \end{cases} \tag{4.22b}$$

$$\begin{cases} \boldsymbol{I}_1 = [\frac{T_{\max}-T_1}{T_{\max}-T_{\min}} \frac{T_{\max}-T_2}{T_{\max}-T_{\min}} \cdots \frac{T_{\max}-T_N}{T_{\max}-T_{\min}} 0 \cdots 0]^T \\ \boldsymbol{I}_2 = [\frac{T_1-T_{\min}}{T_{\max}-T_{\min}} \frac{T_2-T_{\min}}{T_{\max}-T_{\min}} \cdots \frac{T_N-T_{\min}}{T_{\max}-T_{\min}} 0 \cdots 0]^T \end{cases} \tag{4.22c}$$

$$\begin{cases} \boldsymbol{I}_3 = [0 \cdots 0 \frac{T_{N+1}-T_{\min}}{T_{\max}-T_{\min}} \frac{T_{N+2}-T_{\min}}{T_{\max}-T_{\min}} \cdots \frac{T_{2N}-T_{\min}}{T_{\max}-T_{\min}}]^T \\ \boldsymbol{I}_4 = [0 \cdots 0 \frac{T_{\max}-T_{N+1}}{T_{\max}-T_{\min}} \frac{T_{\max}-T_{N+2}}{T_{\max}-T_{\min}} \cdots \frac{T_{\max}-T_{2N}}{T_{\max}-T_{\min}}]^T \end{cases} \tag{4.22d}$$

The aggregated TCL model with control signals is derived as follows.

(i) $-1 \leq u_0(k) \leq 0$: When TCLs receive $u_0(k)$, some of them in on state will be turned off and locked immediately. Then the unlocked TCL population, $\boldsymbol{x}^a(k)$, which is the difference between the total and the locked TCLs, are expressed as:

$$\boldsymbol{x}^a(k) = \boldsymbol{x}(k) - \boldsymbol{x}^L(k) - \boldsymbol{u}(k) \tag{4.23a}$$

$$\boldsymbol{x}^L(k) = \sum_{i=1}^{nL} \boldsymbol{x}_i^L(k) \tag{4.23b}$$

where $\boldsymbol{x}^L(k)$ represents the number of locked TCLs in each state bin. Some TCLs are added to $\boldsymbol{X}^L(k)$ when they respond to $u_0(k)$ and are locked. $\boldsymbol{X}^L$ at the $(k + 1)$th dispatch period is stated as:

$$\boldsymbol{X}^L(k+1) = \boldsymbol{A} \cdot (\boldsymbol{X}^L(k) + \boldsymbol{U}(k)) \tag{4.24a}$$

$$\boldsymbol{U}(k) = [\boldsymbol{B} \cdot \boldsymbol{u}(k), \boldsymbol{0}, \boldsymbol{0}, \cdots, \boldsymbol{0}] \tag{4.24b}$$

where $\boldsymbol{U}(k)$ is a $2N \times nL$ matrix and $\boldsymbol{0}$ represents a $2N \times 1$ null matrix. $\boldsymbol{B}\cdot\boldsymbol{u}(k)$ is the added locked TCLs after responding to $u_0(k)$. $\boldsymbol{B}$ is a $2N \times 2N$ matrix which is expressed as:

$$\boldsymbol{B} = \begin{bmatrix} 0 & \cdots & 0 & 1 \\ \cdots & 0 & 1 & 0 \\ \vdots & \vdots & \vdots & \vdots \\ 1 & 0 & \cdots & 0 \end{bmatrix} \tag{4.25}$$

The locked time of TCLs described by X^L will be reduced by Δt at the $(k + 1)$th period as compared with that at the kth period. Meanwhile, some unlocked TCLs are transferred to the first state bin by default which are locked immediately. Thus, $X^L(k + 1)$ is modified, based on (4.24a) and (4.24b), which is given as:

$$\boldsymbol{X}^L(k+1) = [\boldsymbol{A}' \cdot \boldsymbol{x}^a(k), \boldsymbol{x}_1^L(k+1), \boldsymbol{x}_2^L(k+1), \cdots, \boldsymbol{x}_{nL-1}^L(k+1)] \tag{4.26}$$

where $A' \cdot x^a(k)$ represents the TCLs that are locked by default at the $(k + 1)$th period. A' is a $2N \times 2N$ matrix, in which the elements in the first row are the same as those of $\boldsymbol{A}$ and the other elements are 0.

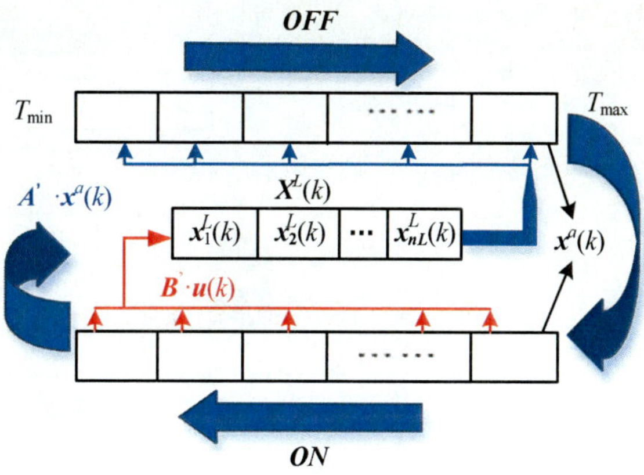

Fig. 4.8 State transition diagram of TCLs when $-1 \le u_0(k) \le 0$ (blue line is the default TCL transition and the red line is the transition process by control signals)

The total TCLs at the $(k+1)$th period, $\boldsymbol{x}$, are updated as:

$$\boldsymbol{x}(k+1) = \boldsymbol{A} \cdot (\boldsymbol{x}(k) + \boldsymbol{B}' \cdot \boldsymbol{u}(k)) \tag{4.27a}$$

$$\boldsymbol{B}' = \boldsymbol{B} - \boldsymbol{I} \tag{4.27b}$$

where $\boldsymbol{B}' \cdot \boldsymbol{u}(k)$ is the sum of added locked TCLs and the reduced TCLs in on state. $\boldsymbol{I}$ is a $2N \times 2N$ identity matrix, which describes the reduced TCLs in on state. The TCL state transition diagram when $-1 \le u_0(k) \le 0$ is shown in Fig. 4.8.

(ii) $0 < u_0(k) \le 1$: When some of the unlocked TCLs in the off state are turned on by responding to $u_0(k)$. the corresponding $\boldsymbol{x}^a(k)$ is calculated as:

$$\boldsymbol{x}^a(k) = \boldsymbol{x}(k) - \boldsymbol{x}^L(k) + \boldsymbol{B}' \cdot \boldsymbol{u}(k) \tag{4.28}$$

where $\boldsymbol{B}' \cdot \boldsymbol{u}(k)$ represents the added TCLs in on state and reduced TCLs in off-state because of $u_0(k)$.

Accordingly, there are no added locked TCLs and $\boldsymbol{X}^L$ at the $(k+1)$th dispatch period is given as:

$$\boldsymbol{X}^L(k+1) = \boldsymbol{A} \cdot \boldsymbol{X}^L(k) \tag{4.29}$$

Considering that some of the unlocked TCLs are locked when they arrive in the first state bin and some of the locked TCLs are released at the $(k+1)$th dispatch period, $\boldsymbol{X}^L$ $(k+1)$ is modified according to (4.26).The total TCL population at the $(k+1)$th period, $\boldsymbol{x}(k+1)$, is updated in (4.27a). The power output of the TCL population, $y(k+1)$, is expressed as:

$$y(k+1) = \boldsymbol{C} \cdot \boldsymbol{x}(k+1) \tag{4.30a}$$

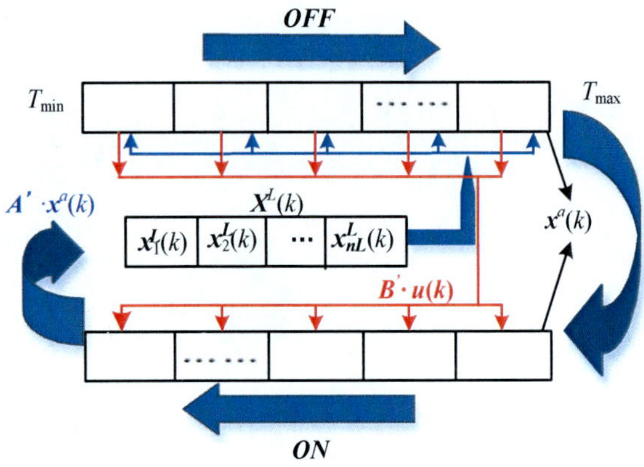

Fig. 4.9 State transition diagram when $0 < u_0(k) \leq 1$ (blue line is the default TCL transition process and the red line is the transition process by control signals)

$$C = P \cdot [\underbrace{0 \cdots 0}_{N}\ \underbrace{1 \cdots 1}_{N}] \tag{4.30b}$$

Accordingly, unlocked TCLs at the $(k + 1)$th dispatch period, are described as:

$$\boldsymbol{x}^a(k+1) = \boldsymbol{x}(k+1) - \boldsymbol{x}^L(k+1) \tag{4.31}$$

The state transition diagram of TCLs when $0 < u_0(k) \leq 1$ is shown in Fig. 4.9.

In summary, $\boldsymbol{u}(k)$ in (4.22) is the bridge between $u_0(k)$ and the aggregated model which describes the effect of $u_0(k)$ on the collective TCL behavior. The aggregated model with control signals in (4.23)–(4.31) show the operating dynamics of TCLs when end users respond to $u_0(k)$, which will guide aggregators to generate control signals in the next dispatch period. The procedure for the aggregated TCL model with control signals is shown in Fig. 4.10.

4.5 Case Studies

In this section, the proposed aggregated TCL model is validated. There are 26,532 heterogeneous TCLs [32] connected to a utility and ten heterogeneous buildings/communities take part in DR programs. TCLs in the same building/community are controlled by an aggregator and the corresponding parameters and number of TCLs are shown in Table 4.1. To simulate more realistic TCL dynamics, a second-order thermodynamic model [10] is employed to describe the actual operating behaviors of TCLs by individual simulations. C_M and R_M are mass thermal capacitance and mass thermal resistance of the second-order thermodynamic model, respectively. C_M is uniformly distributed within [0,4] kWh/°C and R_M is uniformly distributed within [2,6] °C/kW for TCLs managed by an aggregator, which helps characterize

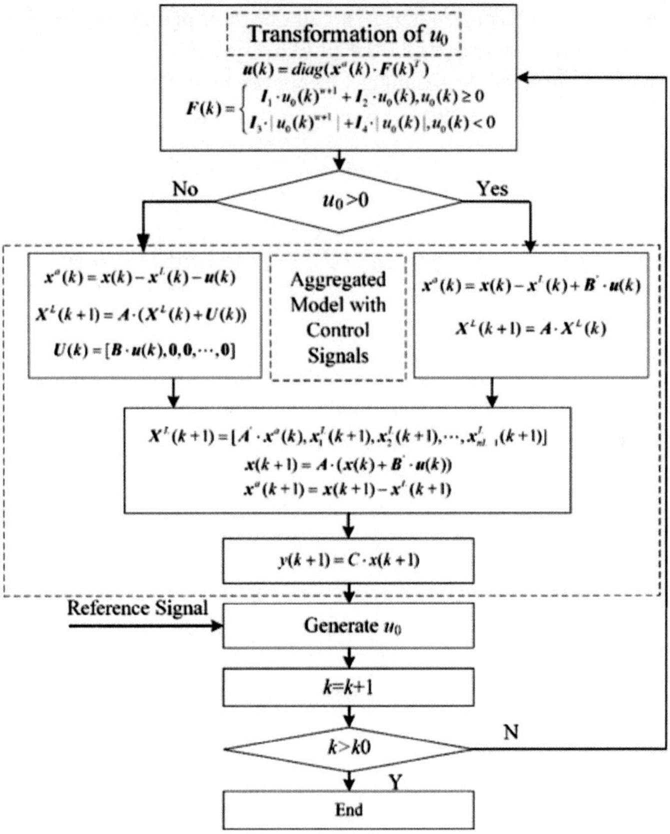

Fig. 4.10 Flowchart for the aggregated TCL model with control signals

Table 4.1 TCL parameters

No	n	R	C_a	P	η	T_{set}	δ
L1	2312	2.5	1.5	6	2.5	23.0	1
L2	3128	3.2	1.8	4	2.5	23.5	1
L3	3874	3.1	2.5	3	2.5	24.0	1
L4	1982	2.7	2.2	5	2.5	25.0	1
L5	2620	2.8	1.8	6	2.5	24.5	1
L6	3435	2.3	2.5	7	2.5	23.0	1
L7	1250	2.5	1.9	8	2.5	24.0	1
L8	2100	2.8	2.3	5	2.5	24.5	1
L9	2543	2.7	2.1	4	2.5	24.5	1
L10	3288	2.6	1.6	6	2.5	24.0	1

the heterogeneity of TCL parameters. The outdoor temperature is 36 °C. w in (4.8) is 3. The simulation time is 1 h and the time delay of TCL compressors is 5 min. Matlab R2014a is utilized to simulate the operating behaviors of the aggregated model and TCLs. The CPU model of the desktop is Intel(R) Core(TM) i7-6700 and the frequency is 3.40 GHz.

4.5.1 Performance of Aggregated TCL Model Without Control Signal

In order to validate the modeling accuracy of the proposed aggregated model in Sect. 4.3.2, two aggregated models of TCLs are compared when there are no control signals. One (M1) is considered using [9, 36], in which $\boldsymbol{A}$ in (4.7) is calculated according to (4.4)–(4.6), (4.8). The other one (M2) is described in Sect. 4.3.2, in which $\boldsymbol{A}$ in (4.7) is derived based on (4.9)–(4.19). The initial TCL indoor temperatures are uniformly distributed in $[T_{\text{set}} - \delta/2, T_{\text{set}} + \delta/2]$.

Δt is set to 15 s, 30 s and 60 s, respectively, considering that the regulation service is dispatched in seconds. In Sect. 4.3, the maximum N in M1 is limited by Δt and the modeling accuracy of M1 is the highest when N is maximum. The maximum N for L1–L10 is given in Table 4.2 when Δt is 15, 30 and 60 s. N in M2 is unlimited which is set as 50. The power in individual simulations of M1 and M2 is compared using different Δt as shown in Fig. 4.11.

According to Table 4.2 and Fig. 4.11, compared to the individual simulations, the maximum N of M1 decreases and the corresponding power error increases as Δt increases. When Δt is 60 s, M1 cannot follow the collective TCL power because of the limited number of state bins. Although the power errors in M2 compared with individual simulations also increase as Δt increases, they are smaller than those in M1, which means M2 is better at modeling accuracy especially when Δt is longer. Additionally, we compare the power among individual simulations for M1 and M2

Table 4.2 Maximum N of 10 TCL groups with different Δt

No	Δt (15 s)	Δt (30 s)	Δt (60 s)
L1	36	18	9
L2	69	34	17
L3	148	74	37
L4	61	30	15
L5	39	19	9
L6	49	24	12
L7	29	14	7
L8	64	32	16
L9	85	42	21
L10	36	18	9

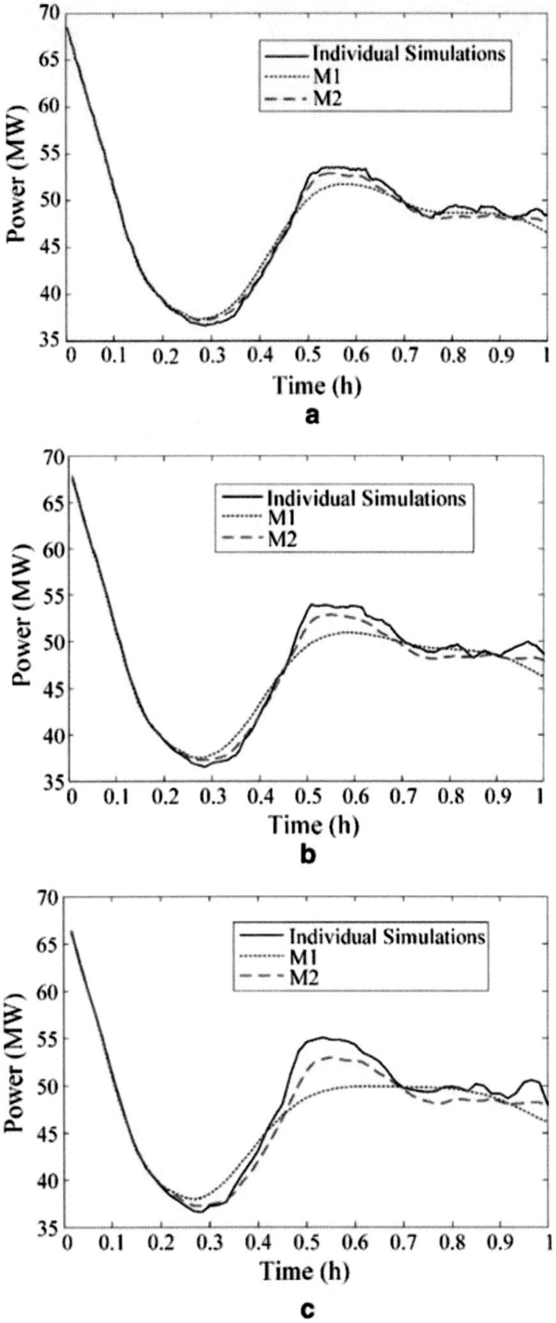

Fig. 4.11 **a** Power comparisons among individual simulations, M1 and M2 when Δt is 15 s (N in M2 is 50). **b** Power comparisons among individual simulations, M1 and M2 when Δt is 30 s (N in M2 is 50). **c** Power comparisons among individual simulations, M1 and M2 when Δt is 60 s (N in M2 is 50)

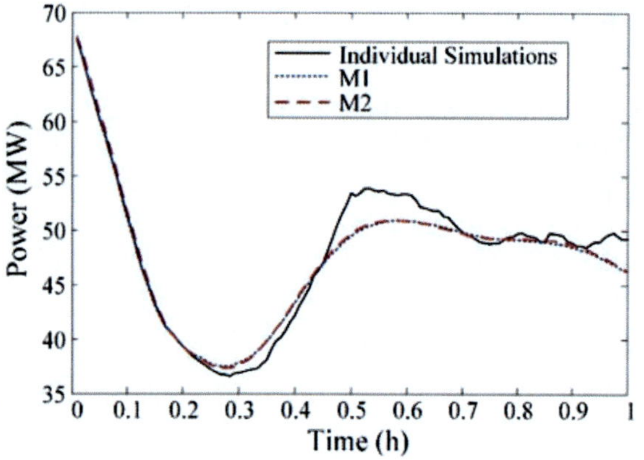

Fig. 4.12 Power comparisons among individual simulations, M1 and M2 when Δt is 30 s (N in M2 is the same as that in M1)

when Δt is 15, 30 and 60 s, in which N in M2 is the same as that in M1. For simplicity, power comparisons when Δt is 30 s are shown in Fig. 4.12. The results indicate that M1 and M2 demonstrate almost the same modeling accuracy. In effect, M2 has a higher modeling accuracy because it possesses a higher number of state bins.

4.5.2 *Performance of Aggregated TCL Model with Control Signals*

This section focuses on the collective TCL behavior followed by the proposed aggregated model with control signals presented in Sect. 4.4. The dispatch period is 30 s. N in the aggregated model is 50. From (4.2) to (4.3) and Table 4.1, the power baseline of 10 TCL groups is 44.5 MW and the regulation service with the upward capacity of 40 MW and downward capacity of 40 MW is provided by TCLs. The desired power profile with a power magnitude that is scaled to keep each regulation signal within upward and downward capacities of TCLs is given in [38]. Considering that the power allocation strategy of the utility is not the focus of this chapter, we assume that regulation signals are proportionally distributed among aggregators based on their adjustable capacity.

The regulation signal provided by the utility, $r(k+1)$, is the error between the desired power and the TCL power baseline [6]. Thus, the desired power, $P_d\ (k+1)$, is calculated as:

$$P_d(k+1) = P_{baseline} + r(k+1) \tag{4.32}$$

The following algorithm is applied to generate TCL control signals at every dispatch period.

Algorithm 1: Generation of Control Signals
1. Calculate the TCL power output according to (2.41)-(2.42), and compare it with $P_d(k+1)$. If $y(k+1) \geq P_d(k+1)$, set $u01$ as -1 and $u02$ as 0. Otherwise, $u01$ is 0 and $u02$ is 1. $u_0(k+1) \in [u01, u02]$.
2. Divide $[u01, u02]$ into l groups equally in which the boundary point is $u1, u2, \ldots, u(l), u(l+1)$, where $(u1 = u01, u(l+1) = u02)$, respectively. According to the aggregated model described in Section 3.4, calculate the corresponding aggregated TCL power output population, $y(1), y(2), \ldots, y(l+1)$, where $u_0(k+1)$ is $u1, u2, \ldots, u(l+1)$, respectively;
3. Compare $P_d(k+1)$ to $y(1), y(2), \ldots, y(l+1)$. If $y(h) < P_d(k+1) < y(h+1)$ $(1 \leq h \leq l+1)$, $u_0(k)$ is given as $u_0(k+1) = 0.5(u(h)+u(h+1))$ If $P_d(k+1) \leq y(1)$ when $u01$=-1, then $u_0(k+1)$ =-1; If $P_d(k+1) \geq y(l+1)$ when $u02$=1, then$u_0(k+1)$= 1;
4. $k=k+1$, and then go to step 1

When aggregators distribute control signals $u_0(k)$ among TCLs, each local controller will generate a random number within [0,1] and calculate the response probability, $f(T_i(k), S(k), u_0(k))$, according to (4.21). If the random number is less than $f(T_i(k), S(k), u_0(k))$, TCL will change its switch state. Otherwise, TCL will retain its current status. Compared to the control strategies given in [6, 7, 14], in which a multitude of TCL control signals are generated, there is only one global control signal in every dispatch period according to the proposed control strategy which reduces the communication requirements between the central controller and TCLs [8].

There are three scenarios to be compared.

Scenario 1 (S1): The aggregated TCL model with control signals is the same as that described in Sect. 4.4. This is the benchmark scenario.

Scenario 2 (S2): The compressor time delay is set as 0 (i.e., $t_{lock} = 0$) [19, 20, 22] in the aggregated TCL model with control signals. The other parts are the same as those in S1.

Scenario 3 (S3): Aggregators send a global control signal to all TCLs to regulate their aggregated power. TCLs change their switch state using the same probability [8, 9, 23]. The other parts are the same as those in S1. The desired power, actual TCL power in S1 (i.e., the collective power of TCLs by individual simulations when they respond to control signals) and actual TCL power in S2 are shown in Fig. 4.13. The control signals u_0 in S1 are given in Fig. 4.14. Figures 4.15 and 4.16 show the corresponding indoor air and mass TCL temperatures in S1. Figure 4.17 depicts the comparison of unlocked TCL (off) numbers in the aggregated model and individual simulations in S2.

The following observations are based on our simulation results.

(1) TCLs in S1 can follow the desired power closely as shown in Fig. 4.13, which means that the proposed aggregated TCL model with control signals can track the collective TCL behavior. When some TCLs need to be turned off to follow the desired power (e.g., about 0.15 h in Fig. 4.14), the control signals are negative; otherwise, the control signals are positive when some of the TCLs are turned on (e.g., about 0.95 h in Fig. 4.14). The proper control signals can

Fig. 4.13 Power comparisons among desired power, actual power in S1 andactual power in S2

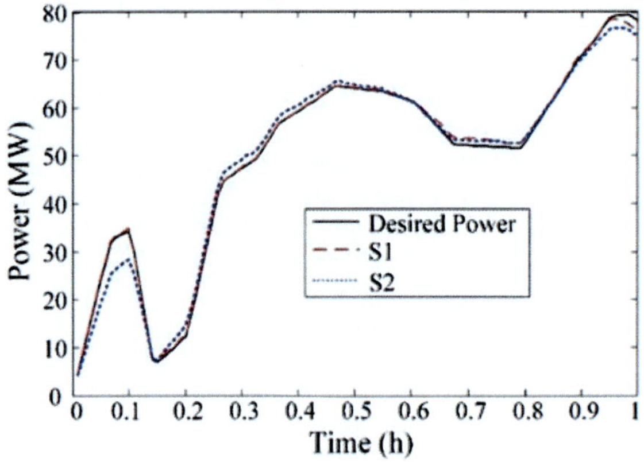

Fig. 4.14 Control signals in S1

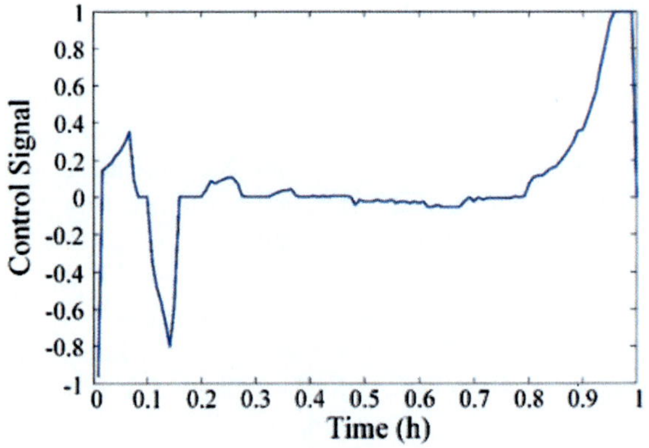

Fig. 4.15 Indoor air temperature profiles of TCLs in S1

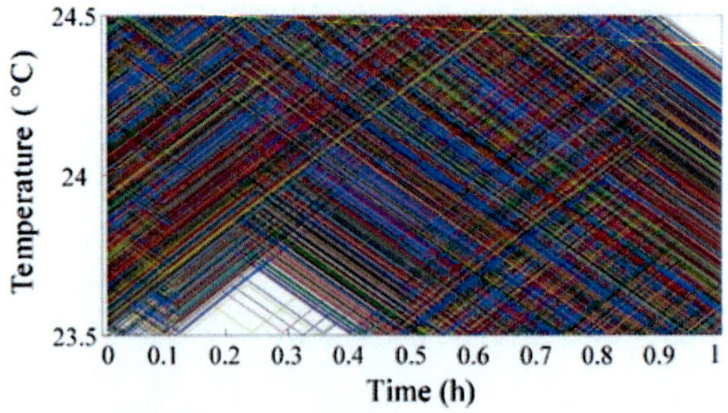

be generated based on the aggregated TCL model. So, when TCLs respond to control signals according to their operating states, their indoor temperatures are always kept within [23.5, 24.5] in Figs. 4.15 and 4.16 and end users' comfort requirements are guaranteed;

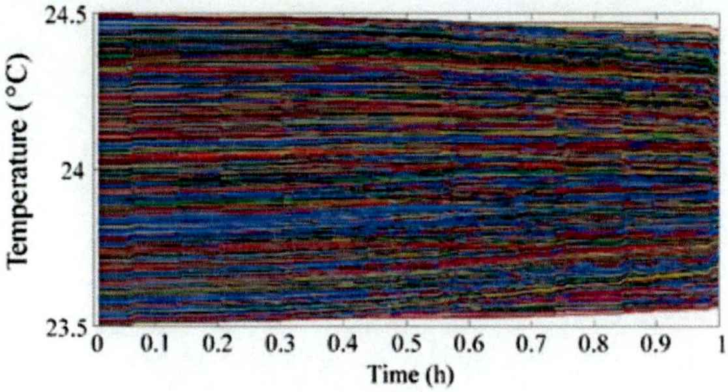

Fig. 4.16 Indoor mass temperature profiles of TCLs in S1

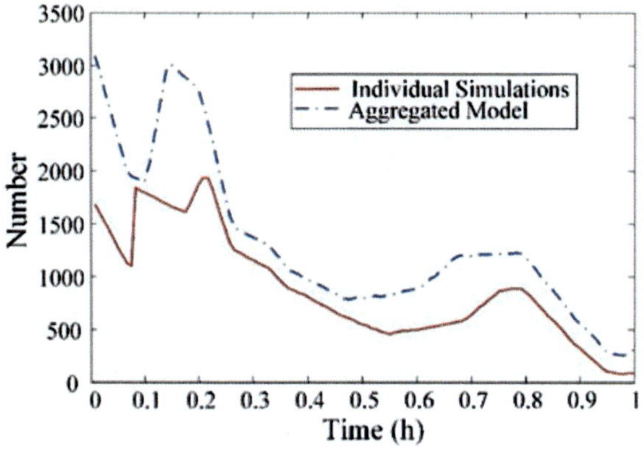

Fig. 4.17 Comparison of unlocked TCL (off) numbers in the aggregated model and individual simulations in S2

(2) Compared to the simulation results of S1, TCLs in S2 have worse performances and there are larger power errors between the desired power and the actual power. This is because TCLs will be locked for a certain time in individual simulations when they are turned off. However, the locked TCLs are not taken into account in the aggregated TCL model. Thus, the unlocked TCL number in individual simulations is quite different from that of the aggregated model in Fig. 4.17. It demonstrates that the aggregated TCL model with control signals cannot describe the actual operating TCL behavior if the compressor time delay is ignored in the aggregated model.

The average number of switching in 10 TCL groups is given in Table 4.3 in which the average number of TCL switching in S3 is more than that in S1. This is because TCLs in S3 respond to control signals with the same probability, which makes it possible for some TCLs to be switched on/off quickly. Especially when indoor TCL temperatures are around $T_{set} - \delta/2$ or $T_{set} + \delta/2$, TCLs can be turned on/off (by default and by responding to control signals) several times in a short time.

Table 4.3 Number of switching of 10 TCL groups in S1 and S3

	L1	L2	L3	L4	L5	L6	L7	L8	L9	L10
S1	3.3	3.8	3.3	3.3	2.9	3.7	3.4	3.1	3.4	2.8
S3	5.1	4.4	3.8	3.9	3.3	5.1	3.9	3.5	3.9	3.1

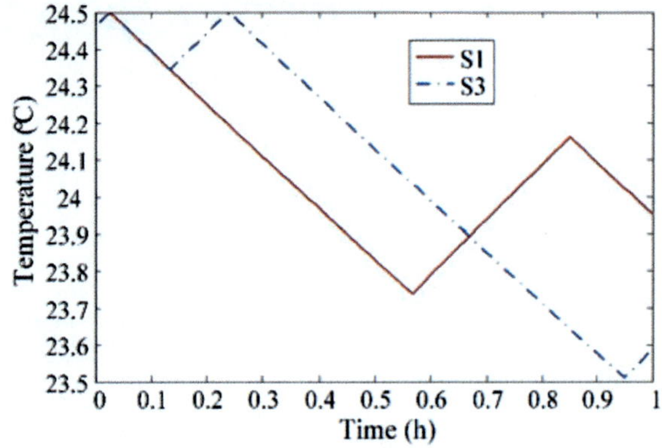

Fig. 4.18 Indoor air temperature profiles of a TCL in S1 and S3

Figure 4.18 shows indoor air temperature profiles of a TCL in S1 and S3, respectively. The TCL in S3 is turned off (by responding to control signals) shortly after it is turned on by default and then it is turned on by default again. Thus, the number of TCL switching is increased. However, TCLs in S1 respond to control signals with a different probability when the indoor temperature, an important parameter, is considered. And when they are turned off/on with less probability in a short period of time when the TCLs are to be turned on/off by default, and they are to be turned on/off by default at $T_{set} + \delta/2/T_{set} - \delta/2$, they will respond to control signals with a larger probability to change their switching states,.This also helps to reduce the number of TCL switching with a less adverse effect on the durability of the compressor [6, 14].

4.6 Conclusions

In this chapter, a modified state-space model of aggregated TCLs is proposed which improves the modeling accuracy for regulation services. The corresponding aggregated control model of TCLs is also presented to describe the collective TCL behavior with control signals. Meanwhile, a probability control strategy is given to reduce the adverse control effects on the durability of compressors and communication requirements between the central controller and TCLs. Simulation results are summarized as: (1) the state-space model accuracy of aggregated TCLs in most studies (e.g., [9, 36]) is aggravated as the dispatch period is increased, which is because of the limited number of state bins. The traditional model even fails to track the collective TCL behavior when the number of state bins is improperly chosen; (2) the modified state-space model can improve the modeling accuracy of aggregated TCLs by releasing the limitation on the number of state bins especially when the dispatch period is longer; (3) the proposed aggregated control model of TCLs with control signals is valid which can track the TCL dynamic behavior more closely as compared to the aggregated model proposed in [19, 20, 22] in which the compressor time delay is not taken into account; (4) The proposed control strategy can relieve communication requirements for implementing control actions and reduce the number of TCL switching with less adverse effects on the durability of compressors as compared to the models provided in [8, 9, 23].

References

1. Song, M., Gao, C., Shahidehpour, M., Li, Z., Yang, J., & Yan, H. (2018). State space modeling and control of aggregated TCLs for regulation services in power grids. *IEEE Transactions on Smart Grid, 10*(4), 4095–4106.
2. Wu, H., Shahidehpour, M., Alabdulwahab, A., & Abusorrah, A. (2015). Demand response exchange in the stochastic day-ahead scheduling with variable renewable generation. *IEEE Transactions on Sustainable Energy, 6*(2), 516–525.
3. Sajjad, I. A., Chicco, G., & Napoli, R. (2016). Definitions of demand flexibility for aggregate residential loads. *IEEE Transactions on Smart Grid, 7*(6), 2633–2643.
4. Kim, Y.-J., Norford, L. K., & Kirtley, J. L. (2015). Modeling and analysis of a variable speed heat pump for frequency regulation through direct load control. *IEEE Transactions on Power Systems, 30*(1), 397–408.
5. Lu, N., & Zhang, Y. (2013). Design considerations of a centralized load controller using thermostatically controlled appliances for continuous regulation reserves. *IEEE Transactions on Smart Grid, 4*(2), 914–921.
6. Lu, N. (2012). An evaluation of the HVAC load potential for providing load balancing service. *IEEE Transactions on Smart Grid, 3*(3), 1263–1270.
7. Sanandaji, B. M., Vincent, T. L., & Poolla, K. (2016). Ramping rate flexibility of residential HVAC loads. *IEEE Transactions on Sustainable Energy, 7*(2), 865–874.
8. Zhang, W., Lian, J., Chang, C.-Y., & Kalsi, K. (2013). Aggregated modeling and control of air conditioning loads for demand response. *IEEE Transactions on Power Systems, 28*(4), 4655–4664.
9. Sanandaji, B. M., Hao, H., & Poolla, K. (2014). Fast regulation service provision via aggregation of thermostatically controlled loads. In *2014 47th Hawaii International Conference on System Sciences* (pp. 2388–2397).
10. Mahdavi, N., Braslavsky, J. H., Seron, M. M., & West, S. R. (2017). Model predictive control of distributed air-conditioning loads to compensate fluctuations in solar power. *IEEE Transactions on Smart Grid, 8*(6), 3055–3065.
11. Liu, M., Shi, Y., & Liu, X. (2016). Distributed MPC of aggregated heterogeneous thermostatically controlled loads in smart grid. *IEEE Transactions on Industrial Electronics, 63*(2), 1120–1129.
12. Song, M., Gao, C., Yan, H., & Yang, J. (2017). Thermal battery modeling of inverter air conditioning for demand response. *IEEE Transactions on Smart Grid, 9*(6), 5522–5534.
13. Trovato, V., Tindemans, S. H., & Strbac, G. (2016). Leaky storage model for optimal multi-service allocation of thermostatic loads. *IET Generation, Transmission and Distribution, 10*(3), 585–593.

14. Hao, H., Sanandaji, B. M., Poolla, K., & Vincent, T. L. (2015). Aggregate flexibility of thermostatically controlled loads. *IEEE Transactions on Power Systems, 30*(1), 189–198.
15. Mathieu, J. L., Kamgarpour, M., Lygeros, J., Andersson, G., & Callaway, D. S. (2015). Arbitraging intraday wholesale energy market prices with aggregations of thermostatic loads. *IEEE Transactions on Power Systems, 30*(2), 763–772.
16. Zhao, L., Zhang, W., Hao, H., & Kalsi, K. (2016). A geometric approach to aggregate flexibility modeling of thermostatically controlled loads. *IEEE Transactions on Power Systems, 32*(6), 4721–4731.
17. Song, M., Gao, C., Shahidehpour, M., Li, Z., Lu, S., & Lin, G. (2018). Multi time-scale modeling and parameter estimation of TCLs for smoothing out wind power generation variability. *IEEE Trans. Sustain. Energy, 10*(1), 105–118.
18. Hughes, J. T., Domínguez-García, A. D., & Poolla, K. (2016). Identification of virtual battery models for flexible loads. *IEEE Transactions on Power Systems, 31*(6), 4660–4669.
19. Liu, M., & Shi, Y. (2016). Model predictive control of aggregated heterogeneous second-order thermostatically controlled loads for ancillary services. *IEEE Transactions on Power Systems, 31*(3), 1963–1971.
20. Lu, N., Chassin, D. P., & Widergren, S. E. (2005). Modeling uncertainties in aggregated thermostatically controlled loads using a state queueing model. *IEEE Transactions on Power Systems, 20*(2), 725–733.
21. Chang, C.-Y., Zhang, W., Lian, J., & Kalsi, K. (2013, February). Modeling and control of aggregated air conditioning loads under realistic conditions. In *IEEE PES Innovative Smart Grid Technologies Conference* (pp. 1–6).
22. Mathieu, J. L., Koch, S., & Callaway, D. S. (2013). State estimation and control of electric loads to manage real-time energy imbalance. *IEEE Transactions on Power Systems, 28*(1), 430–440.
23. Zhang, W., Kalsi, K., Fuller, J., Elizondo, M., & Chassin, D. (2012). Aggregate model for heterogeneous thermostat-ically controlled loads with demand response. In *IEEE Power and Energy Society General Meeting* (pp. 1–8).
24. Song, M., Gao, C., Qiao, L., Guo, W., & Wang, J. (2017). A Novel aggregated modeling method of air conditioning loads. In *2016 3rd International Conference on Systems and Informatics* (pp. 251–257).
25. Hussain, A., Bui, V. H., & Kim, H. M. (2018). A resilient and privacy-preserving energy management strategy for networked microgrids. *IEEE Transactions on Smart Grid, 9*(3), 2127–2139.
26. Lu, S., Samaan, N., Diao, R., Elizondo, M., Jin, C., Mayhorn, E., Zhang, Y., & Kirkham, H. (2011, January). Centralized and decentralized control for demand response. In *Proceedings of Innovative Smart Grid Technologies* (pp. 1–8).
27. Hu, J., Cao, J., Guerrero, J. M., Yong, T., & Yu, J. (2017). Improving frequency stability based on distributed control of multiple load aggregators. *IEEE Transactions on Smart Grid, 8*(4), 1553–1567.
28. Tindemans, S., Trovato, V., & Strbac, G. (2015). Decentralized control of thermostatic loads for flexible demand response. *IEEE Transactions on Control Systems Technology, 23*(5), 1685–1700.
29. Molina-Garcia, A., Bouffard, F., & Kirschen, D. S. (2011). Decentralized demand-side contribution to primary frequency control. *IEEE Transactions on Power Systems, 26*(1), 411–419.
30. Meng, K., Wang, D., & Dong, Z. (2017). Distributed control of thermostatically controlled loads in distribution network with high penetration of solar PV. CSEE. *Journal of Power and Energy System, 3*(1), 53–62.
31. Zheng, L., & Cai, L. (2014). A distributed demand response control strategy using Lyapunov optimization. *IEEE Transactions on Smart Grid, 5*(4), 2075–2083.
32. Hu, J., et al. (2017). Load following of multiple heterogeneous TCL aggregators by centralized control. *IEEE Transactions on Power Systems, 32*(4), 3157–3167.
33. Kondoh, J., Lu, N., & Hammerstrom, D. J. (2011). An evaluation of the water heater load potential for providing regulation service. *IEEE Transactions on Power Systems, 26*(3), 1309–1316.

34. Lu, N., & Chassin, D. P. (2004). A state-queueing model of thermostatically controlled appliances. *IEEE Transactions on Power Systems, 19*(3), 1666–1673.
35. Koch, S., Mathieu, J. L., & Callaway, D. S. (2011). Modeling and control of aggregated heterogeneous thermosta-tically controlled loads for ancillary services. In *Proceedings of 17th Power System Computation Conference* (pp. 1–7).
36. Zhang, W., Lian, J., Chang, C. Y., Kalsi, K., & Sun, Y. (2012). Reduced-order modeling of aggregated thermostatic loads with demand response. In *IEEE Conference on Decision and Control* (pp. 5592–5597).
37. Ilic, S. M., Bullard, C. W., & Hrnjak, P. S. (2001). Effect of Shorter Compressor On/Off Cycle Times on A/C System Performance. Air Conditioning and Refrigeration Center, College of Engineering, University of Illinois at Urbana-Champaign.
38. PJM Regulation Data. [Online]. Retrieved from http://www.pjm.com/markets-and-operations/ancillary-services/mkt-based-regulation/fast-response-regulation-signal.aspx

Chapter 5
Uniform-Time State Bin Model of Aggregated TCLs for Regulation Services

Abstract TCLs can respond to control signals quickly by on–off switches to provide fast regulation service. A uniform-time state bin aggregate model (UTIM) with higher accuracy is introduced and the corresponding modelling constraints are first presented. Then, based on the UTIM, a novel aggregated control model that takes into account both the external control signals and the time delay of TCL compressor is proposed to follow the dynamic collective behavior of TCLs, which will make the UTIM more practical and accurate. To relieve the communication burden and reduce the number of switching times, a generalized control method is also proposed to control TCLs through the global signal. The simulation results show that the novel aggregate control model can track the collective performances of TCLs more closely, taking the time delay of the compressor into account. Additionally, the proposed control method requires fewer control signals and switching times.

Keywords Uniform-time state bin · Aggregate control model · Time delay · Communication burden

5.1 Introduction

As the increasing renewable energy generation is integrated into the grid, the discrepancy between supply and demand becomes more conspicuous [1], which causes substantial load fluctuations and enormous challenges to the stability and security of the power system operation [2]. To address this problem, fast regulation [3, 4], as an important ancillary service, is employed to maintain the balance between supply and demand. DR has been widely applied as a new solution. It can change the power consumption of end-users to follow the generation [5, 6] and alleviate the tension of supply-demand imbalance. As one of the major DR resources, TCLs have attracted considerable attention for their ability to store thermal energy in the buildings within a certain time frame[7, 8].

TCLs can be controlled by on-off switches [9–12] and thermostat set-point changes [13, 14] to provide different time-scale ancillary services to the power system. It has been stated [15] that compared with thermostat set-point control, on-off switch control can instantaneously change the power consumption of TCLs but

M. Song and C. Gao, *Integration of Distributed Resources in Smart Grids for Demand Response and Transactive Energy*, https://doi.org/10.1007/978-981-16-7170-8_5

cannot maintain the power adjustment for a long time, whereas thermostat set-point control can maintain power regulation for a long time but cannot take effect immediately. Thus, on-off control is more appropriate for TCLs to provide fast regulation service to the grid.

Due to the large number and wide distribution of TCLs, the aggregate models are necessary for the centralized controllers to develop control strategies from the system perspective. A large body of literature is available on the aggregate model, describing the evolution of the indoor temperature of TCL over time.The aggregate models can be classified into three categories. The first one is the state-queuing model [16, 17], in which the operating states of TCLs are assumed to be uniformly distributed and then the aggregated operations can be characterized by several sequential states. Since those external control signals and the constraints of the devices are not considered in the state-queuing model, the aggregate model is difficult to be applied in dispatch and control problems. The second one is the battery model where the aggregated behaviours of TCLs are modelled as charging/discharging action [12, 18–21]. With this model, a population of TCLs can be very conveniently dispatched. However, since the detailed operating states of TCLs are not available,and the time delay of the compressor tends to be ignored, it is difficult to generate accurate control signals for TCLs. The third one is the state-space model where the operating states of TCLs are classified into several state bins and the aggregated time-varying performances can be derived by the state matrix [4, 22–25]. The state-space model helps not only to embed the other variables into the aggregate model but also to reveal the time-varying behaviours of individual TCL. Accordingly, the state-space model will be intensively studied in this paper. However, the state bin in the state-space model has an identical temperature range [22–25] in most reported studies and we refer to this model as the uniform-temperature state bin aggregate model (UTEM). In contrast, the uniform-time state bin aggregate model (UTIM), which makes the transport time of TCL in each state bin identical and is better than the UTEM regarding modeling accuracy [4], is rarely studied. The associated aggregate control model considering both control signals and the constraints of the devices are also rarely investigated, which may cause failure in the application of UTIM in dispatching and controlling a large population of TCLs.

When the regulation trajectory is tracked, the controller will generate control signals and send them to TCLs according to their current operating states. Generally, there are two main factors to be considered in generating the control signals. One is the lifetime of the compressor of TCLs which is affected by the switching frequency. The priority-stack control method was proposed [10, 11, 18] to schedule TCLs according to the indoor temperature, which reduces the switch control frequency of TCLs. The other one is the communication burden between the centralized controller and the population of devices. A probability control method is proposed, in which the centralized controller sends TCLs a probability signal and each TCL decides locally whether to change its switch state with the received probability [23, 26].Accordingly, there is no need for the centralized controller to control TCLs individually, which relieves the communication burden between the controller and the remote devices. Although the above factors have been mentioned in some studies, rarely both are considered. A control method was proposed [4] where TCLs get different probability signals according to the descending/ascending indoor temperature for less switching frequency and communication burden. However, the controller needs to identify and match TCLs with the right signals in advance, which will waste a substantial amount of time.

In general, the time delay of TCL compressor is not modelled [16–18] or regarded as a simple temperature constraint [4, 19]. However, in this way, the actual dynamic behaviour of TCLs cannot be tracked by the aggregate model. Thus, UTIM is used to model the basic performance of TCLs and a novel aggregate control model taking into account both external control signals and the time delay of the compressor is proposed to capture collective operations of TCLs accordingly. The proposed aggregate control model can accurately describe the operating evolution of TCLs with time and make the UTIM more practical to be dispatched in the fast regulation service.

The main work of this chapter is summarized as follows:

(1) We clearly defined the transport time constraint of TCLs in each state bin of the UTIM. From the simulations, we find that the transport time of TCLs in each state bin cannot be shorter than the dispatch period. Otherwise, the UTIM cannot track the operating behaviour of TCLs accurately and the aggregated power given by UTIM will appear divergent.

(2) A UTIM based aggregate control model of TCLs is proposed, considering the external control signals and the time delay of the compressors. It facilitates the function of the centralized controller to conveniently and accurately dispatch and control a large population of TCLs. The expression of the dynamic performance of TCLs is derived analytically when they are controlled to turn on/off considering the time delay of the compressor.

(3) Based on the aggregate control model, the capacity evaluation of TCLs is derived analytically. And a generalized control method, in which a global signal is distributed to all TCLs and each TCL responds locally, is proposed to relieve the communication burden between the controller and end-users. This will also greatly cut down the complexity of control. To decrease the number of switching times without affecting the lifetime of the compressor, the temperature priority is also used in the control method.

5.2 Problem Formulation

5.2.1 The Thermodynamic Model of TCL System

A simplified ETP model [10] of a residential TCL system is formulated by (5.1). The ETP model is suitable to simulate the heat transfer between TCL unit and the surroundings in residential and small commercial buildings. For simplicity, only the cooling mode of TCL is discussed in this chapter.

$$C_a \frac{dT_i}{dt} = \frac{(T_o - T_i)}{R} + Q^{'} - S(t) \cdot Q \tag{5.1}$$

Additionally, η is given as:

$$\eta = \frac{Q}{P} \tag{5.2}$$

TCL can only work at its rated power (on) or 0 power (off), which makes the indoor temperature change periodically within the range [$T_{\min}$, $T_{\max}$] ($T_{\min} = T_{\text{set}}$-1/2·δ, $T_{\max} = T_{\text{set}}$ + 1/2·δ). When TCL is on, it operates at its rated power and then the indoor temperature begins to decline until it reaches $T_{\min}$. Then TCL is turned off and the indoor temperature begins to rise. When the indoor temperature is higher than $T_{\max}$, TCL will be turned on. The switch operating mode is written as

$$S(t + \Delta t) = \begin{cases} 1 \ T_i(t + \Delta t) > T_{\max} \\ 0 \ T_i(t + \Delta t) < T_{\min} \\ S(t) \ T_{\min} \le T_i(t + \Delta t) \le T_{\max} \end{cases} \tag{5.3}$$

Considering (5.1), t_{on}, is expressed as

$$t_{on} = -RC_a \ln \frac{T_o + RQ^{'} - \eta PR - T_{\min}}{T_o + RQ^{'} - \eta PR - T_{\max}} \tag{5.4}$$

Similarly, t_{off}, is given as

$$t_{off} = -RC_a \ln \frac{T_o + RQ^{'} - T_{\max}}{T_o + RQ^{'} - T_{\min}} \tag{5.5}$$

Accordingly, the duty cycle of TCL is described as

$$d_y = \frac{t_{on}}{t_{on} + t_{off}} \tag{5.6}$$

The average electric power consumption of TCL during a cycle is calculated as

$$P_0 = P \cdot d_y \tag{5.7}$$

When a population of TCLs operates without external control signals, the average power consumption (power baseline), P_{baseline}, is expressed as [16]

$$P_{baseline} = \sum_{i=1}^{N} P_0^i \tag{5.8}$$

5.2.2 Utim

Generally, the operating states of TCL are divided into several state bins in terms of the indoor temperature and switch state for modelling the aggregated behaviours of TCLs. Most published reports tend to make the temperature range of each state bin identical [22–25]. However, as TCLs in a state bin have different initial temperatures, they are transported to the next state bin at different rates (the rate is defined as the inverse of the time when TCL moves from a state bin to the next). Then, the average rate between two state bins cannot accurately describe the actual transport behaviours of TCLs for UTEM. In contrast, the transport time of TCLs in each state bin is kept identical in the UTIM so that TCLs can be transported between two neighbouring bins at the same rate. Therefore, UTIM is superior to UTEM in terms of modeling accuracy [4]. A schematic diagram of the UTIM is shown in Fig. 5.1. There are two kinds of states in the aggregate model, namely the OFF state (where the state bins are denoted by 1, 2, …, N_{off} and N_{off} is the number of the state bins) and ON state (where the state bins are denoted by $N_{off}+1, \ldots, N_{off}+N_{on}$ and N_{on} is the number of the state bins). Generally, N_{off} is different from N_{on} for the UTIM.

In case the state-space model is employed to express the aggregated operation of TCLs, the vector $\boldsymbol{x}$ defines the states of TCLs. If no control signal is given to TCLs, the dynamic transition behaviour of TCLs is formulated as [23, 27]:

$$\dot{\boldsymbol{x}} = G \cdot \boldsymbol{x}(t) \tag{5.9a}$$

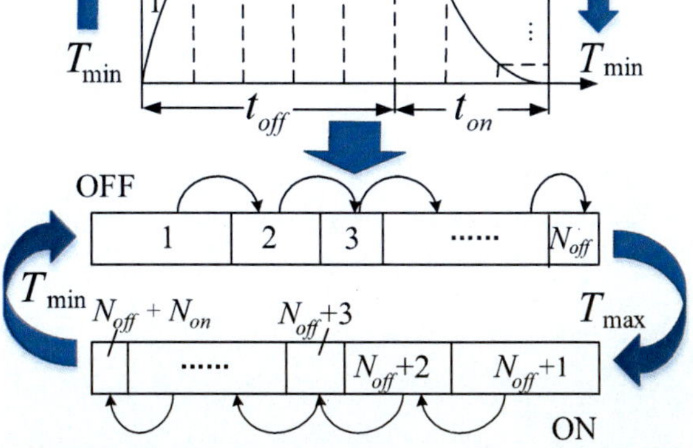

Fig. 5.1 UTIM

$$x(t) = [n_1(t)\ n_2(t) \cdots n_{N_{off}}(t)\ n_{N_{off}+1}(t) \cdots n_{N_{off}+N_{on}}(t)]^T \tag{5.9b}$$

$\boldsymbol{G}$ is a $(N_{off} + N_{on}) \times (N_{off} + N_{on})$ matrix; g_{lq} denotes the (l, q) in $\boldsymbol{G}$, which is the speed at which TCLs in the state bin q are transported to the state bin l. When the time interval is very short, TCLs in a state bin will only move to the next state bin or stay in the same state bin. Therefore, $\boldsymbol{G}$ is a large sparse matrix and g_{lp} is approximately given as [27]:

$$g_{lq} = \begin{cases} -\frac{1}{\Delta t_q'} & \text{if } l = q, 1 \le q \le N_{on} + N_{off} \\ \frac{1}{\Delta t_q'} & \text{if } l = q + 1, 1 \le q \le N_{on} + N_{off} - 1 \\ & or\ l = 1, q = N_{on} + N_{off} \end{cases} \tag{5.10}$$

where $\Delta t_q{}'$ is the time when a TCL is transported from the start point to the end point of the qth state bin.

Discretise (5.9a) as:

$$\boldsymbol{x}(k+1) = \boldsymbol{A} \cdot \boldsymbol{x}(k) \tag{5.11a}$$

$$\boldsymbol{A} = 1 + \boldsymbol{G} \cdot \Delta t \tag{5.11b}$$

where $\boldsymbol{x}(k + 1)$ and $\boldsymbol{x}(k)$ are the state vectors which contain the number of TCLs in each state bin at the $(k + 1)$th and the kth dispatch period, respectively; $\boldsymbol{A}$ is a $(N_{off} + N_{on}) \times (N_{off} + N_{on})$ matrix and is written as:

$$\boldsymbol{A} = \begin{bmatrix} 1 - \frac{\Delta t}{\Delta t_1'} & 0 & \cdots & \cdots & 0 & \frac{\Delta t}{\Delta t_{N_{off}+N_{on}}'} \\ \frac{\Delta t}{\Delta t_1'} & 1 - \frac{\Delta t}{\Delta t_2'} & 0 & \cdots & \cdots & 0 \\ \vdots & \vdots & \vdots & \vdots & \vdots & \vdots \\ \cdots & 0 & \frac{\Delta t}{\Delta t_{q-1}'} & 1 - \frac{\Delta t}{\Delta t_q'} & 0 & \cdots \\ \vdots & \vdots & \vdots & \vdots & \vdots & \vdots \\ 0 & \cdots & \cdots & 0 & \frac{\Delta t}{\Delta t_{N_{off}+N_{on}-1}'} & 1 - \frac{\Delta t}{\Delta t_{N_{off}+N_{on}}'} \end{bmatrix} \tag{5.12}$$

Each element in $\boldsymbol{A}$ is the transition probability of the TCLs when they are transported between the two neighbouring state bins. In the UTIM, the transport time of the TCLs in each state bin is identical ($\Delta t'$). Then (5.12) can be simplified as:

$$A = \begin{bmatrix} 1-\frac{\Delta t}{\Delta t'} & 0 & \cdots & \cdots & 0 & \frac{\Delta t}{\Delta t'} \\ \frac{\Delta t}{\Delta t'} & 1-\frac{\Delta t}{\Delta t'} & 0 & \cdots & \cdots & 0 \\ \vdots & \vdots & \vdots & \vdots & \vdots & \vdots \\ \cdots & 0 & \frac{\Delta t}{\Delta t'} & 1-\frac{\Delta t}{\Delta t'} & 0 & \cdots \\ \vdots & \vdots & \vdots & \vdots & \vdots & \vdots \\ 0 & \cdots & \cdots & 0 & \frac{\Delta t}{\Delta t'} & 1-\frac{\Delta t}{\Delta t'} \end{bmatrix} \tag{5.13}$$

Equation (5.11a) can describe the fundamental operating performances of TCLs without control signals. When TCLs receive the control signal to provide service to the grid, the on/off states of a certain number of TCLs will be switched and then the on/off distribution will be changed accordingly. Equation (5.11a) will not accurately track the collective behaviour of TCLs. Meanwhile, too frequent on/off switching can cause mechanical damage to the compressors. Then the time delay [11, 23] of the compressors needs to be carefully considered to avoid shortening the lifetime of the TCL units. When the time delay is manipulated, some TCLs will be forced to keep off for at least a certain amount of time, which will shape the power capacity of TCLs thus affecting the generation of control signals. Furthermore, from the perspective of the system, an aggregate control model that can describe the aggregated performances of TCLs under more considerations is of vital importance for the aggregator/system operator to conveniently and effectively dispatch and control a large number of TCLs. Thus, modelling the operations of both the external control signals and the time delay of the compressors in the UTIM is necessary and important.

Accordingly, the output power of TCLs at the kth dispatch period, $y(k)$, is given as:

$$y(k) = \boldsymbol{C} \cdot \boldsymbol{x}(k) \tag{5.14a}$$

$$\boldsymbol{C} = P \cdot [\underbrace{0, 0, \cdots, 0}_{N_{off}}, \underbrace{1, 1, \cdots, 1}_{N_{on}}] \tag{5.14b}$$

When it comes to the heterogeneity of TCLs, k-means algorithms [23, 27] can be used to address the problem and achieve a good modelling result. For simplicity, only the aggregate model of homogeneous TCLs is studied in this chapter.

5.3 Aggregate Control Model of TCLs

This section focuses on the transition performance of TCLs based on the UTIM when fast regulation service is provided. To make the aggregate model more practical, both the external control signals and the time delay of the compressor are carefully considered. Additionally, the transient behaviour of the TCLs when they are controlled to change their switch state is also analytically built in the aggregate control model.

When TCL is turned off, a certain amount of time is required to release the air pressure [23]. To avoid shortening the lifetime of the TCL, the time delay of the compressor is designed to keep the compressor off for a minimum amount of time (i.e. the locked time, t_{lock}). If the time during which the TCL remains off is less than t_{lock}, the TCL will not respond to any control signals and remain locked until it is unlocked. Thus, the whole TCLs, $\boldsymbol{x}(k)$, can be divided into two groups: the unconstrained TCLs, $\boldsymbol{x}^{\mathrm{a}}(k)$, which can be controlled freely and respond to control signals immediately, and the locked TCLs, $\boldsymbol{X}^{L}(k)$, which has been locked and cannot respond to control signals. $\boldsymbol{X}^{\mathrm{a}}(k)$ is a $(N_{off}+N_{on})\times 1$ matrix. $\boldsymbol{X}^{L}(k)$ is a $(N_{off}+N_{on})\times n_L(\,nL=t_{lock}/\Delta t)$ matrix, which is given as

$$\boldsymbol{X}^{L}(k)=[\boldsymbol{x}_1^{L}(k)\ \boldsymbol{x}_2^{L}(k)\cdots\boldsymbol{x}_p^{L}(k)\cdots\boldsymbol{x}_{nL}^{L}(k)] \tag{5.15a}$$

$$\begin{gathered}\boldsymbol{x}_p^{L}(k)=[n_{p,1}^{L}(k)\ n_{p,2}^{L}(k)\cdots n_{p,N_{off}}^{L}(k)\ 0\cdots 0]^T\\(1\le p\le nL)\end{gathered} \tag{5.15b}$$

where $\boldsymbol{x}_p^{L}(k)$ is a $(N_{off}+N_{on})\times 1$matrix. It represents the locked TCLs that will be unlocked after the nL-p + 1dispatch periods. $n_{p,h}^{L}(k)$ $(1\le h\le N_{off})$ is the number of locked TCLs in the hth state bin that have been locked for $p-1$ dispatch periods at the kth dispatch period.

$\boldsymbol{U}(k)$ is the control signal, which is expressed as

$$\boldsymbol{u}(k)=[n_1^{c}(k)\ n_2^{c}(k)\cdots n_w^{c}(k)\cdots n_{N_{off}+N_{on}}^{c}(k)]^T(1\le w\le N_{off}+N_{on}) \tag{5.16}$$

where $n_{\mathrm{w}}^{\mathrm{c}}(k)$ is the number of the TCLs in the wth state bin that will be controlled to change their switch status (on → off or off → on) when the dispatch period is k. The aggregate control model of TCLs will be derived below according to different control signals.

(1) The control signal is to turn off some TCLs

When the TCLs get $\boldsymbol{u}(k)$, some of them will be turned off and locked immediately. Then $\boldsymbol{x}^{a}(k)$, the unconstrained TCLs, is calculated as

$$\boldsymbol{x}^{a}(k)=\boldsymbol{x}(k)-\boldsymbol{x}^{L}(k)-\boldsymbol{u}(k) \tag{5.17a}$$

$$\boldsymbol{x}^{L}(k)=\sum_{i=1}^{nL}\boldsymbol{x}_i^{L}(k) \tag{5.17b}$$

where $\boldsymbol{x}^L(k)$ is the total number of the TCLs that are locked at the kth dispatch period.

The TCLs that are locked at the kth period because of $\boldsymbol{u}(k)$ are added to $\boldsymbol{X}^L(k)$. Then $\boldsymbol{X}^L$ at the $(k+1)$th dispatch period is updated as

$$\boldsymbol{X}^L(k+1) = \boldsymbol{A} \cdot (\boldsymbol{X}^L(k) + \boldsymbol{U}(k)) \tag{5.18a}$$

$$\boldsymbol{U}(k) = [\boldsymbol{B} \cdot \boldsymbol{u}(k), \boldsymbol{0}, \boldsymbol{0}, \cdots, \boldsymbol{0}] \tag{5.18b}$$

where $\boldsymbol{U}(k)$ is a $(N_{off} + N_{on}) \times nL$ matrix and $\boldsymbol{0}$ represents a $(N_{off} + N_{on}) \times 1$ null matrix; $\boldsymbol{B}{\cdot}\boldsymbol{u}(k)$ is the added locked TCLs as a result of responding to the control signals; $\boldsymbol{B}$ is a $(N_{off} + N_{on}) \times (N_{off} + N_{on})$ matrix that denotes the (m, n) in matrix $\boldsymbol{B}$ by b_{mn}; b_{mn} is the probability of the TCLs falling into the mth state bin from the nth state bin when the TCLs are controlled to change their switch state. Generally, the mth and nth state bins are located at different states (ON/OFF). Therefore, only if $N_{off} + 1 \leq n \leq N_{off} + N_{on}$ and $1 \leq m \leq N_{off}$ are both satisfied, b_{mn} can be nonzero. Otherwise, b_{mn} is 0.

The derivation of b_{mn} is shown as follows.

Obviously, the nth state bin in which the TCLs are turned off by the control signals is located at the ON state. These are the definitions of some variables: The initial temperature point of the nth state bin is T_n and its corresponding state bin in OFF state is $m1$. Similarly, the end temperature point of the nth state bin is T_{n+1} and its corresponding state bin in the OFF state is $m2$.

- If $m1 = m2$ is satisfied, then the TCLs in the nth state bin will totally fall into the $m1$th (or $m2$th) state bin when they respond to the control signal. Then the probability of the TCL falling into the $m1$th state bin from the nth state bin is 1, which is given as

$$b_{m1n} = b_{m2n} = 1 \tag{5.19}$$

- If $m1$–$m2 = 1$ is satisfied, then the TCLs in the nth state bin will fall into the $m1$th and $m2$th state bins. Since that the TCLs are prone to distribute uniformly with time (i.e. there will be a higher probability for the TCL to be in the temperature range in which it takes more time to be transported), the corresponding probabilities of the TCLs falling into $m1$th and $m2$th state bin from the nth state bin are related with the transport time of the TCL in the shadow of the $m1$th and $m2$th state bin on the nth state bin. According to (5.1), $\Delta t_{m1,n}$ and $\Delta t_{m2,n}$ are given as

$$\begin{cases} \Delta t_{m1,n} = -RC_a \ln \dfrac{T_o + RQ' - \eta PR - T_{m1}}{T_o + RQ' - \eta PR - T_n} \\ \Delta t_{m2,n} = -RC_a \ln \dfrac{T_o + RQ' - \eta PR - T_{n+1}}{T_o + RQ' - \eta PR - T_{m2+1}} \end{cases} \tag{5.20}$$

where T_{m1} and T_{m1+1} are the initial and end-point temperature of the $m1$th state bin, °C; T_{m2} and T_{m2+1} are the initial and end-point temperature of the $m2$th state bin,

°C; $\Delta t_{m1,n}$ is the transport time when the TCL is on and being transferred from T_n to T_{m1}; $\Delta t_{m2,n}$ is the transport time when the TCL is on and being transferred from T_{m2+1} to T_{n+1}.

Therefore, b_{m1n} and b_{m2n} are described as

$$\begin{cases} b_{m1n} = \frac{\Delta t_{m1,n}}{\Delta t} \\ b_{m2n} = \frac{\Delta t_{m2,n}}{\Delta t} \end{cases} \tag{5.21}$$

- If $m1–m2 > 1$is satisfied, then the TCLs in the nth state bin will fall into more than two state bins and they are specifically $m2$, ma ($m2 \leq ma \leq m1$), $m1$. The corresponding transport time of the TCL in the shadow of the $m1$th, math and $m2$th state bin on the nth state bin, $\Delta t_{m1,n}$, $\Delta t_{ma,n}$ and $\Delta t_{m2,n}$ are described as

$$\begin{cases} \Delta t_{m1,n} = -RC_a \ln \frac{T_o+RQ'-\eta PR-T_{m1}}{T_o+RQ'-\eta PR-T_n} \\ \Delta t_{ma,n} = -RC_a \ln \frac{T_o+RQ'-\eta PR-T_{ma}}{T_o+RQ'-\eta PR-T_{ma+1}} \\ \Delta t_{m2,n} = -RC_a \ln \frac{T_o+RQ'-\eta PR-T_{n+1}}{T_o+RQ'-\eta PR-T_{m2+1}} \end{cases} \tag{5.22}$$

where T_{ma} and T_{ma+1} are the initial and end-temperature points of the math state bin, °C.

The probabilities of the TCLs in the nth state bin falling into the $m2$th, math and $m1$th state bin are given as

$$\begin{cases} b_{m1n} = \frac{\Delta t_{m1,n}}{\Delta t} \\ b_{man} = \frac{\Delta t_{ma,n}}{\Delta t} \\ b_{m2n} = \frac{\Delta t_{m2,n}}{\Delta t} \end{cases} \tag{5.23}$$

The derivation diagram of b_{mn} is shown in Fig. 5.2.

It is noted that the locked time of the TCLs described by $\boldsymbol{X}^L$ will be reduced by Δt at the (k + 1)th dispatch period compared with that at the kth dispatch period. Then the column number of $\boldsymbol{x}_p^L(k)$ should be reduced by a time step. At the same

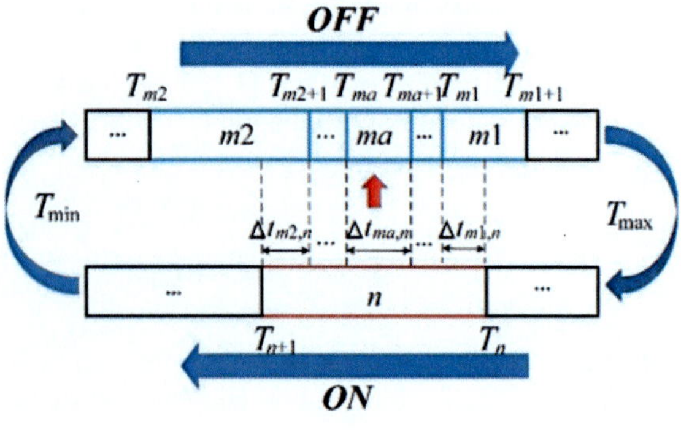

Fig. 5.2 The derivation diagram of b_{mn}

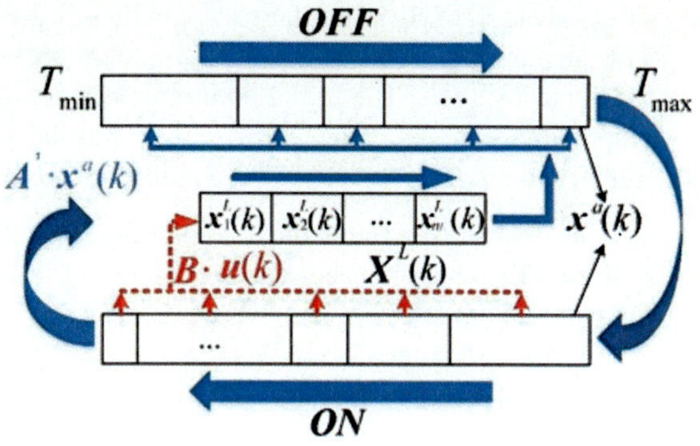

Fig. 5.3 The state transition diagram of the TCLs when the control signals are to turn off some TCLs (the blue solid line represents the natural transition process of the TCLs and the red dashed line is control signal-induced transition process caused by the control signals)

time, some unconstrained TCLs are naturally transported into the first state bin and are locked. Thus, $\boldsymbol{X}^L(k+1)$ should be further modified based on (5.18a) and (5.18b) and it is expressed as

$$\boldsymbol{X}^L(k+1) = [\boldsymbol{A}' \cdot \boldsymbol{x}^a(k), \boldsymbol{x}_1^L(k+1), \boldsymbol{x}_2^L(k+1), \ldots, \boldsymbol{x}_{nL-1}^L(k+1)] \tag{5.24}$$

where $\boldsymbol{A}' \cdot \boldsymbol{x}^a(k)$ are the TCLs that are naturally locked at the $(k+1)$th dispatch period. $\boldsymbol{A}'$ is a $(N_{\text{off}} + N_{\text{on}}) \times (N_{\mathit{off}} + N_{\text{on}})$ matrix. Its elements in the first row are the same as $\boldsymbol{A}$ and the other elements are 0.

The state transition diagram of $\boldsymbol{x}^a(k)$ and $\boldsymbol{X}(k)$ is shown in Fig. 5.3.

Given the control signals, $\boldsymbol{x}$ is updated as

$$\boldsymbol{x}(k+1) = \boldsymbol{A} \cdot (\boldsymbol{x}(k) + \boldsymbol{B}' \cdot \boldsymbol{u}(k)) \tag{5.25a}$$

$$\boldsymbol{B}' = \boldsymbol{B} - \boldsymbol{I} \tag{5.25b}$$

where $\boldsymbol{B}' \cdot \boldsymbol{u}(k)$ represents the added locked TCLs and the reduced TCLs in the ON state; $\boldsymbol{I}$ is a unit matrix, which is used to describe the reduced TCLs in the ON state.

(2) The control signal is to turn on some TCLs

When some TCLs are turned on after responding to $\boldsymbol{u}(k)$, $\boldsymbol{x}^a(k)$ is calculated as:

$$\boldsymbol{x}^a(k) = \boldsymbol{x}(k) - \boldsymbol{x}^L(k) + \boldsymbol{D} \cdot \boldsymbol{u}(k) \tag{5.26}$$

where $\boldsymbol{D} \cdot \boldsymbol{u}(k)$ represents the added TCLs in the ON state and the reduced TCLs in the OFF state because of $\boldsymbol{u}(k)$; $\boldsymbol{D}$ is a $(N_{\text{off}} + N_{\text{on}}) \times (N_{\mathit{off}} + N_{\text{on}})$ matrix and denotes the (i, j) in the matrix $\boldsymbol{D}$ by $d_{\text{ij.}}$ D_{ij} is the probability of TCLs falling into the ith state bin from the jth state bin. If $i = j (1 \le j \le N_{\mathit{off}})$ is satisfied, then d_{ij} is -1, which represents the reduced TCLs in the OFF state; if $N_{\text{off}} + 1 \le i \le N_{\text{off}} + N_{\text{on}}$ and $1 \le j \le N_{\mathit{off}}$ are both satisfied, then d_{ij} will be calculated as follow; otherwise, d_{ij} is 0.

The derivation of d_{ij} is similar to b_{mn}, which is given briefly below.

The jth state bin is located at the OFF state, in which the TCLs are turned on by the control signals. The initial point temperature of the jth state bin is T_j and its corresponding state bin in the ON state is $i1$. Similarly, the end-point temperature of the jth state bin is T_{j+1} and its corresponding state bin in the OFF state is $i2$.

- If $i1 = i2$ is satisfied, thenthe TCLs in the jth state bin will totally fall into the $i1$th (or $i2$th) state bin. Then the probability of the TCL falling into the $i1$th state bin from the jth state bin is 1, which is given as:

$$d_{i1j} = d_{i2j} = 1 \tag{5.27}$$

- If $i1–i2 = 1$is satisfied, the corresponding time when the TCL is transported through the shadow of the $i1$th and $i2$th state bin on the j thstate bin, $\Delta t_{i1,j}$ and $\Delta t_{i2,j}$ are given as:

$$\begin{cases} \Delta t_{i1,j} = -RC_a \ln \frac{T_o + RQ' - T_{i1}}{T_o + RQ' - T_j} \\ \Delta t_{i2,j} = -RC_a \ln \frac{T_o + RQ' - T_{j+1}}{T_o + RQ' - T_{i2+1}} \end{cases} \tag{5.28}$$

where T_{i1} and T_{i1+1} are the initial and end-point temperature of the $i1$th state bin, °C; T_{i2} and T_{i2+1} are the initial and end-point temperature of the $i2$th state bin, °C.

Then b_{i1j} and b_{i2j} are described as:

$$\begin{cases} d_{i1j} = \frac{\Delta t_{i1,j}}{\Delta t} \\ d_{i2j} = \frac{\Delta t_{i2,j}}{\Delta t} \end{cases} \tag{5.29}$$

- If $i1–i2 > 1$ is satisfied, the TCLs in the jth state bin will fall into more than two state bins and they are specifically $i2$, ia ($i2 \le ia \le i1$), $i1$. The corresponding operating time of the TCL in the shadow of the $i1$th, iath and $i2$th state bin on the jth state bin, $\Delta t_{i1,j}$, $\Delta t_{ia,j}$ and $\Delta t_{i2,j}$ are given as:

$$\begin{cases} \Delta t_{i1,j} = -RC_a \ln \frac{T_o + RQ' - T_{i1}}{T_o + RQ' - T_j} \\ \Delta t_{ia,j} = -RC_a \ln \frac{T_o + RQ' - T_{ia}}{T_o + RQ' - T_{ia+1}} \\ \Delta t_{i2,j} = -RC_a \ln \frac{T_o + RQ' - T_{j+1}}{T_o + RQ' - T_{i2+1}} \end{cases} \tag{5.30}$$

where T_{ia} and T_{ia+1} are the initial and end-point temperatures of the iath state bin, °C.

The probability of the TCLs in the jth state bin falling into the ith ($i2 \le i \le i1$) state bin is given as:

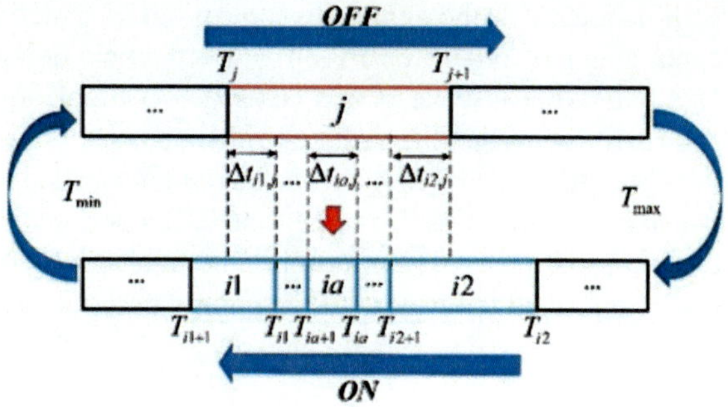

Fig. 5.4 Diagram of the derivation of d_{ij}

$$\begin{cases} d_{i1j} = \frac{\Delta t_{i1,j}}{\Delta t} \\ d_{iaj} = \frac{\Delta t_{ia,j}}{\Delta t} \\ d_{i2j} = \frac{\Delta t_{i2,j}}{\Delta t} \end{cases} \tag{5.31}$$

The derivation of d_{ij} is shown in Fig. 5.4.

There are no added locked TCLs and $\boldsymbol{X}^L$ at the (k + 1)th period is given as:

$$\boldsymbol{X}^L(k+1) = \boldsymbol{A} \cdot \boldsymbol{X}^L(k) \tag{5.32}$$

Considering that some unconstrained TCLs are naturally locked and some locked TCLs are released at the $(k+1)$th dispatch period, $\boldsymbol{X}^L$ $(k+1)$ needs to be modified as (5.24).

The state transition diagram of $\boldsymbol{x}^a(k)$ and $\boldsymbol{X}(k)$, when the control signals are to turn TCLs on, is shown in Fig. 5.5.

$\boldsymbol{x}$ is Updated as:

$$\boldsymbol{x}(k+1) = \boldsymbol{A} \cdot (\boldsymbol{x}(k) + \boldsymbol{D} \cdot \boldsymbol{u}(k)) \tag{5.33}$$

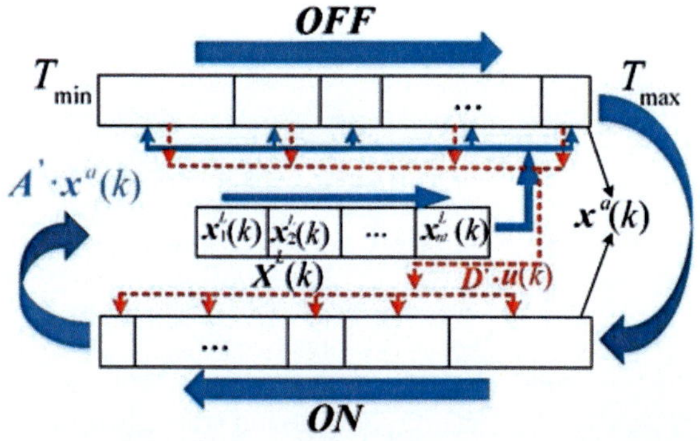

Fig. 5.5 The state transition diagram of the TCLs when the control signals are to turn on TCLs ($\boldsymbol{D}' = \boldsymbol{D} + \boldsymbol{I}$, $\boldsymbol{D}' \cdot \boldsymbol{u}(k)$ represents the reduced TCLs in the OFF state)

It is noted that the aggregate control model is derived according to the different control signals (on–off or off–on) when the time delay of the compressor is considered. This is because the locked TCLs, $\boldsymbol{X}^L(k)$, is updated in different ways [see (5.18a) and (5.32)] with the different control signals andrequires careful consideraation. In this way, the locked TCLs will be analytically modelled, which is helpful to distinguish the locked TCLs from the total TCLs and makes the aggregate model follow the behaviour of the TCLs closely. The transient behaviour of TCLs when they are being controlled to turn on/off are mainly described by the derivations of $\boldsymbol{B}$ and $\boldsymbol{D}$.

5.4 Control of TCLs for Fast Regulation Service

The aggregate control model of TCLs under external control signals and the time delay of the compressors was built in Sect. 5.3. Then the detailed operating information of TCLs (e.g. the switch states, the locked time, indoor temperature, etc.) can be acquired and provided to the aggregator/system operator to make further decisions. This section focuses on evaluating the potential (power capacity) of TCLs in the next dispatch period and generating control signals to be sent to the population of TCLs.

The unconstrained TCLs at the $(k+1)$th dispatch period, $\boldsymbol{x}^a(k+1)$, are described as:

$$\boldsymbol{x}^a(k+1) = \boldsymbol{x}(k+1) - \boldsymbol{x}^L(k+1) \tag{5.34}$$

where $\boldsymbol{x}^L(k+1)$ denotes the locked TCLs at the $(k+1)$th dispatch period. It can be calculated according to the update of $\boldsymbol{X}^L$ in (5.18a) and (5.32); $\boldsymbol{x}^a(k+1)$ is used to describe the controllable TCLs at the $(k+1)$th dispatch period and provides the power capacity information that will guide the centralized controller to develop the appropriate control strategy.

The up power is the collective power of the TCLs that are in the OFF state but not locked. Then the up power at the $(k+1)$th dispatch period, $M^{\mathrm{up}}(k+1)$ is calculated as follows.

$$M^{up}(k+1) = \boldsymbol{H}^{up} \cdot \boldsymbol{x}^a(k+1) \tag{5.35a}$$

$$\boldsymbol{H}^{up} = P \cdot [\underbrace{1, 1, \ldots, 1}_{N_{off}}, \underbrace{0, 0, \ldots 0}_{N_{on}}] \tag{5.35b}$$

The down power is the collective power of the TCLs in the ON state. Considering the time delay of the compressor, the indoor temperature will be out of the range $[T_{\min}, T_{\max}]$ after a dispatch period when the indoor temperature of the controlled TCLs is very close to $T_{\max}$. Therefore, there is an insensitive region in the ON state where the TCLs will not be turned off.

From (5.1), the temperature point where it will take TCL in the OFF state a dispatch period to be transported to $T_{\max}$, T_s, is calculated as:

$$T_s = T_o + RQ' - (T_o + RQ' - T_{\max})e^{\frac{t_{lock}}{RC_a}} \tag{5.36}$$

Accordingly, the indoor temperature of the TCLs in the insensitive region should be higher than T_s. Assuming that the end state bin in the insensitive region is $N_{\text{off}} + nl$ (nl is the number of state bins in the insensitive region), the initial and end-point temperature of the $(N_{off} + nl)$th state bin, $T_{Noff+nl}$ and $T_{Noff+nl+1}$ should satisfy:

$$T_{N_{off}+nl} > T_s \ \&\ T_{N_{off}+nl+1} \leq T_s \tag{5.37a}$$

$$\begin{aligned} T_{N_{off}+nl} = {} & T_o + RQ' - \eta RP \\ & + (T_{\max} - T_o - RQ' + \eta RP)e^{-\frac{1}{RC_a}(N_{off}+nl-1)\Delta t'} \end{aligned} \tag{5.37b}$$

$$\begin{aligned} T_{N_{off}+nl+1} = {} & T_o + RQ' - \eta RP \\ & + (T_{\max} - T_o - RQ' + \eta RP)e^{-\frac{1}{RC_a}(N_{off}+nl)\Delta t'} \end{aligned} \tag{5.37c}$$

From (5.37a)–(5.37c), nl is expressed as:

$$nl = \{nl \in N_+ | l \leq nl < h\} \tag{5.38a}$$

$$l = -\frac{RC_a}{\Delta t'} \ln \frac{\eta RP - (T_o + RQ' - T_{\max})e^{\frac{t_{lock}}{RC_a}}}{T_{\max} - T_o - RQ' + \eta RP} - N_{off} \tag{5.38b}$$

$$h = -\frac{RC_a}{\Delta t'} \ln \frac{\eta RP - (T_o + RQ' - T_{\max})e^{\frac{t_{lock}}{RC_a}}}{T_{\max} - T_o - RQ' + \eta RP} - N_{off} + 1 \tag{5.38c}$$

Then the insensitive region in ON state will be determined. Considering that the operating states of the TCLs in a state bin are regarded as the same, the whole $(N_{\text{off}} + nl)$th state bin is included in the insensitive region even though there are some TCLs whose indoor temperature is lower than T_s. The diagram of the insensitive region is shown in Fig. 5.6.

Thus, the down power of the TCLs at the $(k + 1)$th dispatch period, $M^{\text{down}}(k + 1)$, is calculated as:

$$M^{down}(k+1) = \boldsymbol{H}^{down} \cdot \boldsymbol{x}^a(k+1) \tag{5.39a}$$

$$\boldsymbol{H}^{down} = P \cdot [\underbrace{0, 0, \ldots, 0}_{N_{off}+nl}, \underbrace{1, 1, \ldots 1}_{N_{on}-nl}] \tag{5.39b}$$

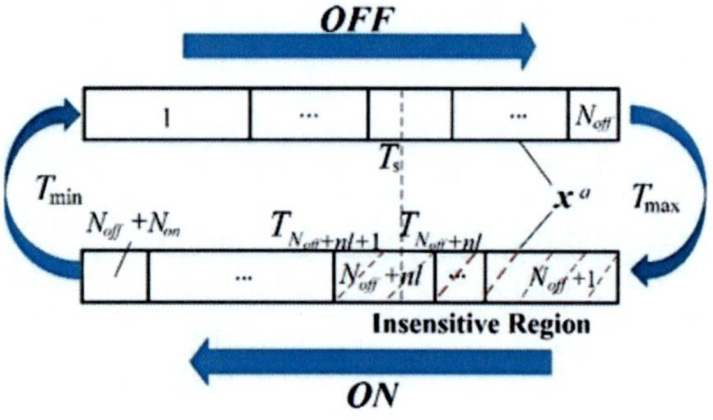

Fig. 5.6 Diagram of the insensitive region (the state bins in red line represent the insensitive region)

When the fast regulation service is served ($k = k + 1$), the aggregated power of the TCLs will be controlled to follow the regulation signal $P_f(k)$ which is a series of power deviations (compared with P_{baseline}) at a 4-s interval [28]. Then the desired aggregated power, $P_d(k)$, is calculated as:

$$P_d(k) = P_f(k) + P_{baseline} \tag{5.40}$$

The desired regulation power obtained by the on/off control of the TCLs, $e(k)$, is expressed as:

$$e(k) = P_d(k) - y(k) \tag{5.41}$$

Due to the limit of the actual power capacity, the centralized controller needs to determine the actual regulation power, $\Delta P(k)$, which is calculated as:

$$\Delta P(k) = \begin{cases} -M^{down}(k) \quad \text{if } e(k) < 0\&\&|e(k)| \geq M^{down}(k) \\ e(k) \quad \text{if } e(k) < 0\&\&|e(k)| < M^{down}(k) \\ M^{up}(k) \quad \text{if } e(k) \geq 0\&\&e(k) \geq M^{up}(k) \\ e(k) \quad \text{if } e(k) \geq 0\&\&e(k) < M^{up}(k) \end{cases} \tag{5.42}$$

Then the number of the TCLs that are turned on/off, z^c, is given as:

$$z^c = \frac{\Delta P(k)}{P} \tag{5.43}$$

To reduce the switching times, a priority queue control method [5] is applied here. When $\Delta P(k) \geq 0$ is satisfied, it means that there are some TCLs to be turned on to track the regulation signal. The unconstrained TCLs in the OFF state will be turned on in descending indoor temperature order until the increased power most closely approaches $\Delta P\ (k)$. When $\Delta P(k) < 0$ is satisfied, it means that some TCLs need to be turned off to shed the load. The TCLs with lower indoor temperature in the ON state are a bigger priority to be controlled and the TCLs in the $N_{\text{off}} + N_{\text{on}}$ state bin are at the top of the queue to be turned off. Then $\boldsymbol{u}(k)$, which is the input of the aggregate control model, can be given accordingly.

When the TCLs are turned on, the element of $\boldsymbol{u}(k)$ is expressed as:

$$n_w^c(k) = \begin{cases} \boldsymbol{x}^a(k)(w,1), \text{ if } \sum_{j=w}^{N_{off}} \boldsymbol{x}^a(k)(j,1) \le z^c \\ z^c - \sum_{j=w+1}^{N_{off}} \boldsymbol{x}^a(k)(j,1), if \sum_{j=w+1}^{N_{off}} \boldsymbol{x}^a(k)(j,1) \le z^c \\ \quad \& \sum_{j=w}^{N_{off}} \boldsymbol{x}^a(k)(j,1) > z^c \\ 0 \quad , \text{otherwise} \end{cases} \tag{5.44a}$$

where $\boldsymbol{x}^a(k)(j,1)$ is the jth element of $\boldsymbol{x}^a(k)$.

When the TCLs are turned off, the element of $\boldsymbol{u}(k)$ is calculated as:

$$n_w^c(k) = \begin{cases} \boldsymbol{x}^a(k)(w,1), if \sum_{j=w}^{N_{off}+N_{on}} \boldsymbol{x}^a(k)(j,1) \le z^c \\ z^c - \sum_{j=w+1}^{N_{off}+N_{on}} \boldsymbol{x}^a(k)(j,1), \text{ if } \sum_{j=w+1}^{N_{off}+N_{on}} \boldsymbol{x}^a(k)(j,1) \le z^c \\ \quad \& \sum_{j=w}^{N_{off}+N_{on}} \boldsymbol{x}^a(k)(j,1) > z^c \\ 0 \quad , otherwise \end{cases} \tag{5.44b}$$

$\boldsymbol{u}(k)$ as the input of the aggregate control model will be applied to update the operating states of the TCLs. And from (5.44a) and (5.44b), the critical w^*th state bin, in which, only some TCLs are turned on/off, will also be highlighted.

Evidently, the controller will need plenty of time to transmit control signals to each TCL, and the communication between the controller and the TCLs will be heavy when the control signals are distributed to the TCLs one by one according to $\boldsymbol{u}(k)$. Accordingly, a generalized control method is designed in which each TCL gets the same signal and changes its operating state based on its local information to help relieve the communication burden on the controller and make the system work effectively. The global signal in this chapter contains two variables, the desired indoor temperature, T^*, and the desired switch state, s^* (only two states, 1 and 0). If $\Delta P(k) \ge 0$ is satisfied, s^* is 1. If $\Delta P(k) < 0$ is satisfied, s^* is 0. When TCL gets the control signal (T^*, s^*), it will decide whether to change the switch state locally. Let s^* be 0, and the TCL will turn off only when the indoor temperature of the TCL that is on is lower than T^*. Let s^* be 1, and the TCL will turn on only when the indoor temperature of the TCL that is off and not locked is higher than T^*. Otherwise, the TCL will remain in its current state. T^* can be approximately calculated as:

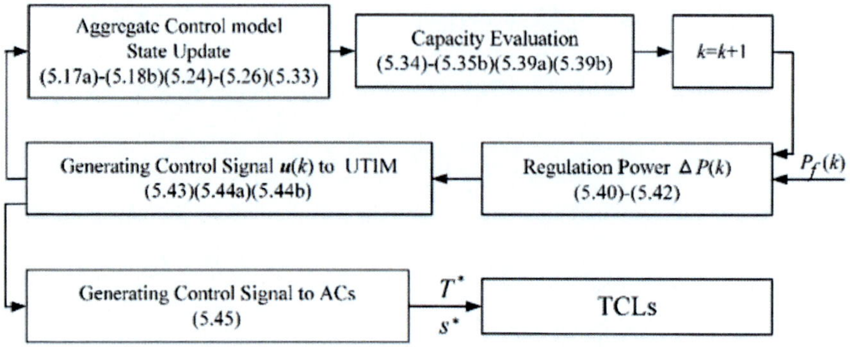

Fig. 5.7 The control framework of TCLs

$$T^* = \begin{cases} T_{w^*+1} - \frac{n^c_{w^*}(k)(T_{w^*+1} - T_{w^*})}{x^a(k)(w^*,1)}, & \text{if } s^* = 1 \\ T_{w^*+1} + \frac{n^c_{w^*}(k)(T_{w^*} - T_{w^*+1})}{x^a(k)(w^*,1)}, & \text{if } s^* = 0 \end{cases} \tag{5.45}$$

where T_{w*} and T_{w*+1} are the start and end-point temperature of the w^*th state bin.

The control framework of the population of TCLs is shown in Fig. 5.7.

5.5 Case Studies

In this case, the parameters of TCLs shown in Table. 5.1 are from a previously reported study [4] and the dispatch period is 4 s. The number of TCLs is 10000 and the default value of the compressor time delay is 10 min. The proposed aggregate control model is applied to follow the fast regulation signals previously reported [28], whose power magnitude has been scaled to match the power deviation capacity of 10,000 TCLs to keep each regulation signal within the upward and downward

Table 5.1 Parameters of TCLs

Parameter	Value	Unit
R	2	°C/kW
C_a	2	kWh/°C
Q'	0.2	kW
P	5.6	kW
$T_{\min}$	19	°C
$T_{\max}$	21	°C
H	2.5	/
T_{out}	32	°C
Δt	4	S

capacity of TCLs. Individual simulations [i.e. the results obtained by the simulation of the individual TCL model described in (5.1)] are regarded as the actual operating states of TCLs.

5.5.1 Performances of UTIM

In this section, the initial temperatures of TCLs are stochastically distributed within the range $[T_{\min}, T_{\max}]$ and the switch state (on/off) of TCLs are also given randomly. The simulation time is 4 h. The behaviour of the UTIM with different transport times of the TCL in each state bin is studied in this section. When the transport time of the TCL in each state bin is shorter than the dispatch time (4 s), the power error between the UTIM and individual simulations when the transport time is 3.5, 3 and 2 s are shown in Fig. 5.8. And the power error at the specific time is also given in Table. 5.2. Clearly, the magnitude of the power errors is very large (e.g. the magnitude is

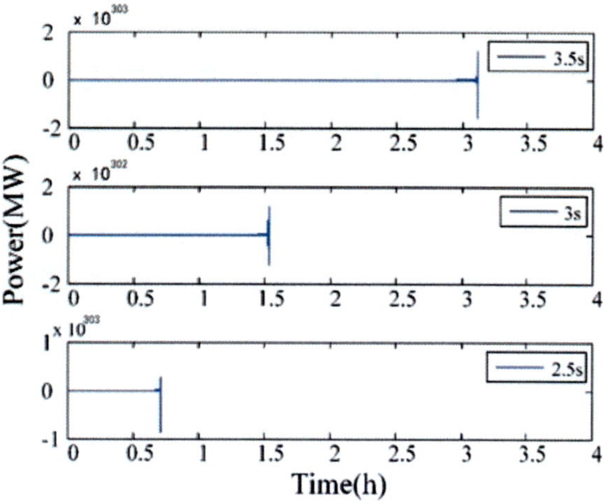

Fig. 5.8 The power errors between the UTIM and individual simulations when the transport time of the TCL in each state bin is 3.5, 3 and 2 s

Table 5.2 The power errors at the specific time (MW)

Specific time	The transport time in each state bin		
	3.5 (s)	3 (s)	2 (s)
0.5 (h)	1.2×10^{46}	-1.9×10^{96}	-2.9×10^{211}
1 (h)	4.5×10^{95}	4.1×10^{196}	\
1.5 (h)	2.2×10^{143}	\	\
2 (h)	1.6×10^{193}	\	\
2.5 (h)	-7.3×10^{242}	\	\
3 (h)	-5.2×10^{291}	\	\
3.5 (h)	\	\	\

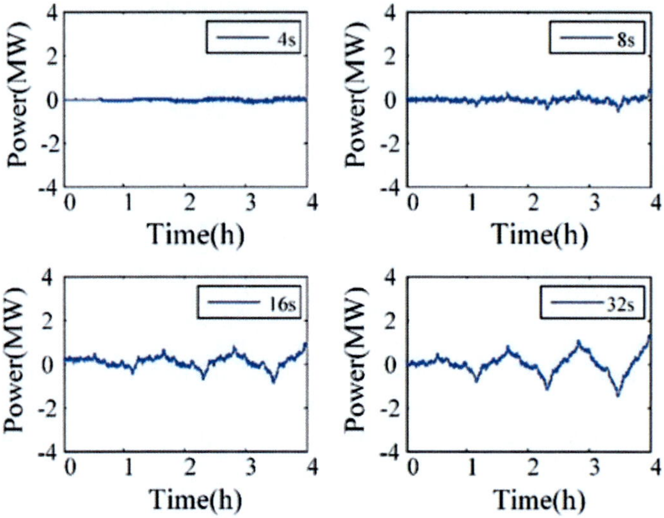

Fig. 5.9 The power errors between UTIM and individual simulations when the transport time of the TCL in each state is 4, 8, 16 and 32 s

10^{303} when the transport time is 3.5 s in Fig. 5.8, which means the power errors are divergent and the UTIM fails in following the dynamic behaviour of TCLs. Moreover, it will take less time to make the UTIM completely unfeasible as the transport time becomes shorter. In contrast, when the transport time is 4,8,16 and 32 s, the power errors, as shown in Fig. 5.9, indicate that the UTIM can closely track the behaviours of TCLs when the transport time is not shorter than the dispatch period. In addition, the modelling accuracy gradually improves when the transport time is shorter. A detailed explanation is given below.

The matrix $\boldsymbol{A}$ in (5.12) is used to describe the transition performance of TCLs and each element in $\boldsymbol{A}$ is the transition probability of TCLs from a state bin to another one. When the dispatch period is very short, the aggregate model described by (5.9a)–(5.13) will work and TCLs in a state bin can only stay in the same bin or be transported to the next one during a dispatch period. Accordingly, all elements in $\boldsymbol{A}$ should be kept within the range [0, 1]. If the transport time is shorter than the dispatch period, $\Delta t/\Delta t'_{q-1}$ and $1 - \Delta t/\Delta t'_q$ in $\boldsymbol{A}$ will be out of the range [0, 1]. At this time, TCLs in a state bin will be passed to the next state bin and transported to a farther one during a dispatch period, which prevents (5.11a) from accurately capturing the aggregation performance of TCLs. When the transport time is not shorter than the dispatch period, the aggregate model will be more accurate with more state bins [23]. In other words, the shorter the transport time, the better the model performance. Therefore, to ensure the good operation of the UTIM, the transport time of TCLin each state bin should not be shorter than the dispatch period.

5.5.2 *Performance of TCLs Following the Fast Regulation Signals*

The aggregate control model of TCLs is applied to track the fast regulation signals. The simulation time is 60 min. The transport time of TCLs in each state bin is equal to the dispatch period. The initial indoor temperature of TCLs is randomly distributed in the range $[T_{\min}, T_{\max}]$ and the ratio of the number of TCLs in the OFF state to that in the ON state is equal to t_{off}/t_{on}. According to the parameters in Table. 5.1, the power baseline of 10,000 TCLs is 24.8 MW and the aggregated rated power is 56 MW. Thus, we design a scenario where TCLs are dispatched to provide the fast regulation service with an upward and downward capacity of 20 MW.

The comparisons of the power with the aggregate control model, individual simulations and the desired power (i.e. the sum of the power baseline and the fast regulation signals) are shown in Fig. 5.10. It is evident that the three power profiles are almost coincident and the errors among them are very small.It demonstrates that the aggregate control model can precisely track the collective behaviours of TCLs and TCLs can follow the fast regulation signals very well. The comparisons of the locked TCL number of the aggregate control model and individual simulations are given in Fig. 5.11. The locked TCL number of the aggregate control model is nearly the same as that of individual simulations, which implies that the time delay of the compressor is accurately considered in the aggregate control model. This also means that the detailed transition behaviours of TCLs can be captured well using aggregate control model when TCLs are controlled to provide fast regulation service. The indoor temperature profiles of the population of TCLs are presented in Fig. 5.12. The indoor temperature of all TCLs fluctuates within the range [19 °C, 21 °C] ensuring the end-users' comfort. This indicates that the proposed control method is functioning well and has excellent accuracy.

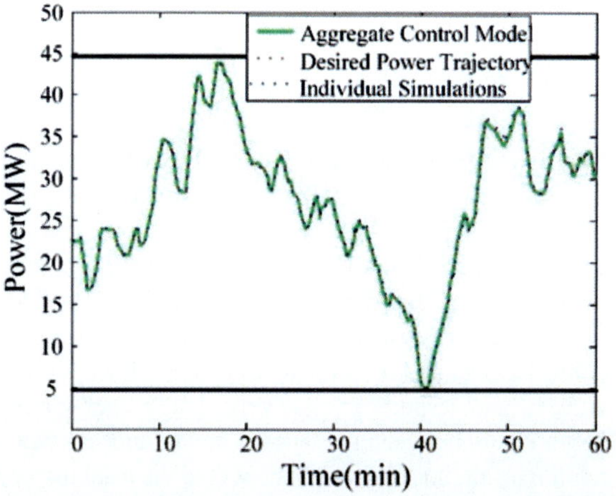

Fig. 5.10 Comparisons of power with the aggregate control model, individual simulations and the desired power

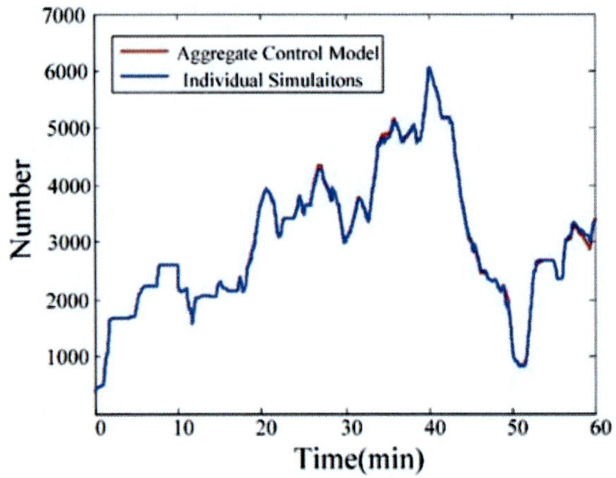

Fig. 5.11 The number of the locked TCLs of the aggregate control model and individual simulations

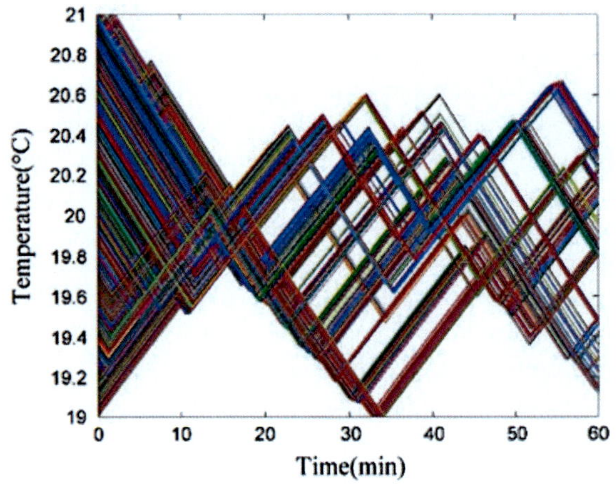

Fig. 5.12 The indoor temperature profiles of TCLs

5.5.3 *Time Delay Impact of the Compressor*

In the aggregate control model of TCLs described in Sect. 5.3, the time delay of the compressor is taken into account for the sake of prolonging the lifetime of the compressor. When the time delay of the compressor is not considered in the aggregate control model of TCLs and only treated as a temperature constraint [4, 19], the comparisons of the desired power trajectory and individual simulations are shown in Fig. 5.13. The corresponding number of TCLs in the OFF state of the aggregate control model and individual simulations is given in Fig. 5.14 when the locked time is 5, 10 and 15 min, respectively. The longer locked time obviously can result in larger errors between the desired power and individual simulations. In other words, the longer the locking time, the worse the tracking performance. This is because the time delay of some TCLs cannot be described by the simple design of the temperature

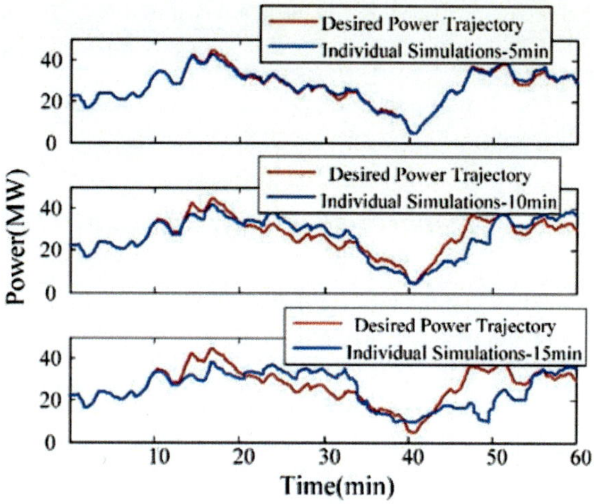

Fig. 5.13 The aggregated behaviours of TCLs when the time delay of the compressor is only modelled as a temperature constraint (the locked time is 5, 10 and 15 min, respectively)

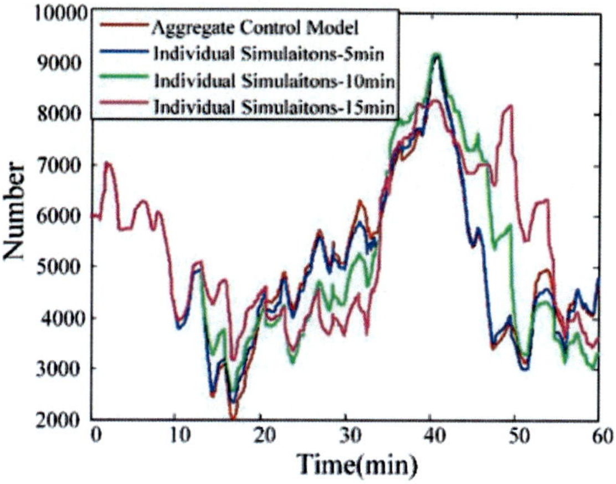

Fig. 5.14 The number of TCLs in the OFF state of the aggregate control model and individual simulations when the time delay of the compressor is only modelled as a temperature constraint (the locked time is 5, 10 and 15 min, respectively)

constraint when they are controlled to be turned off, which will lead to the incorrect signals being sent to TCLs. Moreover, the time delay is essentially a physical parameter of the compressor of each TCL. When the delay time of the compressor is longer, TCLs will be locked for a longer time and more TCLs will be out of the model causing larger errors, which may eventually produce economic losses or even operational risks to the power system. Therefore, the locked time is a key factor influencing the fast regulation signals following effect, which also means that proper modelling of the locked time of the compressor in the aggregate control model is very critical.

Table 5.3 ASTT and AST in the four scenarios

	S1	S2	S3	S4
ASTT	43.1	1	2	1
AST	3.8	24.1	3.8	3.8

5.5.4 Comparisons of Different Control Methods

To verify that the proposed generalized control method will relieve the communication burden and reduce the number of switching times, we compared the following four scenarios.

Scenario1 (S1): The centralized controller centrally distributes diverse control signals to individual TCL according to the operating states of TCLs [10, 18]. When the power consumption of TCLs needs to be adjusted, the controller will send massive signals to the controlled TCLs separately in descending/ascending indoor temperature order.

Scenario 2 (S2): The centralized controller generates one probability signal to all TCLs at each dispatch period [23, 26]. All TCLs locally respond to the common signal according to their operating states.

Scenario 3 (S3): The centralized controller generates two probability signals to the controlled TCLs at each dispatch period [4]. The controller will not only match TCLs with the signals but also distribute two different probability signals to the proper TCLs. Then TCLs will locally respond to the signals.

Scenario 4 (S4): The proposed control method in Sect. 5.4 is applied to control TCLs. The centralized controller only needs to generate one global signal to the population of TCLs at each dispatch period.

To quantify the communication burden and number of switching times of the four scenarios, the average signal transmission times (ASTT) of the controller per dispatch period and the average switch times (AST) of each TCL during the 60 min are shown in Table 5.3. The indoor temperature profiles of TCLs in Scenario 1, 3 and 4 are quite similar as shown in Fig. 5.12.

Noticeably, compared with the other two scenarios, it takes less average signal transmission times in the S2 and S4 during the simulation sending only one control signal per dispatch period. Meanwhile, the temperature priority is considered in S1, S3 and S4, so that fewerswitching times are required. This is because the TCLs in S1, S3 and S4, having indoor temperature closer to the natural temperature switch point ($T_{\min}$ /$T_{\max}$), will have higher priority to change their switch states when the aggregated power needs to be reduced/increased. In this way, the controlled TCLs will change their switch states before their natural switch time.Moreover, the indoor temperatures of the TCLs with higher priorities are closer to the last natural temperature switch point ($T_{\min}$/$T_{\max}$) and farther from the next one ($T_{\max}$/$T_{\min}$) after they are controlled, which means a longer time is needed to change the switching state naturally. Thus, the indoor temperature of very few TCLs' reaches $T_{\min}$/$T_{\max}$

and fewer switching times are performed in Fig. 5.12. In contrast, each TCL has the same probability (priority) to be controlled to change the operating states in S2. The TCLs that just naturally switched their on/off states can be controlled to change their switch states again. Moreover, their switch states will naturally change in a short time. Therefore, there will be more switching times and the indoor temperature of many more TCLs reaches T_{min}/T_{max} in S2. In conclusion, the proposed control method can considerably relieve the communication burden and reduce the number of switching times.

5.6 Conclusions

A novel aggregate control model that takes into account both the control signals and time delay of the compressor is proposed via UTIM. From the numerical simulation, we found that (1) the transport time of TCLs in each state bin of the UTIM should not be shorter than the dispatch period. Otherwise, the UTIM will fail to follow the performance of TCLs; (2) the aggregate control model can closely track the dynamic behaviour of TCLs and enable TCLs to follow fast regulation signals well, which demonstrates that the aggregate control model is accurately built; (3) TCLs cannot closely follow the frequency regulation signals when the locked time of the compressor is modelled as a simple temperature constraint. Additionally, the tracking behaviour will get worse with a longer locked time. (4) The proposed control method can relieve the communication burden between the centralized controller and TCLs and reduce the number of switching times.

References

1. Jad, I.A., Chicco, G., & Napoli, R. (2016). definitions of demand flexibility for aggregate residential loads. *IEEE Transactions on Smart Grid, 7*(6), 2633–2643.
2. Hui, H., Ding, Y., Liu, W., et al. (2017). Operating reserve evaluation of aggregated air conditioners. *Applied Energy, 196*, 218–228.
3. Kim, Y. J., Norford, L. K., & Kirtley, J. L. (2015). Modeling and analysis of a variable speed heat pump for frequency regulation through direct load control. *IEEE Transactions on Power Systems, 30*(1), 397–408.
4. Song, M., Ciwei, G., Yang, J., Liu, Y., & Cui, G. (2017). Novel aggregate control model of air conditioning loads for fast regulation service. *IET Generation, Transmission & Distribution, 11*(17), 4391–4401.
5. Lu, N., Du, P., & Makarov, Y. V. (2012, July). The potential of thermostatically controlled appliances for intra-hour energy storage applications. In *2012 IEEE Power and Energy Society General Meeting* (pp. 1–6).
6. Mai, W., & Chung, C. Y. (2015). Economic MPC of aggregating commercial buildings for providing flexible power reserve. *IEEE Transactions on Power Systems, 30*(5), 2685–2694.
7. Zhang, Y., & Lu, N. (2013). Parameter selection for a centralized thermostatically controlled appliances load controller used for intra-hour load balancing. *IEEE Transactions on Smart Grid, 4*(4), 2100–2108.

8. Saleh, S. A., Pijnenburg, P., & Castillo-Guerra, E. (2017). Load Aggregation from generation-follows-load to load-follows-generation: Residential loads. *IEEE Transactions on Industry Applications, 53*(2), 833–842.
9. Lu, N., & Zhang, Y. (2013). Design considerations of a centralized load controller using thermostatically controlled appliances for continuous regulation reserves. *IEEE Transactions on Smart Grid, 4*(2), 914–921.
10. Lu, N. (2012). An evaluation of the HVAC load potential for providing load balancing service. *IEEE Transactions on Smart Grid, 3*(3), 1263–1270.
11. Sanandaji, B. M., Vincent, T. L., & Poolla, K. (2016). Ramping rate flexibility of residential hvac loads. *IEEE Transactions on Sustainable Energy, 7*(2), 865–874.
12. Trovato, V., Tindemans, S. H., & Strbac, G. (2016). Leaky storage model for optimal multi-service allocation of thermostatic loads. *IET Generation, Transmission & Distribution, 10*(30), 585–593.
13. Zhang, B., & Baillieul, J. (2013). A packetized direct load control mechanism for demand side management. *Computer Science, 23*(1), 3658–3665.
14. Hong, Y. Y., Lin, J. K., Wu, C. P., & Chuang, C. C. (2012). Multi-objective air-conditioning control considering fuzzy parameters using immune clonal selection programming. *IEEE Transactions on Smart Grid, 3*(3), 1603–1610.
15. Wai, C. H., Beaudin, M., Zareipour, H., et al. (2015). cooling devices in demand response: A comparison of control methods. *IEEE Transactions on Smart Grid, 6*(1), 249–260.
16. Lu, N., Chassin, D. P., & Widergren, S. E. (2005). Modelling uncertainties in aggregated thermostatically controlled loads using a State queuing model. *IEEE Transactions on Power Systems, 20*(2), 725–733.
17. Lu, N., & Chassin, D. P. (2004). A state-queueing model of thermostatically controlled appliances. *IEEE Transactions on Power Systems, 19*(3), 1666–1673.
18. Hao, H., Sanandaji, B. M., Poolla, K., et al. (2015). Aggregate flexibility of thermostatically controlled loads. *IEEE Transactions on Power Systems, 30*(1), 189–198.
19. Sanandaji, B. M., Hao, H., & Poolla, K. (2014). Improved battery models of an aggregation of thermostatically controlled loads for frequency regulation. In *2014 American Control Conference* (pp. 38–45).
20. Zhao, L., & Zhang, W. (2016). A geometric approach to virtual battery modeling of thermostatically controlled loads. In *American Control Conference (ACC)* (pp. 1452–1457).
21. Trovato, V., Teng, F., & Strbac, G. (2017). Role and benefits of flexible thermostatically controlled loads in future low-carbon systems. *IEEE Transactions on Smart Grid, 9*(5), 5067–5079.
22. Bashash, S., & Fathy, H. K. (2013). Modeling and control of aggregate air conditioning loads for robust renewable power management. *IEEE Transactions on Control Systems Technologies, 21*(4), 1318–1327.
23. Zhang, W., Lian, J., Chang, C. Y., et al. (2013). Aggregated modeling and control of air conditioning loads for demand response. *IEEE Transactions on Power Systems, 28*(4), 4655–4664.
24. Liu, M., & Shi, Y. (2016). Model predictive control of aggregated heterogeneous second-order thermostatically controlled loads for ancillary services. *IEEE Transactions on Power Systems, 31*(3), 1963–1971.
25. Mathieu, J. L., Kamgarpour, M., Lygeros, J., et al. (2015). Arbitraging intraday wholesale energy market prices with aggregations of thermostatic loads. *IEEE Transactions on Power Systems, 30*(2), 763–772.
26. Kara, E. C., Bergés, M., & Hug, G. (2015). Impact of disturbances on modeling of thermostatically controlled loads for demand response. *IEEE Transactions on Smart Grid, 6*(5), 2560–2568.
27. W. Zhang, K. Kalsi, J. Fuller, M. Elizondo & D. Chassin. (2012). Aggregate model for heterogeneous thermostatically controlled loads with demand response. In *2012 IEEE Power and Energy Society General Meeting* (pp. 1–8).
28. PJM regulation data. http://www.pjm.com/markets-and-operations/ancillary-services/mkt-based-regulation/fast-response-regulation-signal.aspx.

Chapter 6
Thermal Battery Modeling of TCLs for Demand Response

Abstract To achieve the compatibility of the inverter TCLs with the current dispatch models, this chapter attempts to model an inverter TCL system as a thermal battery (TB). The comparisons between the TB and the lithium-ion battery are conducted. A hierarchical control framework is designed and an aggregated TB model is proposed to handle the heterogeneity of the inverter TCLs, which can protect the end-users' privacy and relieve the computational burden of the centralized control. A finite-horizon optimization model is used to compare the operating performances of the aggregated TBs and the lithium-ion batteries. Simulation results demonstrate that the aggregated TB model works well with the power dispatch model developed for lithium-ion batteries. In other words, with the TB modeling of TCLs, lithium-ion batteries can be replaced in current dispatch models providing service for the grid.

Keywords Inverter TCL · Thermal battery · Hierarchical control

6.1 Introduction

Direct load control (DLC) is an important control method of DR [1–4]. As a main category of DLC, centralized control is often used to control loads to take part in demand response programs because of its controllability and predictability [5]. Arora et al. [6] proposes an optimal dispatching mechanism for centralized TCLs to reduce the customer's electricity bills. A novel dynamic parameter selection method is presented to optimize the operating behaviors of a centralized controller and all uncertainties are taken into consideration [2]. The rate of the variation of temperature is calculated and communicated to the centralized controller for a performance enhancement [7]. A centralized control architecture- priority stacks [8] is proposed to select TCLs to turn on/off for frequency regulation, which helps to reduce the impacts of the centralized control on end-users' comfort. However, when a large number of loads are managed, the centralized control fails to be applied due either to the end-users' privacy protection or to the heavy computational burden [9, 10]. Thus, how to solve the two problems of centralized control is an important research point.

M. Song and C. Gao, *Integration of Distributed Resources in Smart Grids for Demand Response and Transactive Energy*, https://doi.org/10.1007/978-981-16-7170-8_6

Thermal energy generated by TCLs can be stored in the building, which is similar to the conventional battery. There have been several pieces of literatures [8, 11–14] that propose the battery modeling of the TCL system. A battery model of the TCL system is presented in [8] for the frequency regulation service to the grid. The battery model is used to aggregate the flexibility of TCLs and provide the adjustment potential information (i.e. the power limit and energy capacity of TCLs) for the operator. Based on [8], clustering of heterogeneous TCLs and no-short-cycling are taken into account in the improved battery model [11, 12]. A time-varying battery is proposed to keep track of the energy state of the TCLs and calculate the optimal control trajectories [13, 14]. Although the battery models of TCL systems have been investigated in some literature, these models above are all about fixed frequency TCLs and not very detailed. The energy storage modeling of inverter TCL systems has rarely been studied. Besides, there are no related literature about how to build a battery model compared with the conventional battery.

A large number of heterogeneous TCLs are hard to be controlled individually and an aggregated model is necessary for the centralized control. In most literature, as for heterogeneous parameters of loads, k-means algorithms [15, 16] are used to handle the problem for the aggregated models. The heterogeneous loads are classified into several clusters and each cluster is associated with a center. The parameters of the loads in a cluster are regarded as homogeneous [17–19]. Then the modeling problem of heterogeneous loads is simplified as the sum of several aggregated models of homogeneous loads. And there have been many aggregated modeling methods for homogeneous TCLs [13, 20–22]. Although the k-means algorithms are effective to solve the problem of heterogeneity, the associated computational burden is huge. Apparently, more clusters will make the aggregated model better at following the operating performances of the heterogeneous loads. However, more clusters indicate more computational burden. Therefore, a succinct aggregated model of heterogeneous TCLs is necessary.

The compressors of TCLs have evolved from fixed speed to variable speed units. Inverter TCLs have gained a noticeably growing market share for their higher energy efficiency [23]. However, fixed frequency TCLs are the main highlights in the majority of present literature and there are many differences between fixed frequency and inverter TCLs. It means that it is impossible to model and control inverter TCLs in the way of fixed frequency TCLs. As a result, residential and small commercial inverter TCLs are considered fully with regards to the thermal battery (TB) modeling and dispatch.

This chapter focuses on the following points.

(1) In the majority of literature, DR of the fixed frequency TCLs are thoroughly studied, while inverter TCLs are rarely studied in DR from the perspective of the system.
(2) Full derivation of the TB model for individual inverter TCL system is given. And the comparisons between the TB model and lithium-ion battery are also provided. Then a general inverter TCL system is encapsulated as a TB model,

which helps to protect the privacy of the end-users and be compatible with the existing dispatch model.

(3) A hierarchical control framework is designed to solve the two problems of centralized control (i.e. end-users' privacy disclosure and the computational burden). An aggregated battery model is proposed, which solves the parameter heterogeneity problem of TBs and relieves the computational burden to some extent.

6.2 Fundamental Model of the Inverter TCL System

6.2.1 Thermodynamic Model of the Inverter TCL System

The thermodynamic model of the inverter TCL system is used to describe the thermal dynamic transition process in a room. The simplified ETP model [4] of a residential TCL system described in (6.1) is common and it shows the dynamic behaviors of T_{i} with Q. There are two operating modes (i.e. heating and cooling mode) for the TCL unit and the cooling mode is discussed in this context.

$$C_a \frac{dT_i}{dt} = \frac{(T_o - T_i)}{R} + Q^{'} - Q \tag{6.1}$$

6.2.2 Electrical Model of the Inverter TCL System

The electrical model of the inverter TCL system describes the relationship between P and Q. The majority of power consumption of the inverter TCL occurs during the compressor operation, whose coefficient of performance is not constant and varies with f. According to [24, 25], by further simplifications, the relationship between f and P, is expressed approximately as

$$P = k_1 f + l_1 \tag{6.2}$$

Similarly, Q is given as

$$Q = k_2 f + l_2 \tag{6.3}$$

From (6.2) and (6.3), the relationship between P and Q is described as.

$$Q = \frac{k_2}{k_1} P + \frac{k_1 l_2 - l_1 k_2}{k_1} \tag{6.4}$$

The fundamental model of an inverter TCL system is composed of (6.1) and (6.4), which bridges the relationship between T_{i} and P. Thus, it is possible to adjust P continuously via controlling f to change T_{i} according to (6.1)–(6.4). This makes the inverter TCL be better at flexible control and have more various operating performances than fixed frequency TCLs [16–19] that can only operate at two states: rated power and off. In this context, f as the control input is used to adjust the electricity consumption of the inverter TCL system.

6.3 Thermal Battery Modeling of Individual Inverter TCL System

6.3.1 Lithium-Ion Battery Model

Lithium-ion batteries [26–28] are the most conventional batteries. They usually have the following parameters and constraints.

$x(k)$ is a very important parameter and is regarded as constant during the time interval between k and $k + 1$. $x(k) > 0$ means the battery is charging and $x(k) < 0$ is for discharging. Due to the operating limits of the battery, $x(k)$ is constrained by

$$-x^{d} \leq x(k) \leq x^{c} \tag{6.5}$$

$s(k)$ represents the state of charge (SOC), which is used to describe the energy storage level of the lithium-ion battery. $S(k)$ is defined as

$$s(k) = \frac{o(k)}{o^{\max}} \tag{6.6}$$

Considering the controlled charge/discharge and self-discharge, the update of SOC is expressed as

$$s(k+1) = (1-\sigma)s(k) + \frac{\eta(k)\cdot\Delta t}{o^{\max}}x(k) \tag{6.7}$$

$$\eta(k) = \begin{cases} \eta^{in}, x(k) \geq 0 \\ \frac{1}{\eta^{out}}, x(k) < 0 \end{cases} \tag{6.8}$$

σ is the energy dissipation rate as a result of the self-discharge power of the battery [29]. The self-discharge power of the battery is so small that σ is taken as 0 in some literature [26–28]. $0 < \eta^{\mathrm{in}} < 1$ and $0 < \eta^{\mathrm{out}} < 1$ are both satisfied [26]. Thus, when the battery is charging, $\eta(k) < 1$ while $\eta(k) > 1$ when the battery is discharging.

Considering the capacity limits of the battery, $s(k + 1)$ should be constrained by

$$s_{\min} \leq s(k+1) \leq s_{\max} \tag{6.9}$$

To sum up, a conventional battery model can be modeled by (6.5) (6.7) and (6.9) and its parameters are mainly $o^{\max}$, x^d, x^c, $s_{\min}$ and $s_{\max}$. For the system operator (SO), it might be familiar with the conventional battery models and there have been many dispatch models developed for the batteries [29–31]. However, the model of the inverter TCL system and its parameters are alien to the SO. Thus, some special considerations need to be taken into account and the existing dispatch models may be modified to accommodate the TCLs' participation, which will bring some schedule troubles. To solve the compatibility problem, a TB model of an inverter TCL system is proposed. It shares the same expression with the lithium-ion battery, which will make the inverter TCL dispatched in the same way as the lithium-ion battery. Thus, the modification of the current scheduling model is not required, which will provide great convenience.

6.3.2 TB Modeling of the Inverter TCL System

(1) *Capacity*

The stored energy in the buildings exists in the form of thermal energy and its capacity is related to the difference between indoor and outdoor temperature. Assuming that indoor temperature is kept in $[T_{\min}, T_{\max}]$ to guarantee the end-users' comfort, it is obvious that the stored energy is maximum when the indoor temperature is $T_{\min}$ while it is minimum when the indoor temperature is $T_{\max}$. Make the stored energy when the indoor temperature is $T_{\max}$ as the energy baseline. According to (6.1), $O_a(k)$ is described as:

$$O_a(k) = C_a(T_{\max} - T_i(k)) \tag{6.10}$$

When $T_i(k)$ is equal to $T_{\min}$, the stored energy is maximum and O is given as

$$O = C_a(T_{\max} - T_{\min}) \tag{6.11}$$

(2) *Maximum Charge and Discharge Power*

Generally, the inverter TCL works at the initial temperature set-point T_{set0} if there are no external control signals. The electric power is regarded as P_{baseline} when the TCL works at T_{set0}. From (6.1), the corresponding $Q_{baseline}$, is calculated as.

$$Q_{baseline} = \frac{1}{R}(T_o - T_{set0}) + Q' \tag{6.12}$$

Considering (6.4) and (6.12), P_{baseline} is given as

$$P_{baseline} = \frac{k_1(T_o - Tset_0)}{k_2 R} + \frac{k_1 Q' + k_2 l_1 - k_1 l_2}{k_2} \tag{6.13}$$

It should be noted that $P_{baseline}$ is related with T_o according to (6.13). When T_o changes, $P_{baseline}$ will fluctuate. However, T_o changes slowly with time and it can be seen as a fixed value during a small time interval. Then, it is reasonable to treat $P_{baseline}$ as a fixed value within a small time increment.

When the electric power of the inverter TCL deviates from P_{baseline} as a result of the external control signals, the power deviation is considered as the charge/discharge power [13, 14], $P_{ch/dis}(k)$, which is given as

$$P_{ch/dis}(k) = P(k) - P_{baseline} \tag{6.14}$$

It is obvious that the electric power cannot be increased/ decreased infinitely and it is restricted in $[P_{\min}, P_{\max}]$ because of the f limits. Therefore, in any time intervals, $P_{ch/dis}(k)$ satisfies

$$-P^d \leq P_{ch/dis}(k) \leq P^c \tag{6.15a}$$

$$P^d = P_{baseline} - P_{\min} \tag{6.15b}$$

$$P^c = P_{\max} - P_{baseline} \tag{6.15c}$$

Equation (6.15a)–(6.15c) describes the charge/discharge power limits of a TB. It means that $P_{ch/dis}(k)$ will be limited in $[-P^d, P^c]$ in any time, which is the same as that of the lithium-ion battery.

(3) *The Definition and Update of SOC*

The SOC of a TB is also used to describe the stored energy level and is defined identically to that of a lithium-ion battery. From (6.10) and (6.11), the SOC of the TB is given as

$$soc(k) = \frac{O_g(k)}{O} = \frac{T_{\max} - T_i(k)}{T_{\max} - T_{\min}} \tag{6.16}$$

According to the definition of SOC in (6.6), the update of *SOC* of the TB is formulated as

$$soc(k+1) = soc(k) + \frac{\Delta O(k)}{O} \tag{6.17}$$

$\Delta O(k)$ is the increment of the stored energy during the kth period. Similarly, it is also composed of two parts and they are the dissipation energy as a result of self-discharge and the cooling energy generated by the inverter TCL. The self-discharge energy is used to transfer the internal heat generation and the heat energy from the outdoor surroundings. Thus, the dissipation power at the time t, $P_{diss}(t)$, is calculated as

$$P_{diss}(t) = Q' + \frac{1}{R}(T_o - T_i(t))\ (k\Delta t \le t \le (k+1)\Delta t) \quad (6.18)$$

$\Delta O(k)$ in (6.17) is the sum of the increment of the cooling power during the kth dispatch period, which is written as

$$\Delta O(k) = \int_{k\Delta t}^{(k+1)\Delta t} (Q(t) - P_{diss}(t))dt \quad (6.19)$$

For simplicity, make the initial time point of the kth time interval as the time 0. From (6.18), $\Delta O(k)$ in (6.19) is simplified as

$$\Delta O(k) = \int_{0}^{\Delta t} (Q(t) - Q' - \frac{1}{R}(T_o - T_i(t)))dt \quad (6.20)$$

It should be noted that $Q(t)$ in (6.20) is controlled to be constant during the kth dispatch period and $Q(t) = Q(k)$ is satisfied. According to (6.4) and (6.14), $Q(t)$ is given as

$$Q(t) = Q(k) = \frac{k_2}{k_1}(P_{ch/dis}(k) + P_{baseline}) + \frac{k_1 l_2 - l_1 k_2}{k_1} \quad (6.21)$$

Considering the SOC definition of the TB model in (6.16), P_{baseline} is rewritten as

$$P_{baseline} = \frac{k_1(T_{\max} - T_{\min})}{k_2 R} soc(0) + \frac{k_1(T_o - T_{\max}) + R\left(k_1 Q' + k_2 l_1 - k_1 l_2\right)}{k_2 R} \quad (6.22a)$$

$$soc(0) = \frac{T_{\max} - T_{set0}}{T_{\max} - T_{\min}} \quad (6.22b)$$

$soc(0)$ represents the initial energy storage level and it is only related to T_{set0}. It is seen as a constant parameter in the TB model.

Substitute (6.22a) and (6.22b) to (6.21) and then $Q(t)$ is rewritten as

$$Q(t) = \frac{k_2}{k_1} P_{ch/dis}(k) + \frac{T_{\max} - T_{\min}}{R} soc(0) + \frac{T_o - T_{\max} + RQ'}{R} \quad (6.23)$$

By solving (6.1), $T_{\text{i}}(t)$ in (6.20) is given as

$$T_i(t) = (T_i(0) - T_o - RQ' + RQ(t))e^{-\frac{t}{RC_a}} + T_o + RQ' - RQ(t) \quad (6.24)$$

$T_i(0)$ in (6.24) is the initial point of indoor temperature during the kth time interval and then $T_i(0) = T_i(k)$ is satisfied. According to (6.16), $T_i(0)$ is rewritten as

$$T_i(0) = T_i(k) = T_{\max} - soc(k)(T_{\max} - T_{\min}) \tag{6.25}$$

Substituting (6.11), (6.20) and (6.23)–(6.25) to (6.17), the SOC update of the TB is calculated as:

$$soc(k+1) = \alpha soc(k) + \beta P_{ch/dis}(k) + \gamma \tag{6.26a}$$

$$\alpha = e^{-\frac{\Delta t}{RC_a}} \tag{6.26b}$$

$$\beta = \frac{k_2 R(1 - e^{-\frac{\Delta t}{RC_a}})}{k_1(T_{\max} - T_{\min})} \tag{6.26c}$$

$$\gamma = (1 - \alpha) \cdot soc(0) \tag{6.26d}$$

$soc(k+1)$ is 0 when $T_i(k+1)$ is $T_{\max}$, and $soc(k+1)$ is 1 when $T_i(k+1)$ is $T_{\min}$. Otherwise, $0 < soc(k+1) < 1$ is satisfied. Thus, in any time intervals, $soc(k+1)$ should satisfy

$$0 \leq soc(k+1) \leq 1 \tag{6.27}$$

To sum up, the TB model of an inverter TCL system can be described by (6.15a) (6.26a) and (6.27). Compared with the lithium-ion battery model presented by (6.5) (6.7) and (6.8), the TB model is with the same expressions of the conventional lithium-ion battery model. Then, the inverter TCL can be treated as a conventional battery in regular dispatch.

6.3.3 Comparisons of the Two Battery Models

Specifically, the comparisons between the lithium-ion battery model and the TB model are shown in Table 6.1.

According to Table 6.1, it is clear that the TB model shares identical expressions and constraints with the lithium-ion battery model. However, there are some differences between them, which are summarized as follows.

(1) The maximum stored energy (i.e. capacity) is a deterministic parameter for the lithium-ion battery model while it is a relative parameter for the TB model. In fact, the maximum stored energy is much larger than O for a TB. Likewise, the charge/discharge power is also a relative parameter, which makes the controlled TCL loads the sum of TBs and the uncontrolled TCL loads. In other words,

Table 6.1 Comparisons between the lithium-ion battery and TB model

Model	The lithium-ion battery model	The TB model
Expression	$-x^d \leq x(k) \leq x^c$ $s(k+1) = (1-\tau)\cdot s(k)$ $+\eta(k)\cdot \Delta t \cdot x(k)/o^{\max}$ $s_{\min} \leq s(k+1) \leq s_{\max}$	$-P^d \leq P_{ch/dis}(k) \leq P^c$ $soc(k+1) = \alpha soc(k)$ $+\beta P_{ch/dis}(k) + \gamma$ $0 \leq soc(k+1) \leq 1$
Variables and constraints	$o^{\max}$	O
	$x(k)$	$P_{ch/dis}(k)$
	x^d	P^d
	x^c	P^c
	$s(k)$	$soc(k)$
	$s_{\min}$	0
	$s_{\max}$	1

extra virtual batteries (TBs) are available based on the original system when the TCLs are dispatched by demand response programs.

(2) The dissipation power of the lithium-ion battery is much smaller than that of the TB. The energy dissipation rate of the lithium-ion battery is very small and it can be neglected in this chapter. However, the self-discharge power of the TB is considerable, which leads to distinctive operating performances from a lithium-ion battery. For example, when the SOC keeps unchanged during the kth time interval, according to (6.26a), $P_{ch/dis}(k)$ is calculated as

$$P_{ch/dis}(k) = \frac{(1-\alpha)(soc(k) - soc(0))}{\beta} \tag{6.28}$$

From (6.28), the TB is charging when $soc(k)$ is larger than $soc(0)$ while the TB is discharging when $soc(k)$ is smaller than $soc(0)$. $P_{ch/dis}(k)$ is 0 only when $soc(k)$ is equal to $soc(0)$ $(0 < \alpha < 1)$. It means that the SOC can keep fixed even though the TB is charging/discharging. Especially, the TB can discharge when $soc(k)$ is 0 and charge when $soc(k)$ is 1. However, for the lithium-ion battery, $x(k)$ is roughly 0 as long as $s(k+1)$ is equal to $s(k)$. This is one of the most important operating differences between the two models, which will be verified in Sect. 6.5.

(3) P^c and P^d will change with the varying T_o according to (6.13) (6.15b) and (6.15c) while x^c and x^d are constant. Thus, it is necessary to take T_o into account and update the parameters of the TB when the TB model is dispatched.

6.4 Hierarchical Control Design

6.4.1 Battery Encapsulation and Conversion

The heterogeneous TCLs are at the bottom layer of the hierarchical control framework. A smart controller is installed on each TCL and it can store data, calculate, measure outdoor temperature and receive/send signals from/to its upper layer by the way of broadcasting. Each end-user can choose an indoor temperature comfort range $[T_{\min}, T_{\max}]$ according to their own preference. It is assumed that the T_{set0} is in the range of $[T_{\min}, T_{\max}]$.

To protect the privacy of the end-users, the inverter TCL system is encapsulated as a battery model by the TB modeling described in Sect. 6.2. Then, the parameters (i.e. C_a, R, $Q^{'}$, k_1, k_2, l_1, l_2, $P_{\max}$, $P_{\min}$, $T_{\min}$, $T_{\max}$,T_{set0}) of the inverter TCL system are invisible to others and it appears to be a TB device with the parameters (P^d, P^c, α, β, γ). In other words, the end-user's information about its building design, device type and its preferences is protected well. Thus, it is reasonable to transmit the parameters of the TB model to a centralized controller without leaking end-user' private information.

Before the demand response program runs, each controller of the TB device will broadcast its parameters (P^d, P^c, α, β, γ) to its upper controller. Generally, Δt is fixed during the program, and then α, β, γ are constant. Therefore, the controller of each TB only needs to broadcast α, β, γ to its upper layer once. P^d and P^c are related to T_o and it is necessary to update them when T_o changes.

At each dispatch period, each smart controller will receive an identical broadcast signal $\delta(k)$ (it is $soc(k+1)/\beta$ and this will be explained in Sect. 6.4.2 from the upper controller. The controllers transfer $\delta(k)$ into frequency signals to change their operating state according to their operating behaviors and parameters.

From (6.26a), $P_{ch/dis}(k)$ is calculated as

$$P_{ch/dis}(k) = \delta(k) - \alpha \frac{soc(k)}{\beta} - \frac{\gamma}{\beta} \tag{6.29}$$

According to (6.2) (6.14) (6.15c) and (6.29), $f(k)$ should be given as

$$f(k) = \frac{P_{\max} - P^c + \delta(k) - l_1}{k_1} - \frac{\alpha soc(k) + \gamma}{\beta k_1} \tag{6.30}$$

Then $\delta(k)$ is converted to a frequency signal that will be implemented by the TCL device.

At the same time, $soc(k)$ is updated as:

$$soc(k) = \beta \cdot \delta(k) \tag{6.31}$$

The design diagram of the smart controller is shown in Fig. 6.1.

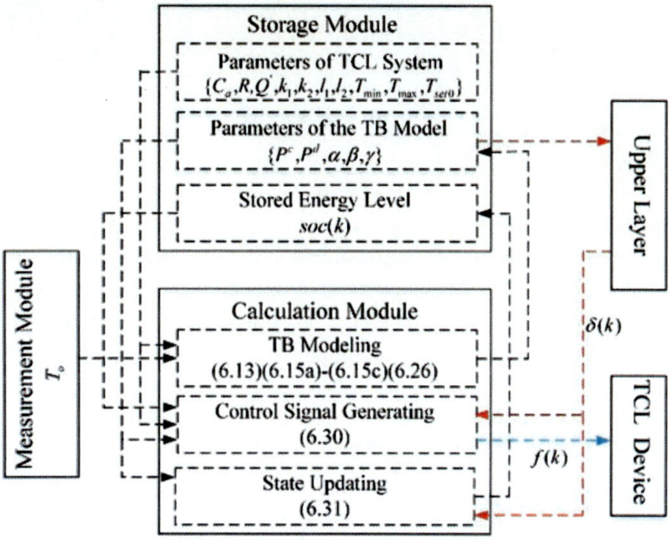

Fig. 6.1 The design diagram of the smart controller (the red line is the communication between the controller and the upper layer. The blue line is the control signal generated by the smart controller to the TCL device to adjust electric power. The black line is the communication among different modules in the smart controller)

6.4.2 *Aggregation and Control*

The aggregator is at the middle layer and its main aim is to provide a succinct aggregated TB model, get the dispatch signals from the top layer, and broadcast control signals to the TBs in the bottom layer.

For the population of heterogeneous TBs, the update of SOC is expressed as

$$\begin{cases} soc_1(k+1) = \alpha_1 soc_1(k) + \beta_1 P_{ch/dis\_1}(k) + \gamma_1 \\ soc_2(k+1) = \alpha_2 soc_2(k) + \beta_2 P_{ch/dis\_2}(k) + \gamma_2 \\ \qquad \vdots \\ soc_n(k+1) = \alpha_n soc_n(k) + \beta_n P_{ch/dis\_n}(k) + \gamma_n \end{cases} \tag{6.32}$$

It is impractical to control each TB individually and an aggregated model is necessary. However, the heterogeneous β makes it hard to add $P_{ch/dis}(k)$ of each TB in (6.32) for the power aggregation. To solve the problem, β_i is divided by the ith ($1 \le i \le n$) equation and add all the equations in (6.32). Thus, (6.32) is rewritten as

$$\sum_{i=1}^{n} \frac{soc(k+1)}{\beta_i} = \sum_{i=1}^{n} \frac{soc(k)}{\beta_i} + p_{agg}(k) + \gamma_{agg} \tag{6.33a}$$

$$P_{agg}(k) = \sum_{i=1}^{n} P_{ch/dis\_i}(k) \tag{6.33b}$$

$$\gamma_{agg} = \sum_{i=1}^{n} \frac{\gamma_i}{\beta_i} \tag{6.33c}$$

Equation (6.33a) can describe the relationship between the aggregated power of the TBs and the SOC of each TB. Unfortunately, the aggregated stored energy level of the TBs cannot be obtained because of heterogeneous α. To address this problem, $soc(k)/\beta$ is treated as a whole to describe the stored energy level of the TB. Then soc/β of each TB is controlled to be identical and $soc(k+1)/\beta = soc_1(k+1)/\beta_1 = soc_2(k+1)/\beta_2 = \cdots = soc_n(k+1)/\beta_n$ is satisfied. Generally, the initial soc/β *of* each TB is diverse and the aggregator must send the TBs a signal to synchronize them before the demand response. (6.33a) is simplified as

$$n\frac{soc(k+1)}{\beta} = \frac{soc(k+1)}{\beta}\sum_{i=1}^{n}\alpha_i + P_{agg}(k) + \gamma_{agg} \tag{6.34}$$

$SOC_{agg}(k)$ is defined to describe the stored energy level of the aggregated TBs and given as

$$SOC_{agg}(k) = n\frac{soc(k)}{\beta} \tag{6.35}$$

Considering (6.35), (6.34) is formulated as

$$SOC_{agg}(k+1) = \xi \cdot SOC_{agg}(k) + P_{agg}(k) + \gamma_{agg} \tag{6.36a}$$

$$\xi = \frac{\sum_{i=1}^{n}\alpha_i}{n} \tag{6.36b}$$

Then the update of the stored energy level of the aggregated TBs is expressed by (6.36a).

Accordingly, $SOC_{agg}(k+1)$ should be constrained by

$$0 \le SOC_{agg}(k+1) \le \tau \tag{6.37a}$$

$$\tau = n \cdot \min\left\{\frac{1}{\beta_i}, 1 \le i \le n\right\} \tag{6.37b}$$

$P_{agg}(k)$ is the sum of the charge/discharge power of TBs, which should be constrained by

$$-P_{agg}^{d} \leq P_{agg}(k) \leq P_{agg}^{c} \tag{6.38a}$$

$$P_{agg}^{d} = \sum_{i=1}^{n} P_i^{d} \tag{6.38b}$$

$$P_{agg}^{c} = \sum_{i=1}^{n} P_i^{c} \tag{6.38c}$$

To sum up, the aggregated model of the TBs can be expressed by (6.36a) (6.37a) and (6.38a) and it is expressed with a battery model well, namely, a large number of TBs can be represented with an aggregated TB model. The aggregator only sends the parameters $\{Pd\ agg, Pc\ agg, \xi, \gamma_{agg}, \tau\}$ to the top layer and then the heterogeneous TBs will be dispatched as a whole.

When the aggregator receives dispatch signal $B(k)$ from the top layer, it will generate control signal $\delta(k)$ according to (6.39) and broadcast the signal to the TBs.

$$\delta(k) = \frac{soc(k)}{\beta} = \frac{\delta \cdot SOC_{agg}(k) + B(k) + \gamma_{agg}}{n} \tag{6.39}$$

At the same time, the stored energy level of the aggregated TB is updated ($k = k + 1$) as:

$$SOC_{agg}(k) = \xi \cdot SOC_{agg}(k) + B(k) + \gamma_{agg} \tag{6.40}$$

The centralized controller design of the aggregator is shown in Fig. 6.2.

Compared with other centralized control methods and aggregated models, the proposed aggregated model will help the aggregator relieve the computation burden, which is illustrated by the following two points. (1) When the control signals to the devices are generated, the centralized controller tends to generate different signals according to the operating state of each device [4, 8, 12], which will bring huge calculations and even communication delays. But in this dispatch framework, the identical control signal δ(k) for all the devices will be generated by (6.39), which is very concise and practical. (2) For the aggregated models, the traditional approach is with the k-means algorithms [15, 16], which are often used to handle the problem of heterogeneous parameters. The TBs will be divided into several groups according to the similarities of the parameters and the TBs in a group are aggregated analytically. Therefore, the heterogeneity of the TBs is expressed by forming several aggregated models, which will complicate the dispatch model and bring calculation trouble to the aggregator. By contrast, the heterogeneous TBs can be represented with one aggregated model in our framework, and the heterogeneity of the parameters is also fully considered. The aggregated model is simplified with much less computation burden.

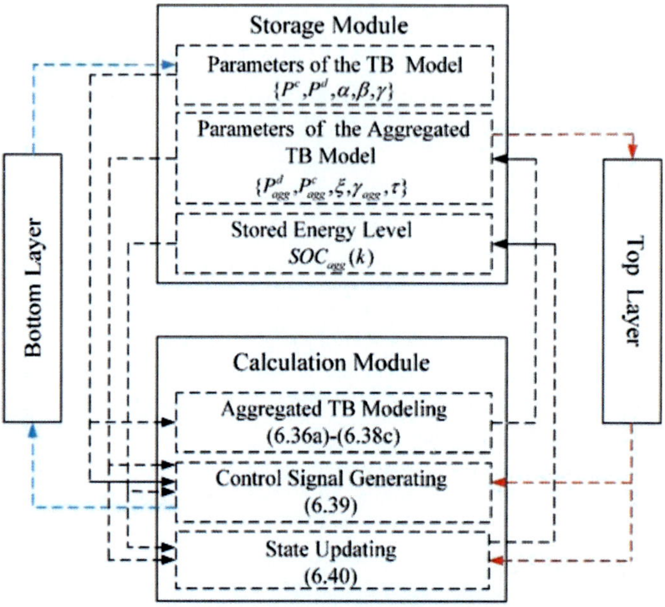

Fig. 6.2 The design diagram of the centralized controller. The red line is the communication between the controller and the upper layer. The blue line is the control signal generated by the centralized controller to the TBs at the bottom layer. The black line is the communication among different modules in the centralized controller

6.4.3 Optimal Dispatch

The main object of SO at the top layer is to develop an optimal dispatch strategy. There have been extensive literature focusing on the dispatch strategies where batteries are scheduled to participate in unit commitment [29, 32], peak shifting [33, 34], output smoothing of wind generation [31, 35, 36] or economic optimization in the electricity market [30, 37]. Among the above literature, batteries as conventional devices are dispatched for better system operating performance or more economic profit because of their flexible performances.

To show the compatibility of the TBs with current dispatch models, a conventional scheduling strategy for lithium-ion batteries is chosen from the extensive literature. SO participates in the real-time market to purchase/sell power to achieve supply–demand balance [38] and wind generation is also considered. Prior to a market period, market participants are informed to submit their schedules, and then market-clearing price is worked out by market rules. SO optimizes charge/discharge behaviors of TBs before submitting its offer according to wind generation, load, and electricity price forecast to maintain supply–demand balance and minimize cost. Then it submits its schedule using the optimization results. Two hypotheses are made in this context: (1) the offer of SO is too small for the whole market and SO is regarded as a price taker; (2) the load is inelastic to electricity price.

The objective function of SO is minimal energy cost in the real-time market. For the lithium-ion batteries, the dispatch model is:

$$\min F_c = \sum_{k=i}^{i+w} [(L(k) + B(k) - G(k))p_{rt}(k)]$$

Subject to:

$$-x^d \leq B(k) \leq x^c$$

$$SOC(k+1) = SOC(k) + \chi(k) \cdot B(k) \tag{6.41}$$

$$\chi(k) = \begin{cases} \frac{\eta^{in} \cdot \Delta t}{o^{\max}}, & B(k) \geq 0 \\ \frac{\Delta t}{\eta^{out} \cdot o^{\max}}, & B(k) < 0 \end{cases}$$

$$s_{\min} \leq SOC(k+1) \leq s_{\max}$$

For the aggregated TB model, the dispatch model is modified as:

$$\min F_c = \sum_{k=i}^{i+w} [(L(k) + B(k) - G(k))p_{rt}(k)]$$

Subject to:

$$-P_{agg}^d \leq B(k) \leq P_{agg}^c \tag{6.42}$$

$$SOC(k+1) = \xi \cdot SOC(k) + B(k) + \gamma_{agg}$$

$$0 \leq SOC(k+1) \leq \tau$$

It is a finite-horizon optimization problem [38, 39]. A mixed-integer linear programming algorithm is employed to solve the problem.

From (6.41) (6.42), the TB aggregated model quite fit the dispatch model of lithium-ion batteries. The only difference between the two dispatch models of the TBs and lithium-ion batteries lies in the coefficients of the SOC update. Thus, the aggregated TB model is compatible with the current dispatch model. In this way, the heterogeneous inverter TCLs are dispatched as a conventional battery device and thermal constraints of TCL are considered with electrical constraints of the SOC limit.

In general, the controllers at the bottom layer encapsulate the inverter TCL systems to protect the privacy of the end-users and undertake the main calculation burden.

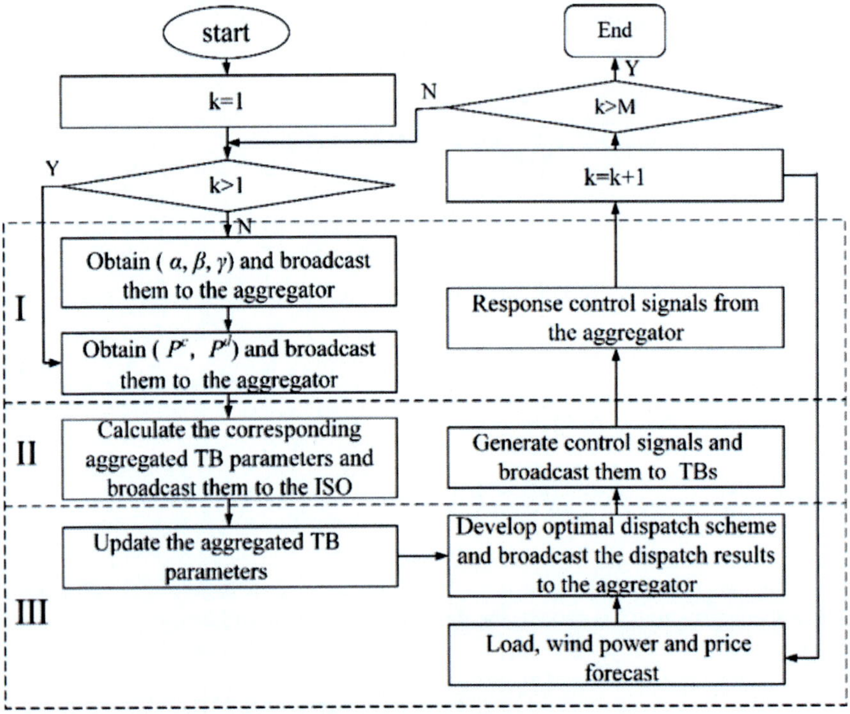

Fig. 6.3 Procedure of the hierarchical control framework. I–III stand for the bottom, middle and top layer, respectively. M is the total number of dispatch periods, which is determined by the SO. It should be noted that the smart controllers at *I* will calculate and broadcast *w* forecast values of (P^c, P^d) at each dispatch period according to the outdoor temperature forecast

The aggregator that coordinates the bottom and top layers aggregate heterogeneous TBs and generates control signals to the TBs. The optimal dispatch strategy with other system constraints is provided by the SO at the top layer. The procedure of the hierarchical control framework is summarized in Fig. 6.3.

6.5 Case Studies

In this case, we apply the proposed hierarchical control framework to verify the coordination of the three layers and the operating differences between the aggregated TBs and the lithium-ion batteries. The number of inverter TCLs is 2000. The related parameters of the TCL systems are shown in Table 6.1. R, C_a, Q' and T_{set0} are all distributed randomly in their ranges. The dispatch period is 0.5 h and the simulation time is 24 periods. The receding optimization horizon period is 4. Wind generation, load, and electricity price are from [38–40]. We use the outdoor temperature provided by [41], which is a typical summer outdoor temperature profile in the region of PJM.

During each dispatch period, the outdoor temperature is regarded as a fixed value. Because forecast is not the research point, the forecast method is from [42], which is shown as

$$V_f = V_{actual}(1 + h \cdot \Delta E \cdot Z) \quad (6.43)$$

where V_f, V_{actual} are the forecast and actual value of wind generation, load,electricity price and outdoor temperature, respectively. ΔE is the unit forecast error and it is 0.05, 0.01 0.02, and 0.01 for wind generation, load, electricity price and outdoor temperature, respectively. Z is a rand number that is distributed uniformly in $[-1, 1]$. h $(1 \leq h \leq w)$ represents the hth receding optimization period, which is used to show the fact that the forecast error will become larger when the forecast time is longer.

6.5.1 Optimization Results of TBs Compared with Lithium-Ion Batteries in Real-Time Market

Three dispatch strategies are designed to analyze the dynamic behaviors of TBs and lithium-ion batteries in the real-time market.

Strategy 1: The TCL loads as a part of the total load are not controlled and there are no any conventional batteries. SO purchases/sells electricity in the real-time market according to the power mismatches between load and wind generation;

Strategy 2: The lithium-ion batteries are equipped and the TCL loads are not controlled.SO optimizes the charge/discharge power of the aggregated TBs in the real-time market for minimal cost according to the dispatch model presented in Sect. 6.4;

Strategy 3: There are lithium-ion batteries and the TCL loads are not controlled. SO optimizes the charge/discharge power of lithium-ion batteries in a real-time market for minimal cost according to the dispatch model presented in Sect. 6.4. And the total cost in a day is similar to that of strategy 2.

The maximum charge and maximum discharge power profiles of the aggregated TBs with varying outdoor temperature are shown in Fig. 6.4, in which, the maximum charge power decreases when the outdoor temperature rises and it will increase when the outdoor temperature drops. It means that the maximum charge power of TB has an opposite changing tendency with outdoor temperature. Similarly, the maximum discharge power of the TB is with the same varying tendency of outdoor temperature. This is because that the baseline power of the TB [see (6.13)] has a positive linear relationship with outdoor temperature. When the outdoor temperature rises, the baseline power of the TB will increase, which will lead to a smaller charge power limit and a larger discharge limit according to the definitions of the TB in (6.15b) and (6.15c).

The other parameters of the aggregated TB model are fixed and calculated in Table 6.3.

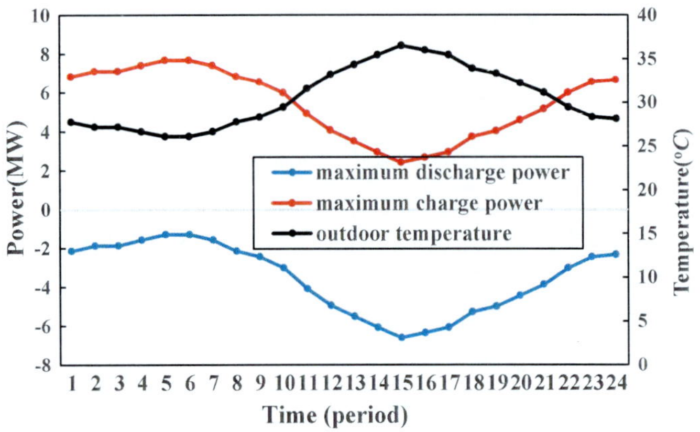

Fig. 6.4 The maximum charge and discharge power profiles of the TB with the changing outdoor temperature

Initially, the aggregator broadcasts the control signal to the TBs to synchronize them and $SOC_{\text{agg}}(0)$ is adjusted to half of τ, 3209.

When the cost of strategy 3 is nearly equal to that of strategy 2, the parameters of lithium-ion batteries are given in Table 6.4. It should be noted that there are many different ways of setting parameters to make the cost of strategy3 similar to that of strategy 2. The parameters of the lithium-ion batteries here are only representative examples and other parameter sets are also available.

According to [43], the capital cost of battery energy storage, C_{capital}, is expressed as:

$$C_{capital} = C_p P^{\max} + C_w W_{\max} \tag{6.44}$$

where $P^{\max}$ is the power capacity of battery energy storage, MW; $W_{\max}$ is the energy capacity, MWh; C_p, C_{w} are power and energy cost coefficients, respectively.

From [43], C_p is 206.7 dollar/kW and C_w is 437.4 dollar/kWh. Then C_{capital} is 4.6 million dollars. It means the service provided by the TBs is similar to that of lithium-ion batteries with a capital cost of 4.6 million dollars. Therefore, replacing lithium-ion batteries with TBs helps to save money. The increased cost is only the expense of extra measurement and control devices investment. In light of energy storage and economy, TBs is a good choice.

The cost of SO with the three strategies in the real-time market is 7464 dollars, 5982 dollars and 6010 dollars, respectively. The actual imported power of SO with the three strategies is shown in Fig. 6.5. Compared with SO without any energy storage devices, it is obvious that SO with energy storage devices (i.e. TBs or lithium-ion batteries) purchases more power at a lower price (e.g. the 3rd, 9th, 17th period) and purchases less power at a higher price (e.g. the 8th, 14th, 18th period). TBs or lithium-ion batteries can make SO operate more flexibly according to the electricity

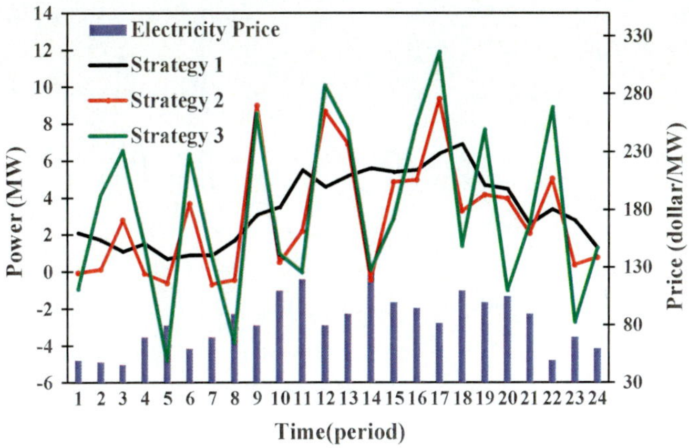

Fig. 6.5 The actual imported power of SO with the three dispatch strategies in real-time market

price variations and spend less money to keep supply–demand balance. It means that TBs have similar operating behaviors with lithium-ion batteries and can be used where the lithium-ion batteries are applied.

Figure 6.6 shows the charge/discharge power comparisons of TBs and lithium-ion batteries in the real-time market. With strategy 3, the conventional batteries are equipped and the maximum charge/discharge power are the constant parameters. Therefore, the peak charge power of each period is equal, which also applies to the peak discharge power. For example, the lithium-ion batteries always charge with 5.5 MW at the 3rd, 6th, 9th, 12th and 22th period. While with strategy 2, we find that the peak charge/discharge power of TBs are various since it is dependent upon the outdoor temperature. The $SOC_{\text{agg}}(k)$ profile of the aggregated TB model is given in Fig. 6.7. From Figs. 6.6 and 6.7, it is seen that the TBs can discharge when the SOC

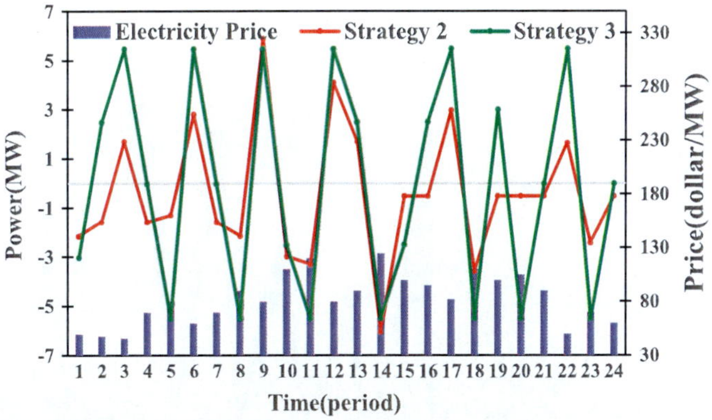

Fig. 6.6 The charge/discharge power of the aggregated TBs and lithium-ion batteries

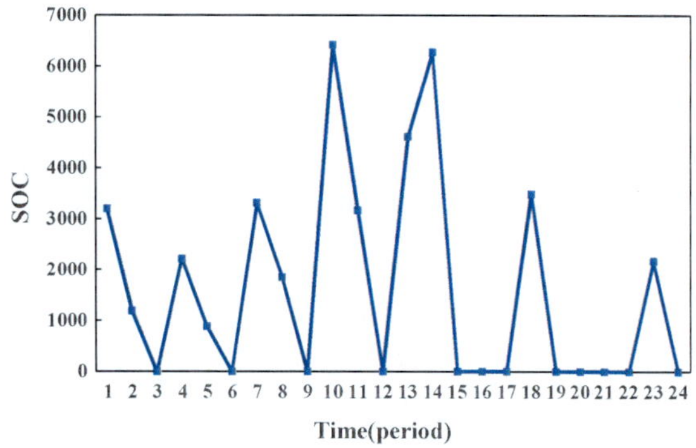

Fig. 6.7 The $SOC_{agg}(k)$ profile of the aggregated TBs

is 0 (e.g. the 15th, 16th, 19th-21th periods), which is not available for the lithium-ion batteries. Although the aggregated TBs are discharging, SOC is always kept 0. It means that the TBs can discharge forever and operate like batteries without capacity limits (this has been explained in Sect. 6.2.3). This is the most important difference between the two battery models.

Figure 6.8 shows the SOC profiles of the TBs when the TBs are dispatched with Strategy 2. The corresponding indoor temperature profiles of TCLs are given in Fig. 6.9. Before the time 0, soc/β of each TB is synchronized by the aggregator and the indoor temperature of TCLs is also adjusted accordingly. The indoor temperature profiles of TCLs develop oppositely compared with the SOC profiles of the TBs, which is consistent with the relationship between the SOC and indoor temperature [see (6.16)]. It demonstrates that the proposed TB model is correct and applicable. Meanwhile, since the upper limit of $SOC_{\text{agg}}(k+1)$ [see (6.37b)] is determined by the minimum $1/\beta$ of the TBs, the SOC of some TBs cannot reach 1 when the aggregated TB model is charged ***at utmost***. For example, the SOC of some TBs is only about 0.6 at the 9th period when the SOC_{agg} reaches the maximum value, τ. It means that

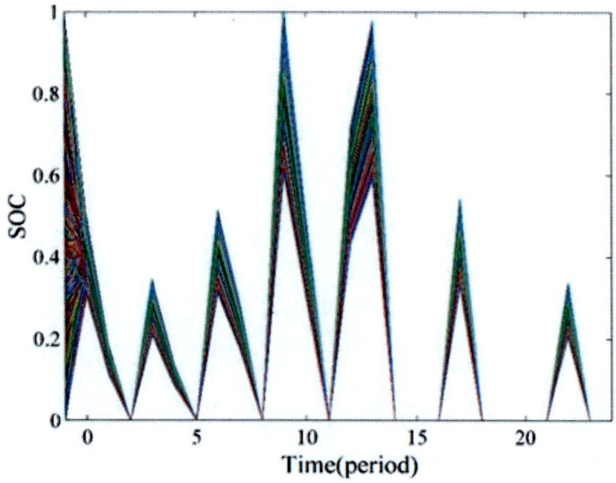

Fig. 6.8 The SOC profiles of the TBs

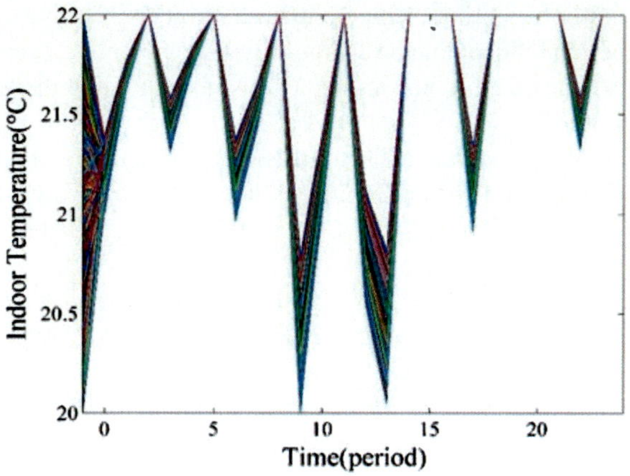

Fig. 6.9 The indoor temperature profiles of TCLs

some TBs cannot be utilized fully, which is the defect of the proposed synchronized control.

To verify that the proposed dispatch framework can relieve the calculation burden of the centralized control, we compare the computation time of the two scenarios with MATAB R2014a. In the first scenario, the proposed method is used to solve the dispatch model and generate the control signals to the TBs. This requires 0.3 s. In the second scenario, the k-means method [15, 16] is applied to cluster the TBs into 10 groups and the control signals are generated by the priority-stack method [4, 8, 12]. It takes about 4.7 s, which is much longer than that in scenario 1.

6.5.2 *Dispatch Period Impact*

According to the definitions of α (6.26b), β (6.26c) and γ (6.26d), it is known that α, β and γ are related with the dispatch period. Thus, this section focuses on analyzing the impact of the dispatch period on the power adjustment potential of the TB. From (6.15a) (6.26a) and (6.27), the maximum discharge and charge power of a TB at the kth dispatch period, $P_{\min}(k)$ and $P_{\max}(k)$, are given as:

$$P_{\min}(k) \leq P_{ch/dis}(k) \leq P_{\max}(k) \tag{6.45a}$$

$$P_{\min}(k) = -\min\left(\frac{\alpha soc(k) + \gamma}{\beta}, P^{d}\right) \tag{6.45b}$$

$$P_{\max}(k) = min\left(\frac{1 - \alpha soc(k) - \gamma}{\beta}, P^{c}\right) \tag{6.45c}$$

Obviously, $P_{\min}(k)$ and $P_{\max}(k)$ are dependent on three variables and they are the dispatch period, outdoor temperature, and the current SOC of TB. In the simulation, firstly, the outdoor temperature is fixed at 36 °C trying to get the insight of the dispatch period influencing the charge/discharge potential. Secondly, the outdoor temperature is considered to be varying as it is in reality and the associated result is thoroughly studied.

A representative TB is chosen to simulate the impact of the dispatch period. R, C, Q', and T_{set0} are 2 °C/kW, 2kWh/°C, 0.5 kW, 21 °C, respectively, and the other parameters are the same as those in Table 6.2. When $soc(k)$ is 0, 0.2, 0.5, 0.8 and 1, the profiles of $P_{\min}(k)$ (<0) and $P_{\max}(k)$ (>0) with different dispatch periods are shown in Fig. 6.10. When $soc(k)$ is 0, 0.5 and 1, the profiles of P^{c} (Pc1), $(1 - \alpha soc(k) - \gamma)/\beta$ (Pc2), P^{d} (Pd1) and $(\alpha soc(\text{k}) + \gamma)/\beta$ (Pd2) with different dispatch periods are shown in Fig. 6.11.

The impact of the dispatch period on $P_{\max}(k)$ is analyzed primarily. From Fig. 6.10, $P_{\max}(k)$ will stay fixed or become smaller as the dispatch period becomes longer. It means the longer dispatch period is disadvantageous for TBs to have larger adjustment potential in demand response. When $soc(k)$ is fixed (e.g. $soc(k) = 0.5$), $P_{\max}(k)$ keeps constant if the dispatch period is shorter than about 0.9 h (see Fig. 6.10). This is because $(1 - \alpha soc(k) - \gamma)/\beta$ is larger than P^{c} [see Fig. 6.11b] when the dispatch period is less than 0.9 h, which makes $P_{\max}(k)$ determined by P^{c}. P^{c} is a constant and only related with outdoor temperature. Then $P_{\max}(k)$ keeps unchanged. As the dispatch period increases, $(1 - \alpha soc(k) - \gamma)/\beta$ becomes smaller than P^{c}. Then $P_{\max}(k)$ will be determined by $(1 - \alpha soc(k) - \gamma)/\beta$. It is obvious that $(1 - \alpha soc(k) - \gamma)/\beta$ becomes less as the dispatch period increases in Fig. 6.11b. Thus, $P_{\max}(k)$ will be smaller with the dispatch period when the dispatch period is longer than 0.9 h.

When $soc(k)$ is 1, $P_{\max}(k)$ is minimum, independent of the dispatch period (see Fig. 6.10). According to Fig. 6.11c, both $(1 - \alpha - \gamma)/\beta$ and P^{c} have nothing to do with the dispatch period. Thus, the profile of $P_{\max}(k)$ in Fig. 6.10 is a flat line. As $soc(k)$ decreases, $(1 - \alpha soc(k) - \gamma)/\beta$ begins to increase (see the green line in

Table 6.2 The parameters of inverter TCLs

Parameter	Value	Unit
R	1.5–2.5	°C/kW
C_a	1.5–2.5	kWh/°C
Q'	0.1–0.5	kW
T_{set0}	20–22	°C
$T_{\min}$	20	°C
$T_{\max}$	22	°C
$P_{\min}$	0.5	kW
$P_{\max}$	5	kW
k_1	0.03	kW/Hz
$l1$	−0.4	kW
k_2	0.06	kW/Hz
l_2	−0.3	kW

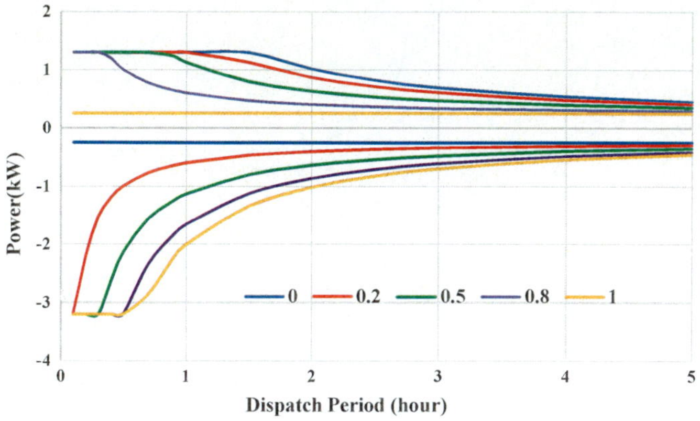

Fig. 6.10 The profiles of $P_{\min}$ and $P_{\max}$ with different dispatch periods when $soc(k)$ is 0, 0.2, 0.5, 0.8 and 1, respectively

Fig. 6.11c–b–a. Then $P_{\max}(k)$ is easier to be determined by P^{c} at the shorter dispatch period.

To sum up, there is a 'critical' dispatch period (it is about 1.5 h in this case) for the TB. When the dispatch period is shorter than the 'critical' value, the charge potential of the TB will be determined by P^{c} and is dependent upon outdoor temperature. When the dispatch period is longer than the 'critical' value, the charge potential will be determined by the dispatch period and will no longer be affected by the outdoor temperature.

Similarly, the 'critical' dispatch period for TB discharge is about 0.6 h. Considering that there are both charging and discharging processes for the TB in a full operation cycle of TB, we take the longer period as the critical value, which is 1.5 h in this case. Once the dispatch period is longer than 1.5 h, the charging/discharging power potential of the TB will not be affected by the outdoor temperature and will become less as the dispatch period increases.

As the outdoor temperature varies, the 'critical' dispatch period profile is given in Fig. 6.12. Apparently, the "critical" dispatch period curve varies with the outdoor temperature. When the outdoor temperature is lower than 32 °C, the 'critical' dispatch period becomes shorter as the outdoor temperature rises. When the outdoor temperature is higher than 32 °C, the 'critical' dispatch period will become longer as the outdoor temperature rises. Thus, when the outdoor temperature is 32 °C, the 'critical' dispatch period is the shortest, and as long as the dispatch period is longer than 0.9 h, the TB model will be independent of the outdoor temperature. Accordingly, the smart controllers of the TBs will judge whether it is necessary to update their parameters, which helps maintain the accuracy of the TB model and reduces the communication burden among the three layers.

Fig. 6.11 The profiles of P^c (Pc1), $(1 - \alpha soc(k) - \gamma)/\beta$ (Pc2), P^d (Pd1) and $(\alpha soc(k) + \gamma)/\beta$ (Pd2) with different dispatch periods

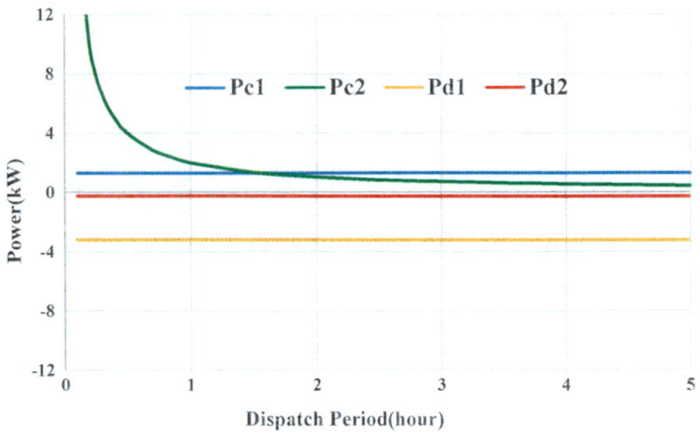

(a) $soc(k)$=0

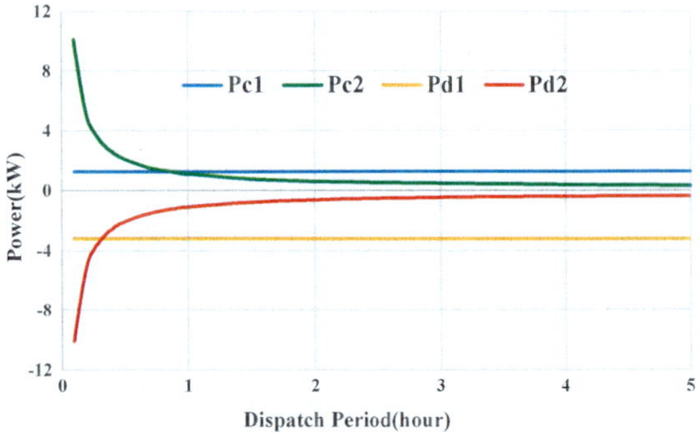

(b) $soc(k)$=0.5

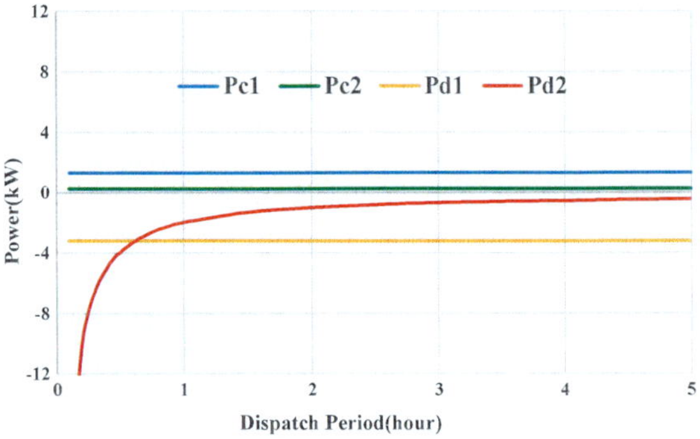

(c) $soc(k)$=1

Table 6.3 The parameters of the aggregated TB model

Parameter	ξ	γ_{agg}	τ	SOC_{agg} (0)
Value	0.879	518.1	6418	3209

Table 6.4 Parameters of the lithium-ion batteries

Parameter	x^d	x^c	η^{in}	η^{out}
Value	5.5	5.5	0.9	0.9
Unit	MW	MW	/	/
Parameter	$s_{\min}$	$s_{\max}$	$o^{\max}$	$SOC(0)$
value	0	1	8	0.5
Unit	/	/	MWh	/

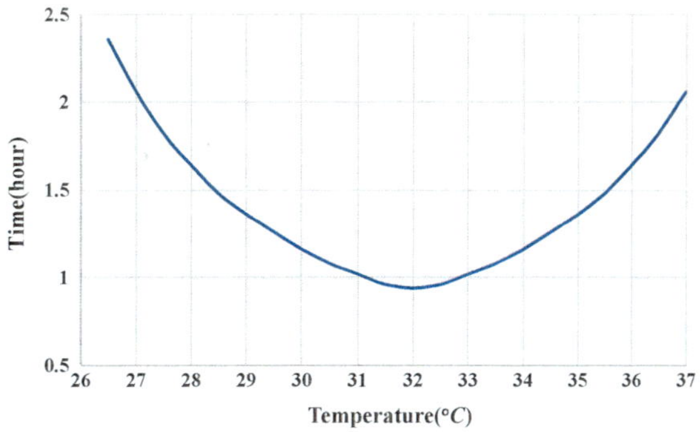

Fig. 6.12 The relationship between outdoor temperature and the 'critical' dispatch period

6.6 Conclusion

Inverter TCLs are modeled as TBs for dispatching compatibility. Firstly, the TB model of an inverter TCL system is proposed and the comparisons between the TB model and the lithium-ion battery are given. Then a hierarchical control framework is designed to protect end-users' privacy and relieve the computational burden of the centralized control. In the end, operating behaviors of TBs in the real-time market are simulated. The results are as follows: (1) the aggregated TB model works well with the power dispatch model developed for the lithium-ion batteries. In other words, with the TB modeling of TCLs, lithium-ion batteries can be replaced in current dispatch models providing service for the grid. (2) TBs can still charge/discharge when the SOC keeps unchanged while the lithium-ion batteries cannot charge/discharge without changing the state of energy storage. Particularly, the TB can also discharge when the SOC is 0; (3) some TBs are utilized with low efficiency (e.g. the maximum value of the SOC is only about 0.6 when the TB is charged at utmost), which is a defect of the proposed synchronizing control; (4) The longer dispatch period is disadvantageous to have a larger power adjustment range. When the dispatch period is longer than the 'critical' value, the TB is required to update its parameters according to the outdoor temperature.

References

1. Song, M., Gao, C., Yan, H., & Yang, J. (2017). Thermal battery modeling of inverter air conditioning for demand response. *IEEE Transactions on Smart Grid, 9*(6), 5522–5534.
2. Zhang, Y., & Lu, N. (2013). Parameter selection for a centralized thermostatically controlled appliances load controller used for intra-hour load balancing. *IEEE Transactions on Smart Grid, 4*(4), 2100–2108.
3. Lu, N., & Zhang, Y. (2012). Design considerations of a centralized load controller using thermostatically controlled appliances for continuous regulation reserves. *IEEE Transactions on Smart Grid, 4*(2), 914–921.
4. Lu, N. (2012). An evaluation of the HVAC load potential for providing load balancing service. *IEEE Transactions on Smart Grid, 3*(3), 1263–1270.
5. Lu, S., Samaan, N., Diao, R., Elizondo, M., Jin, C., Mayhorn, E., Zhang, Y., & Kirkham, H. (2011, January). Centralized and decentralized control for demand response. ISGT 2011 (pp. 1–8).
6. Arora, M., Chanana, S., & Kumar, A. (2014, December). A real time price based optimal scheduling mechanism for centralized air conditioning load. 2014 Eighteenth National Power Systems Conference (NPSC) (pp. 1–5).
7. Vanouni, M., & Lu, N. (2014). Improving the centralized control of thermostatically controlled appliances by obtaining the right information. *IEEE Transactions on Smart Grid, 6*(2), 946–948.
8. Hao, H., Sanandaji, B. M., Poolla, K., & Vincent, T. L. (2013, October). A generalized battery model of a collection of thermostatically controlled loads for providing ancillary service. In *2013 51st Annual Allerton Conference on Communication, Control, and Computing (Allerton)* (pp. 551–558).
9. Mukherjee, J. C., & Gupta, A. (2014). A review of charge scheduling of electric vehicles in smart grid. *IEEE Systems Journal, 9*(4), 1541–1553.
10. Negenborn, R. R., & Maestre, J. M. (2014). Distributed model predictive control: An overview and roadmap of future research opportunities. *IEEE Control Systems Magazine, 34*(4), 87–97.
11. Sanandaji, B. M., Hao, H., Poolla, K., & Vincent, T. L. (2014, June). Improved battery models of an aggregation of thermostatically controlled loads for frequency regulation. In *2014 American Control Conference* (pp. 38–45).
12. Hao, H., Sanandaji, B. M., Poolla, K., & Vincent, T. L. (2014). Aggregate flexibility of thermostatically controlled loads. *IEEE Transactions on Power Systems, 30*(1), 189–198.
13. Kamgarpour, M., Ellen, C., Soudjani, S. E. Z., Gerwinn, S., Mathieu, J. L., Müllner, N., Abate, A., Callaway, D.S., Fränzle, M., & Lygeros, J. (2013, August). Modeling options for demand side participation of thermostatically controlled loads. In *2013 IREP Symposium Bulk Power System Dynamics and Control-IX Optimization, Security and Control of the Emerging Power Grid* (pp. 1–15).
14. Mathieu, J. L., Kamgarpour, M., Lygeros, J., Andersson, G., & Callaway, D. S. (2014). Arbitraging intraday wholesale energy market prices with aggregations of thermostatic loads. *IEEE Transactions on Power Systems, 30*(2), 763–772.
15. Zhang, W., Kalsi, K., Fuller, J., Elizondo, M., & Chassin, D. (2012, July). Aggregate model for heterogeneous thermostatically controlled loads with demand response. In *2012 IEEE Power and Energy Society General Meeting* (pp. 1–8).
16. Zhang, W., Lian, J., Chang, C. Y., & Kalsi, K. (2014). Aggregated modeling and control of air conditioning loads for demand response. In *2014 IEEE PES General Meeting | Conference & Exposition* (pp. 1–1).
17. Zhang, W., Lian, J., Chang, C. Y., & Kalsi, K. (2013). Aggregated modeling and control of air conditioning loads for demand response. *IEEE transactions on power systems, 28*(4), 4655–4664.
18. Zhang, W., Lian, J., Chang, C. Y., Kalsi, K., & Sun, Y. (2012, December). Reduced-order modeling of aggregated thermostatic loads with demand response. In *2012 IEEE 51st IEEE Conference on Decision and Control (CDC)* (pp. 5592–5597).

19. Sanandaji, B. M., Hao, H., & Poolla, K. (2014, January). Fast regulation service provision via aggregation of thermostatically controlled loads. In *2014 47th Hawaii International Conference on System Sciences* (pp. 2388–2397).
20. Kalsi, K., Chassin, F., & Chassin, D. (2011, December). Aggregated modeling of thermostatic loads in demand response: A systems and control perspective. In *2011 50th IEEE Conference on Decision and Control and European Control Conference* (pp. 15–20).
21. Kalsi, K., Elizondo, M., Fuller, J., Lu, S., & Chassin, D. (2012, January). Development and validation of aggregated models for thermostatic controlled loads with demand response. In *2012 45th Hawaii International Conference on System Sciences* (pp. 1959–1966).
22. Bashash, S., & Fathy, H. K. (2012). Modeling and control of aggregate air conditioning loads for robust renewable power management. *IEEE Transactions on Control Systems Technology, 21*(4), 1318–1327.
23. Shah, N., Waide, P., & Phadke, A. (2013). *Cooling the planet: Opportunities for deployment of superefficient room air conditioners (No. LBNL-6164E)*. Berkeley, CA, USA: Lawrence Berkeley National Lab.(LBNL).
24. Kim, Y. J., Norford, L. K., & Kirtley, J. L. (2014). Modeling and analysis of a variable speed heat pump for frequency regulation through direct load control. *IEEE Transactions on Power Systems, 30*(1), 397–408.
25. Kim, Y. J., Fuentes, E., & Norford, L. K. (2015). Experimental study of grid frequency regulation ancillary service of a variable speed heat pump. *IEEE Transactions on Power Systems, 31*(4), 3090–3099.
26. Xing, H., Fu, M., Lin, Z., & Mou, Y. (2015). Decentralized optimal scheduling for charging and discharging of plug-in electric vehicles in smart grids. *IEEE Transactions on Power Systems, 31*(5), 4118–4127.
27. Shao, C., Wang, X., Wang, X., Du, C., & Wang, B. (2015). Hierarchical charge control of large populations of EVs. *IEEE Transactions on Smart Grid, 7*(2), 1147–1155.
28. Wang, D., Guan, X., Wu, J., Li, P., Zan, P., & Xu, H. (2015). Integrated energy exchange scheduling for multimicrogrid system with electric vehicles. *IEEE Transactions on Smart Grid, 7*(4), 1762–1774.
29. Wen, Y., Li, W., Huang, G., & Liu, X. (2016). Frequency dynamics constrained unit commitment with battery energy storage. *IEEE Transactions on Power Systems, 31*(6), 5115–5125.
30. Zareifard, M. T. (2015, September). Model predictive control of electricity market oriented wind farm dispatch with a battery energy storage system. In *2015 IEEE Conference on Control Applications (CCA)* (pp. 727–732).
31. Piphitpattanaprapt, N., & Bangerdpongchai, D. (2015, November). Optimal dispatch strategy of hybrid power generation with battery energy storage system in islanding mode. In *2015 IEEE Innovative Smart Grid Technologies-Asia (ISGT ASIA)* (pp. 1–6).
32. Yudhaprawira, G., & Hadi, S. P. (2012, July). Unit commitment for power generation system including PV and batteries by mixed integer quadratic programming. In *2012 International Conference on Power Engineering and Renewable Energy (ICPERE)* (pp. 1–5).
33. Bao, G., Lu, C., Yuan, Z., & Lu, Z. (2012, August). Battery energy storage system load shifting control based on real time load forecast and dynamic programming. In *2012 IEEE International Conference on Automation Science and Engineering (CASE)* (pp. 815–820).
34. Ning, Y., Li, X., Ma, X., Jia, X., & Hui, D. (2016, June). Optimal schedule strategy of battery energy storage systems for peak load shifting based on interior point method. In *2016 12th World Congress on Intelligent Control and Automation (WCICA)* (pp. 2285–2288).
35. Li, N., Uckun, C., Constantinescu, E. M., Birge, J. R., Hedman, K. W., & Botterud, A. (2015). Flexible operation of batteries in power system scheduling with renewable energy. *IEEE Transactions on Sustainable Energy, 7*(2), 685–696.
36. Heydarian-Forushani, E., Golshan, M. E. H., Shafie-Khah, M., & Catalão, J. P. S. (2015, August). Optimal coordination of battery energy storages and demand response programs with application to wind integration. In *2015 IEEE International Conference on Smart Energy Grid Engineering (SEGE)* (pp. 1–6).

37. Lujano-Rojas, J. M., Dufo-López, R., Bernal-Agustín, J. L., & Catalão, J. P. (2016). Optimizing daily operation of battery energy storage systems under real-time pricing schemes. *IEEE Transactions on Smart Grid, 8*(1), 316–330.
38. Xu, Y., & Singh, C. (2012). Adequacy and economy analysis of distribution systems integrated with electric energy storage and renewable energy resources. *IEEE Transactions on power systems, 27*(4), 2332–2341.
39. Xu, Y., Xie, L., & Singh, C. (2010, September). Optimal scheduling and operation of load aggregator with electric energy storage in power markets. In*North American Power Symposium 2010* (pp. 1–7).
40. Kang, M. S., Chen, C. S., Ke, Y. L., Lin, C. H., & Huang, C. W. (2007). Load profile synthesis and wind-power-generation prediction for an isolated power system. *IEEE Transactions on Industry Applications, 43*(6), 1459–1464.
41. E. Anastasio (2017, January 10). Load Forecasting at PJM [Online]. Available http://www.pjm.com/~/media/committees-groups/committees/oc/20170110/20170110-item-06-load-forecasting-at-pjm.ashx
42. Xu, Y., & Singh, C. (2012, October). Operation strategies of the load aggregator with electric energy storage. In *2012 IEEE International Conference on Power System Technology (POWERCON)* (pp. 1–6).
43. Lo, C. H., & Anderson, M. D. (1999). Economic dispatch and optimal sizing of battery energy storage systems in utility load-leveling operations. *IEEE Transactions on Energy Conversion, 14*(3), 824–829.

Chapter 7
Comparison Analysis on Energy Storage Behaviors of TCLs Under Different Control Methods

Abstract TCLs are utilized as energy storage resources in the real-time market for less economic loss caused by the fluctuations of wind power. Inverter TCL, is constructed as a power type battery model (PTBM) and a capacity type battery model (CTBM) under different control methods. Its energy storage characteristics are described via circuit model and mathematical model. Moreover, a comparisons between PTBM and CTBM are conducted by analyzing their response speed, power and energy capacity, and the cost of control. It provides guidance to the associated operators in choosing the appropriate models to participatein DR. Given the frequent and large fluctuations of wind power generation, the PTBM is chosen to participate in DR for output optimization of wind power generation. The simulation results demonstrate that PTBMs can optimize wind generation in the real-time market the same way as conventional batteries (CBs).

Keywords Electricity market · Power type battery model · Capacity type battery model,wind generation

7.1 Introduction

In recent years, the global wind power installed capacity continues to increase with an annual growth rate of 28% [1]. China's wind power installed capacity has been increasing since 2005 and it has reached 75.3 GW by the end of 2012, ranking first in the world. However, due to the strong fluctuations of wind power, the prediction accuracy is low, which brings a series of problems to the security and stability of the power grid and causes economic loss to the associated aggregators in the electricity market. To promote large-scale wind power consumption and guarantee economic benefits, energy storage technology has become one of the main ways to improve the wind power utilization rate [2–4]. Nevertheless, due to the high investment in energy storage equipment and the long cost recovery period, the further development of wind generation is seriously restricted. As one of the key technologies of the smart grid, DR has been widely utilized for the reason that it can relieve the tension of supply-demand unbalance, strengthen the grid dealing with power fluctuations, and enhance the ability of the system to accommodate renewable resources [5, 6]. Among

M. Song and C. Gao, *Integration of Distributed Resources in Smart Grids for Demand Response and Transactive Energy*, https://doi.org/10.1007/978-981-16-7170-8_7

all flexible loads, TCLs have received extensive attention for its ability to transfer loads at a certain time to provide a variety of ancillary services for the system [7, 8].

The energy consumption of residential and commercial buildings occupies up to 75% of the total energy consumption of America [9], which makes the TCL an important DR resource to take part in ancillary services. TCLs have been massively studied [10–12]. Lu and Chassin [13] proposed a state-queuing model to describe the energy storage state of TCLs. Moreover, a general thermodynamic model has been developed for the evaluation of load shifting potential under different TCL control principles and electricity rate structures [14]. Ref. [15, 16] proposed a time-varying thermal battery model to keep track of the operating state of the TCLs. A simplified battery was developed in [17–19] to aggregate TCLs to provide frequency regulation service for the grid. Although battery models of TCLs have been studied in some literature, these above-mentioned models are all limited to fixed frequency TCLs. The energy storage modeling of inverter TCL has been rarely studied.

The compressors of TCLs have evolved from fixed speed to variable speed units, which have gained noticeably growing market share for their higher energy efficiency [20, 21]. Therefore, this chapter focuses on inverter TCLs. Generally, fixed-speed TCLs can be controlled by on-off switches and the temperature set-point changes. Under different control methods, fixed-speed TCLs will diversely operate in response speed and power capacity [22]. Similarly, inverter TCLs can be controlled by frequency adjustment and temperature set-point changes. There are rare pieces of literatures that refer to the operating characteristics of the inverter TCLs under the two control methods. Therefore, two battery models of the inverter TCLs system are built in this paper. Accordingly, the main work of this chapter includes, (1) The power type battery model (PTBM) of the inverter TCL system with the frequency control (including circuit and mathematical model) is proposed, which shows the thermal energy storage behaviors of the TCL and helps the TCL be compatible with the current dispatch models. Furthermore, the aggregated model of heterogeneous PTBMs is built to facilitate the scheduling of TCLs; (2) The capacity type battery model (CTBM) of the inverter TCL with the temperature set-point change is also proposed. The differences between PTBM and CTBM in response speed, power capacity and the cost of control are analytically given, which will guide the associated operators to choose appropriate battery models to take part in demand response.

7.2 Basic Model of Inverter TCL System

The inverter TCL system model is composed of a thermodynamic model and an electrical model. A simplified ETP model [8] of a residential TCL system is described as (7.1). The ETP model is well-suited for residential and small commercial buildings in the simulation of heat transition between the TCL and its surroundings. Two operating modes of TCLs are available, namely heating and cooling modes, and only the cooling mode is discussed here.

$$C_{\mathrm{a}}\frac{dT_i}{dt} = \frac{1}{R}(T_{\mathrm{o}} - T_{\mathrm{i}}) + Q^{'} - Q \tag{7.1}$$

The majority of power consumption of the inverter TCL occurs during the compressor operation, and its coefficient of performance is not constant but varies with its rotational frequency. Therefore, considering the power consumption of the compressor as the total electrical power of the inverter TCL is reasonable. According to [20, 21], by further simplifications, the relationship between the frequency, f, and electric power, P, is expressed approximately as

$$P = k_1 f + l_1 \tag{7.2}$$

Similarly, Q is expressed as

$$Q = k_2 f + l_2 \tag{7.3}$$

According to (7.2) and (7.3), the relationship between P and Q is described as

$$Q = \frac{k_2}{k_1}P + \frac{k_1 l_2 - l_1 k_2}{k_1} \tag{7.4}$$

The fundamental model of inverter TCL system is composed of (7.1) and (7.4), which bridges the relationship between T_{i} and P. According to (7.2) and (7.3), P can be continuously adjusted by the frequency control to serve the grid while keeping end-users in a comfortable temperature range. In this way, the controller installed on the inverter TCL is able to get power signals from associated dispatch centres and convert them into frequency signals to change the operating state of the compressor.

7.3 Power Type Battery Modeling of Inverter TCLs

In this section, the battery model of the TCL is proposed when it is controlled by the frequency adjustment. The TCL can change the electric power instantaneously and have a large power capacity, which is similar to that of the power type conventional battery (CB). Thus, we name the model PTBM under this control method.

7.3.1 Circuit Model of PTBM

When the inverter TCL operates at the initial temperature set-point, $T_{\mathrm{i}}(0)$, it can be regarded that the TCL is working in a natural state and the corresponding electric power is seen as the power baseline. From (7.1), the cooling power baseline, Q_{baseline}, is expressed as

$$Q_{\text{baseline}} = \frac{1}{R}(T_{\text{o}} - T_{\text{i}}(0)) + Q^{'} \tag{7.5}$$

According to (7.4), the electric power baseline, P_{baseline}, is given as

$$P_{\text{baseline}} = \frac{k_1}{k_2 R}(T_{\text{o}} - T_{\text{i}}) + \frac{k_1 Q^{'} + k_2 l_1 - k_1 l_2}{k_2} \tag{7.6}$$

Essentially, the TCL is an electric device and cannot authentically charge or discharge. However, the indoor temperature that keeps end-users' comfortable is not a fixed value but a range [23], which makes it possible for the actual electric power P to deviate from P_{baseline}. When P is larger than P_{baseline}, the TCL will absorb more electricity and it can be regarded as a charging process from the perspective of the grid. Similarly, when the actual electric power is smaller than P_{baseline}, it can be seen as a discharging process. Thus, the charging/discharging power of the PTBM, P_{cd}, is defined as [16]

$$P_{\text{cd}} = P - P_{\text{baseline}} \tag{7.7}$$

According to the circuit theory, P is given as

$$P = UI \tag{7.8}$$

From (7.4) and (7.8), (7.1) is rewritten as

$$C_{\text{a}} \frac{dU^{'}}{dt} + \frac{1}{R} U^{'} + I^{'} = I_1 \tag{7.9}$$

$$U^{'} = T_{\text{o}} - T_{\text{i}} \tag{7.10}$$

$$I_1 = \frac{k_2 U}{k_1} I \tag{7.11}$$

$$I^{'} = \frac{k_1 Q^{'} - k_1 l_2 + k_2 l_1}{k_1} \tag{7.12}$$

The equivalent circuit model that is used to show the energy storage behaviors of the inverter TCL system is given in Fig. 7.1. The TCL operates at the initial temperature set-point when the time is 0. When P is increased by a larger frequency (i.e. $P_{\text{cd}}>0$), I will accordingly enlarge. From (7.11), I_1 is determined by I and will also increase. Therefore, the electric current through C_{a} and R will go up and then U' will gradually become larger. Then there will be more energy stored in C_{a} and this is the dynamic charge process. Similarly, when P is reduced by a smaller frequency (i.e. $P_{\text{cd}}<0$), the electric current through R will decline and there will be an adverse

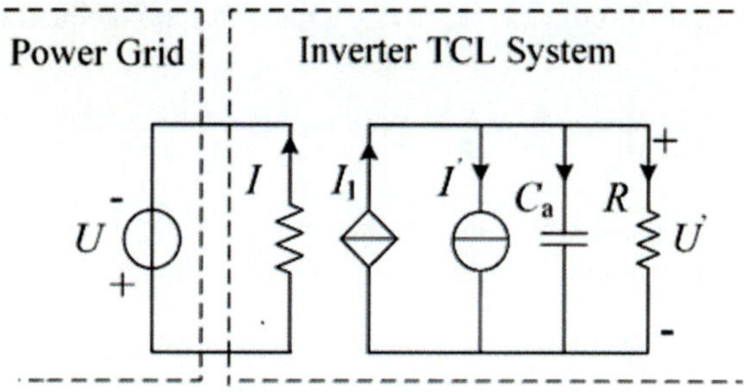

Fig. 7.1 Equivalent energy storage circuit model of the inverter TCL system

electric current through C_a. U' slowly goes down and the stored energy in C_a is cut back. This is the dynamic discharge process.

In fact, according to the circuit model in Fig. 7.1, it is also known that the electric power absorbed by the TCL unit from the grid is consumed in three ways in the form of thermal energy. The first one is to store thermal energy in the buildings ($C_a dU'/dt$). The envelope of the building and internal air work as a "capacitance" to store the thermal energy. The second one is to make up the thermal dissipation because of the difference between the indoor and outdoor temperature (U'/R) and the third one is to take away the internal heat generation (I'). Because the total dissipation power ($U'/R+ I'$) is much larger than that of the CB (the self-discharge power of the CB is very small and can be ignored), the TCL needs to absorb greater amounts of electricity to keep the indoor temperature at the temperature set-point (i.e. make C_a constant). When the actual cooling power is larger than the dissipation power, more energy will be stored in the building (i.e. extra energy is stored in C_a). When the dissipation power is larger than the actual thermal power, the stored energy will decline (i.e. some energy is released from C_a). That is the physical principle of the energy storage of the TCL system.

7.3.2 Mathematical Model of PTBM

CB can store electricity in the form of chemical or mechanical energy while the inverter TCL system can store electricity in the form of thermal energy. To guarantee end-users' comfort, the indoor temperature is kept within $[T_{min}, T_{max}]$. When the indoor temperature is T_{min}, the stored energy is maximal while it is minimal when the indoor temperature is T_{max}. The stored energy is maximized when the room temperature is T_{min} and minimized when the room temperature is T_{max}. Assuming that the room temperature is T_{max}, the stored energy is 0. The state of charge (SOC) of PTBM at time t is defined as

$$SOC(t) = \frac{T_{max} - T_i}{T_{max} - T_{min}} \tag{7.13}$$

Substituting (7.13) to (7.1), the differential equation of SOC is expressed as

$$\frac{dSOC(t)}{dt} = -\frac{1}{RC_{\mathrm{a}}} SOC(t) - \frac{T_o - T_{\max}}{RC_{\mathrm{a}}(T_{\max} - T_{\min})} - \frac{Q' - Q}{C_{\mathrm{a}}(T_{\max} - T_{\min})} \tag{7.14}$$

It is noted that the electric power of the inverter TCL is kept constant during a dispatch period. Solve (7.14) and the formula about the SOC evolution is given by

$$\begin{aligned} SOC(k+1) = e^{-\Delta t/RC_{\mathrm{a}}} \cdot SOC(k) + \frac{R\left(1 - e^{-\Delta t/RC_{\mathrm{a}}}\right)}{T_{\max} - T_{\min}} \cdot Q(k) \\ + \frac{T_{\max} - T_{\mathrm{o}} - RQ'}{T_{\max} - T_{\min}}\left(1 - e^{-\Delta t/RC_{\mathrm{a}}}\right) \end{aligned} \tag{7.15}$$

According to (7.4) and (7.15), the relationship between the SOC and the electric power is described as

$$\begin{aligned} SOC(k+1) = e^{-\Delta t/RC_{\mathrm{a}}} \cdot SOC(k) + \frac{k_2 R\left(1 - e^{-\Delta t/RC_{\mathrm{a}}}\right)}{k_1(T_{\max} - T_{\min})} \cdot P(k) \\ + \frac{k_1(T_{\max} - T_o - RQ') + R(k_1 l_2 - l_1 k_2)}{k_1(T_{\max} - T_{\min})}\left(1 - e^{-\Delta t/RC_{\mathrm{a}}}\right) \end{aligned} \tag{7.16}$$

From (7.7), (7.16) is rewritten as

$$SOC(k+1) = \alpha SOC(k) + \beta P_{\mathrm{cd}}(k) + \gamma \tag{7.17}$$

$$\alpha = e^{-\Delta t/RC_{\mathrm{a}}} \tag{7.18}$$

$$\beta = \frac{k_2 R(1 - e^{-\Delta t/RC_{\mathrm{a}}})}{k_1(T_{\max} - T_{\min})} \tag{7.19}$$

$$\gamma = (1 - \alpha) \cdot SOC(0) \tag{7.20}$$

$$SOC(0) = \frac{T_{\max} - T_{\mathrm{i}}(0)}{T_{\max} - T_{\min}} \tag{7.21}$$

The evolution of the energy storage level with the charging/discharging power of PTBM is described in (7.17), which bridges the thermal energy and the electric energy in the form of the battery. This is a very important characteristic for PTBM. Obviously, SOC should be kept within the range [0, 1] to maintain the end-users' comfort at any time. Then the constraint about SOC is given as

$$0 \le SOC(k+1) \le 1 \tag{7.22}$$

At the same time, $P_{cd}(k)$ should be constrained as

$$-P^{d} \leq P_{cd}(k) \leq P^{c} \tag{7.23}$$

$$P^{d} = P_{baseline} - P_{min} \tag{7.24}$$

$$P^{c} = P_{max} - P_{baseline} \tag{7.25}$$

To sum up, PTBM of the inverter TCL system can be described by (7.17), (7.22) and (7.23), which is identical with the expressions of the CB in [24–26]. Therefore, PTBM can replace the CB in the associated dispatch models, which makes the TCL dispatched in the way of the CB and achieve the compatibility of the TCL with current dispatch models. For the system operator (SO), it might be familiar with the CB models and there have been many dispatch models developed for the batteries [27–29]. However, the model of the inverter TCL system and its parameters are alien to the SO. Special revisions of the current dispatch model should be made to accommodate the TCLs' participation. When the inverter TCL is modeled as a conventional battery, the current dispatch model can be kept as it is, which generates a lot of conveniences.

7.3.3 Aggregation of Heterogeneous PTBMs

As for aggregators, each PTBM is considered difficult to schedule individually from a system perspective, so an aggregation model is required.. The aggregated model of homogeneous PTBMs is built and then the heterogeneity of PTBMs is taken into account. It is obvious that only the heterogeneity of α and β need to be considered when the aggregated model of heterogeneous PTBMs is built according to the model of PTBM. Thus, the homogeneous PTBMs only have the same parameters $\{\alpha, \beta\}$.

The update of SOC of homogeneous PTBMs is calculated as

$$\begin{cases} SOC_1(k+1) = \alpha SOC_1(k) + \beta P_1(k) + \gamma_1 \\ SOC_2(k+1) = \alpha SOC_2(k) + \beta P_2(k) + \gamma_2 \\ \qquad\vdots \\ SOC_m(k+1) = \alpha SOC_m(k) + \beta P_m(k) + \gamma_m \end{cases} \tag{7.26}$$

$SOC_h(k+1)$ and $SOC_h(k)$ are the SOC of the hth ($1 \leq h \leq m$) PTBM at the (k+1)th and kth period, respectively. $P_h(k)$ is the charging/discharging power of the hth PTBM at the kth period, kW. γ_h is the constant coefficient of the hth PTBM.

Aggregate PTBMs and (7.26) is rewritten as

$$\sum_{h=1}^{m} SOC_h(k+1) = \alpha \sum_{h=1}^{m} SOC_h(k) + \beta P_{\text{agg}}(k) + \sum_{h=1}^{m} \gamma_h \quad (7.27)$$

$$P_{\text{agg}}(k) = \sum_{h=1}^{m} P_h(k) \quad (7.28)$$

To simplify the update of *SOC* in (7.27), we introduce SOC_{agg} (k) to describe the aggregated SOC of PTBMs at the kth period. It is defined as follows.

$$SOC_{\text{agg}}(k) = \frac{\sum_{h=1}^{m} SOC_h(k)}{m} \quad (7.29)$$

Then (7.27) is rewritten as

$$SOC_{\text{agg}}(k+1) = \alpha SOC_{\text{agg}}(k) + \kappa P_{\text{agg}}(k) + \sigma \quad (7.30)$$

$$\kappa = \beta / m \quad (7.31)$$

$$\sigma = \sum_{h=1}^{m} \gamma_h / m \quad (7.32)$$

Accordingly, SOC_{agg} (k) in (7.19) should be constrained by

$$0 \leq SOC_{\text{agg}}(k) \leq 1 \quad (7.33)$$

Given the charging/discharging power constraint of PTBM in (7.23), the aggregated charging/discharging power of PTBMs is given as

$$-P_{agg}^{d} \leq P_{agg}(k) \leq P_{agg}^{c} \quad (7.34)$$

$$P_{\text{agg}}^{d} = \sum_{h=1}^{m} P_h^{d} \quad (7.35)$$

$$P_{\text{agg}}^{c} = \sum_{h=1}^{m} P_h^{c} \quad (7.36)$$

The k-means algorithms [11] are employed to classify these heterogeneous parameters $\{\alpha, \beta\}$ of PTBMs into n clusters, and each cluster is associated with a center $\{\alpha^j, \beta^j\}(1 \leq j \leq n)$. It is regarded that PTBMs which fall into the same cluster with the same $\{\alpha^j, \beta^j\}$can be aggregated by the proposed modeling method.

Then the aggregated model of heterogeneous PTBMs is described as

$$-D^{\max} \leq P_n(k) \leq C^{\max}$$

$$\boldsymbol{SOC}(k+1) = \boldsymbol{\alpha} \cdot \boldsymbol{SOC}(k) + \boldsymbol{\kappa} \cdot \boldsymbol{P}_{\mathrm{n}}(k) + \boldsymbol{\sigma} \tag{7.37}$$

$$0 \leq \boldsymbol{SOC}(k+1) \leq 1$$

$C^{\max} = \sum_{j=1}^{n} P_{\mathrm{agg}}^{\mathrm{c}\_j}$, $D^{\max} = \sum_{j=1}^{n} P_{\mathrm{agg}}^{\mathrm{d}\_j}$, $P_n(k) = \sum_{j=1}^{n} P_{\mathrm{agg}}^{j}(k)$, $\boldsymbol{P}_{\mathrm{n}}(k) = [P_{\mathrm{agg}}^{1}(k), P_{\mathrm{agg}}^{2}(k), \ldots, P_{\mathrm{agg}}^{n}(k)]^{T}$, $\boldsymbol{SOC}(k) = [SOC_{\mathrm{agg}}^{1}(k), SOC_{\mathrm{agg}}^{2}(k), \ldots, SOC_{\mathrm{agg}}^{n}(k)]^{T}$, $\boldsymbol{SOC}(k+1) = [SOC_{\mathrm{agg}}^{1}(k+1), SOC_{\mathrm{agg}}^{2}(k+1), \ldots, SOC_{\mathrm{agg}}^{n}(k+1)]^{\mathrm{T}}$. $\boldsymbol{\alpha} = [\alpha^{1}, \alpha^{2}, \ldots, \alpha^{n}]^{\mathrm{T}}$. $\boldsymbol{\kappa}$ is a $n \times n$ matrix and its diagonal matrix is $[\kappa^{1}, \kappa^{2}, \ldots, \kappa^{n}]^{\mathrm{T}}$. $\boldsymbol{\sigma} = [\sigma^{1}, \sigma^{2}, \ldots, \sigma^{n}]^{\mathrm{T}}$. $P_{\mathrm{agg}}^{\mathrm{c}\_j}$ and $P_{\mathrm{agg}}^{\mathrm{d}\_j}$ are the maximal charging/discharging power of PTBMs in the jth cluster, respectively. $P_{\mathrm{agg}}^{j}(k)$ is the aggregated electric power of PTBMs in the jth cluster, kW. $SOC_{\mathrm{agg}}^{j}(k+1)$ and $SOC_{\mathrm{agg}}^{j}(k)$ are the aggregated SOC of PTBMs in the jth cluster at the $(k+1)$th period and kth period. α^{j}, κ^{j} and σ^{j} are the aggregated constant parameters of PTBMs in the jth cluster.

7.4 Capacity Type Battery Modeling of Inverter TCL

In addition to the frequency control, the electricity consumption of TCL can also be changed by the adjustment of the temperature set-point. With the temperature set-point control, TCL cannot instantaneously reach the steady-state but will have a huge energy capacity. Thus, the corresponding battery model is named CTBM.

7.4.1 Circuit Model of CTBM

The charging/discharging process of CTBM can also be described in Fig. 7.1. Different from the frequency control, only the initial and final steady states are focused on and the transient process of the indoor temperature fluctuations will be ignored if the dispatch period is long enough. When U' in Fig. 7.1 reaches a lower set-point after the dynamic discharging process, U' will stay at the lower value for a long time until it is reset; there is no electric current through C_{a}. Similarly, U' can stay at a larger value for a long time. It is obvious that once U' reaches the steady state after a dynamic charging/discharging process, the charging/discharging power will be constant. This means that CTBM will charge/discharge as long as U' is not re-modified. With this control method, CTBM can be treated as a battery with a very large energy capacity.

7.4.2 Mathematical Model of CTBM

When the TCL is controlled by the temperature set-point modification, the frequency will be adjusted according to the difference between the actual and desired temperature until the indoor temperature reaches near the set-point. The frequency evolution is expressed as

$$\Delta T = T_c - T_s \tag{7.38}$$

$$f_{n+1} = \begin{cases} f_{\max}, \Delta T > \Delta T_{\max} \\ f_n + k\Delta T, -\Delta T_{\min} \le \Delta T \le \Delta T_{\max}, f_{\min} \le f_{n+1} \le f_{\max} \\ f_{\min}, \Delta T < -\Delta T_{\min} \end{cases} \tag{7.39}$$

The charging/discharging power and *SOC* of CTBM are defined identically with those of PTBM in (7.7) and (7.13). Thus, the constraints of the charging/discharging power and *SOC* are also described by (7.23) and (7.22).

According to (7.6) and (7.13), when the indoor temperature is kept constant, the relationship between *P* and *SOC* is given as

$$P = \chi SOC + \varsigma \tag{7.40}$$

$$\chi = \frac{k_1(T_{\max} - T_{\min})}{k_2 R} \tag{7.41}$$

$$\varsigma = \frac{-k_1 T_{\max} + k_1 R Q' + k_2 R l_1 - k_1 R l_2 + k_1 T_o}{k_2 R} \tag{7.42}$$

In view of the fact that *SOC* of CTBM stays constant during most of the time of the *k*th period, the charging/discharging power of CTBM at the *k*th period, $P_{\mathrm{CD}}(k)$ is expressed as

$$P_{CD}(k) = P(k) - P_{baseline} = \chi(SOC(k+1) - SOC(0)) \tag{7.43}$$

It should be noted that $P(k)$ in (7.43) is the corresponding electric power when the indoor temperature reaches the steady state during the *k*th dispatch period.

Equation (7.43) is rewritten as

$$SOC(k+1) = \frac{1}{\chi} P_{\mathrm{CD}}(k) + SOC(0) \tag{7.44}$$

To sum up, CTBM of the inverter TCL system can be described by (7.22), (7.23) and (7.44), which is also identical with the CB. The only difference among the three battery models (i.e. the CB, PTBM and CTBM) lies in the coefficients of the *SOC* update.

7.5 Comparisons Between PTBM and CTBM

Since the TCL operate with different behaviors using the two control methods, a comparative study between PTBM and CTBM is conducted to guide aggregators/operators in selecting the appropriate model to participate in DR.

7.5.1 Response Speed

The parameters of the inverter TCL system are given below. R, C_a and Q' are 2 °C/kW, 2kW h/°C and 0.2 kW, respectively. The other parameters are the same as those in Table 7.3. The initial indoor temperature is 21 °C. T_{out} is 32 °C. Accordingly, the parameters of PTBM are calculated in Table 7.1.

PTBM works with the 0 charging/discharging power (i.e. TCL operates at the initial temperature set-point) before the time 0.5 hours. It is then controlled to charge/discharge with corresponding charge/discharge powers of 1.1, 2.1, 2.1, and 2.1 kW, respectively. The dispatch period is 0.5 h. The profiles of the actual electric power, SOC, and indoor temperature are shown in Fig. 7.2. Due to the fact that the compressor of the TCL can instantaneously respond to the frequency signal to change the electricity consumption, the charging/discharging signals will be converted into frequency signals to immediately modify the operating state of PTBM. Thus, the response speed of PTBM is very fast and the charging/discharging power can be

Table 7.1 Parameters of PTBM

Parameter	α	β	γ	$P_{baseline}$	P^d	P^c
Value	0.8825	0.2350	0.0588	2.6 kW	2.1 kW	2.4 kW

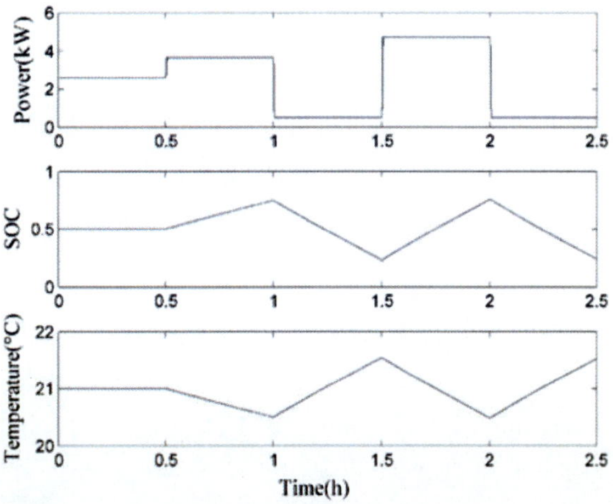

Fig. 7.2 The profiles of the electric power, *SOC* and indoor temperatures of PTBM

Fig. 7.3 The profiles of the electric power, SOC, and indoor temperature of CTBM

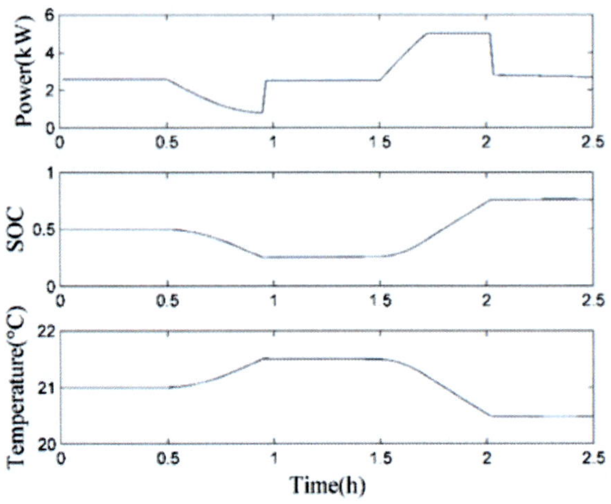

instantaneously obtained if the time delays of the communication and mechanics are ignored.Meanwhile, SOC of PTBM clearly shows an opposite trendcompared with the indoor temperature, which indicates that the modelling method of PTBM is correct.

When the electricity consumption of the TCL is adjusted by changing the temperature set-point, CTBM also stays at the initial temperature point (21 °C) within the initial 0.5 hours. $f_{\min,}f_{\max}$, $\Delta T_{\max}$, $\Delta T_{\min}$ and k are 30Hz, 180Hz, 3 °C, 1 °C and 8, respectively. The dispatch period is 1 hour and the temperature set-point is adjusted to 21.5 and 20.5 °C, respectively. The evolutions of the actual electric power, SOC and indoor temperature are shown in Fig. 7.3. Obviously, the TCL requires a few minutes to make the room temperature operate near the set point. This means that the response speed of CTBM is much slower than that of PTBM and CTBM is not suitable to be applied in such scenarios, in which the dispatch period is short.

It should be noted that there is only one control signal during a dispatch period to change the electricity consumption. For PTBM, the electric power determined by the frequency is fixed and cannot change with the varying outdoor temperature. Thus, the dispatch period should be short so that the variation of the outdoor temperature is small and can be ignored. For CTBM, it can automatically modify the electricity consumption according to the varying outdoor temperature and keep the indoor temperature at the set point. From this point of view, CTBM is also more appropriate in the scenarios where the dispatch period is long.

7.5.2 *Power and Energy Capacity*

According to the definition of PTBM in (7.17), (7.22) and (7.23), the maximum charging and discharging power (i.e. power capacity) of PTBM at the kth dispatch period, $P^{\mathrm{P}}_{\max}(k)$ and $P^{\mathrm{P}}_{\min}(k)$ are given as:

$$-P^{\mathrm{P}}_{\min}(k) \le P_{\mathrm{cd}}(k) \le P^{\mathrm{P}}_{\max}(k) \tag{7.45}$$

$$P^{\mathrm{P}}_{\min}(k) = \min\left(\frac{\alpha SOC(k) + \gamma}{\beta}, P^{\mathrm{d}}\right) \tag{7.46}$$

$$P^{\mathrm{P}}_{\max}(k) = min\left(\frac{1 - \alpha SOC(k) - \gamma}{\beta}, P^{\mathrm{c}}\right) \tag{7.47}$$

Similarly, the maximum charging and discharging power of CTBM at the kth dispatch period, $P^{\mathrm{C}}_{\max}(k)$ and $P^{\mathrm{C}}_{\min}(k)$ are calculated as:

$$-P^{\mathrm{C}}_{\min} \le P_{\mathrm{CD}}(k) \le P^{\mathrm{C}}_{\max} \tag{7.48}$$

$$P^{\mathrm{C}}_{\min} = \min(\chi \cdot SOC(0), P^{\mathrm{d}}) \tag{7.49}$$

$$P^{\mathrm{C}}_{\max} = \min(\chi \cdot (1 - SOC(0)), P^{\mathrm{c}}) \tag{7.50}$$

The maximum charging and discharging power profiles of PTBM and CTBM under different dispatch periods are shown in Fig. 7.4. For PTBM, the maximum charging/discharging power is kept constant when the dispatch period is less than about 0.5 h. As the dispatch period increases, the maximum charging/discharging power becomes smaller. Take the charging capacity as an example to explain this phenomenon. When the dispatch period is short, $[1 - \alpha SOC(k) - \gamma]/\beta$ is larger than P^{c} and $P^{\mathrm{P}}_{\max}(k)$ is determined by the P^{c}. Then $P^{\mathrm{P}}_{\max}(k)$ is kept fixed. The power capacity is large and the energy capacity is small. When the dispatch period becomes

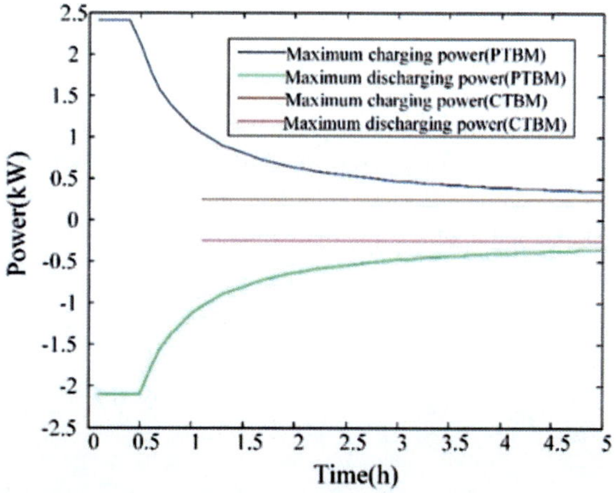

Fig. 7.4 The capacity profiles of PTBM and CTBM under different dispatch periods

longer, $[1 - \alpha SOC(k) - \gamma]/\beta$ becomes less than P^{c} and $P_{\max}^{P}(k)$ is determined by $[1 - \alpha SOC(k) - \gamma]/\beta$. Obviously, $[1 - \alpha SOC(k) - \gamma]/\beta$ will become less with the longer dispatch period. When the dispatch period approaches infinite, $[1 - \alpha SOC(k) - \gamma]/\beta$ infinitely gets closer to $k_1(T_{\max} - T_{\min})(1 - SOC(0))/(k_2 R)$ (i.e. $\chi(1 - SOC(0))$) and the energy capacity is also infinite.

By contrast, the power capacity of CTBM is not affected by the dispatch period and it is kept fixed. It means that CTBM can charge/discharge for a long time and can be seen as a battery with a huge energy capacity. This is because $\chi SOC(0)$ is smaller than P^{d} and $P_{\min}^{C}$ is determined by $\chi SOC(0)$. Similarly, $P_{\max}^{C}(k)$ is determined by $\chi(1 - SOC(0))$. In other words, the power capacity of CTBM is only related to $SOC(0)$. Since the adjustment of the indoor temperature to the new set point takes some time, CTBM is not applied in the case of short scheduling periods. In addition, the power capacity of PTBM is always larger than that of CTBM. When the dispatch period approaches infinite, the power capacity of PTBM infinitely gets closer to that of CTBM. At this time, PTBM can be approximately expressed by CTBM in (7.22), (7.23) and (7.44). The two models share similar operating behaviors and both of them have infinite energy capacity.

7.5.3 *Cost of Control*

Generally, the electricity consumption of the TCL is modified by changing the temperature set-point. Then the frequency of the compressor is adjusted automatically to make the indoor temperature reach the set-point. If the TCL is directly controlled by the frequency, some internal components of the TCL need to be reformed or redesigned, this may cause large expenses. Thus, it is more convenient and cheaper for the TCL to be controlled by changing the temperature set-point compared with adjusting the frequency.

In a conclusion, the comparisons between the two models are summarized in Table 7.2.

Compared with CTBM, PTBM provides a faster response speed and larger power capacity, but generates more cost. Thus, PTBM is more suitable to be applied in some scenarios where the dispatch period is short and the response speed is fast (e.g. load following and frequency regulation) while CTBM is more suitable to be applied in the scenarios where the dispatch period is long and there is no demand needed for the response speed (e.g. the load shaving). Although the power capacity of PTBM is

Table 7.2 Comparisons between PTBM and CTBM

Battery type	Speed	Power capacity	Energy capacity	Cost
PTBM	Fast	Large, related with the dispatch period	Variable	Much
CTBM	Slow	Small	Large	Little

Table 7.3 Parameters of inverter TCL system

Parameter	Value	Unit
R	1.5–2.5	°C/kW
Ca	1.5–2.5	kW h/°C
Q'	0.1–0.5	kW
T_{set0}	20–22	°C
$T_{\min}$	20	°C
$T_{\max}$	22	°C
$P_{\min}$	0.5	kW
$P_{\max}$	5	kW
k_1	0.03	kW/Hz
l_1	− 0.4	kW
k_2	0.06	kW/Hz
l_2	− 0.3	kW

always larger than that of CTBM, CTBM has the advantage of taking part in demand response because of its low control cost and larger energy capacity.

PTBM is dispatched to optimize the output of wind generation due to the frequent and large fluctuations in wind generation. The applications of CTBM will be studied in our future research.

7.6 Dispatch Strategy for Output Optimizing of Wind Generation

In this section, PTBMs are dispatched to participate in a real-time market for reducing the economic loss as a result of the fast fluctuations of wind generation.

The aggregator that manages wind generation and loads are regarded as a single entity, which acts as a generator in the grid. In the day-ahead electricity market, the aggregators submits their power generation schedule of the next day to the system operator (SO), including their wind power, load and electricity price forecasting, etc. The SO receives the schedule information of all market players and the market-clearing price is worked out according to the market rules [30].

Due to the great uncertainties of wind power, the difference between the actual and the planned power of the aggregator is large, which may result in serious punishment and huge economic losses according to the market rules. Therefore, the aggregator needs to purchase/sell power in a real-time market to reduce the losses when the power prediction is more accurate during the day.

In the real-time market, the scheduling of PTBMs is optimized to maximize the benefit of the aggregator. Two assumptions are made here: (1) Aggregators are price takers with little influence on prices in the real-time market because of the limitation of their power; (2) the load is inelastic to electricity price. The MPC is to solve a finite-horizon optimization problem and only the first optimization operation is implemented [30]. The detailed optimization process is given as follows.

(1) Select a receding optimization period w and the forecast information of the wind generation, load and outdoor temperature in the future periods from $i + 1$ to $i + w$ (i is the current dispatch period) are known. The maximum charge/discharge power of PTBMs at the period from $i + 1$ to $i + w$ can be obtained as well;
(2) Solve the optimization problem of the aggregator in a real-time market. The objective function is to maximize the economic income (or minimize the cost) [31, 32], which is a linear programming problem as

$$\max F_{\mathrm{c}} = \sum_{k=i}^{i+w} [(G(k) - P_{\mathrm{n}}(k) - L(k) - Q(k))p_{\mathrm{rt}}(k)] \tag{7.51}$$

Subject to

$$-D^{\max} \le P_n(k) \le C^{\max}$$

$$\boldsymbol{SOC}(k+1) = \boldsymbol{\alpha} \cdot \boldsymbol{SOC}(k) + \boldsymbol{\kappa} \cdot \boldsymbol{P}_n(k) + \boldsymbol{\sigma}$$

$$0 \le \boldsymbol{SOC}(k+1) \le 1$$

(3) Discharge/charge PTBMs in terms of the optimization result of (7.31) at period $i + 1$;
(4) Update the $\boldsymbol{SOC}$ of PTBMs and move to the next period. Repeat the optimization process starting from step (1).

The optimization behaviors of the aggregator in the real-time market are described in Fig. 7.5.

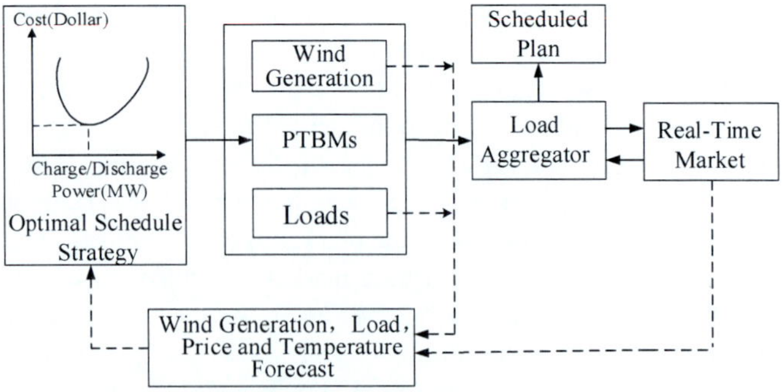

Fig. 7.5 The behaviors of the aggregator in the real-time market

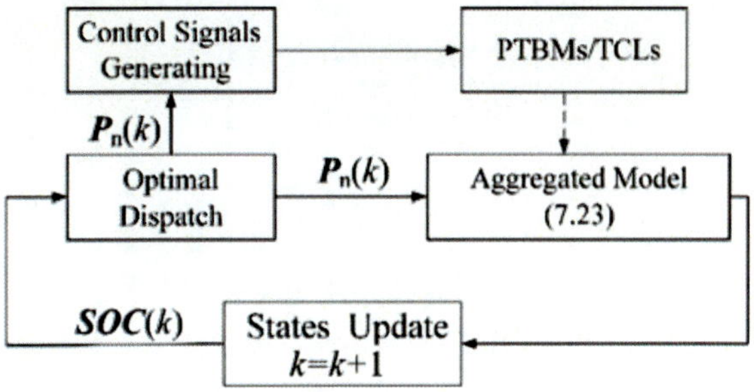

Fig. 7.6 Control framework of PTBMs

The aggregator develops optimal dispatch strategies of PTBMs based on $\boldsymbol{SOC}(k)$ and makes decisions on $\boldsymbol{P}_{\mathrm{n}}(k)$ according to the forecast of the wind generation, load, price and outdoor temperature. The control signals to PTBMs/ACs will be generated by the aggregator to change the electricity consumption. At the same time, $\boldsymbol{P}_{\mathrm{n}}(k)$ as the input of the aggregated model of PTBMs is used to update $\boldsymbol{SOC}(k+1)$ and proceeds to the next period. The control framework is shown in Fig. 7.6.

7.7 Case Studies

In this case, the number of the inverter TCL is 2000 and their parameters are shown in Table 7.3. R, C_{a}, Q' and T_{set0} are all distributed randomly in their ranges. The dispatch period is 1 hour and the receding optimization period is 4. The simulation time is 24 hours. Wind generation, load and electricity price are from [31–33]. The number of clusters is 10. We use the outdoor temperature provided by [34], which is a typical summer outdoor temperature profile in the region of PJM. During each dispatch period, the outdoor temperature is regarded as a fixed value. The forecast method is from [31]. Considering that the fluctuations of wind generation are much larger than those of loads and electricity price, only the uncertainties of wind generation are taken into account.

7.7.1 Optimization Results

The scheduled results of the aggregator in the day-ahead market is not the research point. For simplicity, $Q(k)$ is given as:

$$Q(k) = (G_{\mathrm{a}}(k) - L_{\mathrm{a}}(k)) \times (1 + v_{\mathrm{d}}) \tag{7.52}$$

where $G_{\mathrm{a}}(k)$ and $L_{\mathrm{a}}(k)$ are the actual wind generation and load at the kth period, MW, respectively. v_d is a rand number that is distributed uniformly in [−0.5, 0.5].

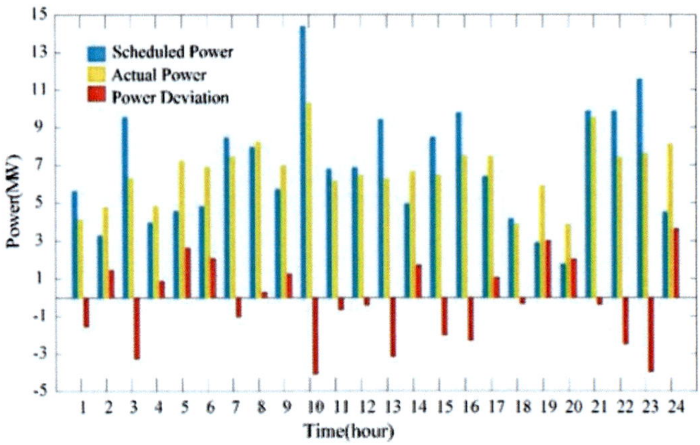

Fig. 7.7 The comparisons of the scheduled power in day-ahead and the actual power

The scheduled power, actual power and the power deviation (actual value-scheduled value) are shown in Fig. 7.7. It can be seen that the difference between the scheduled power in the day-ahead market and actual power is considerable because of the uncertainties of wind generation. If the aggregator purchases/sells power according to the power deviation in the real-time market to provide the scheduled power for the SO and not to be punished heavily, the aggregator will face a huge risk of economic loss.

According to the parameters of PTBM defined in (7.18)–(7.21), (7.24) and (7.25), it can be seen that only P^{c} and P^{d} are related to the outdoor temperature (P_{baseline} varies with the outdoor temperature). The maximum aggregated charge power (i.e. $C^{\max}$ in (7.37)) and maximum aggregated discharge power [i.e. $-D^{\max}$ in (7.37)] with varying outdoor temperatures are shown in Fig. 7.8. It can be seen that the maximum

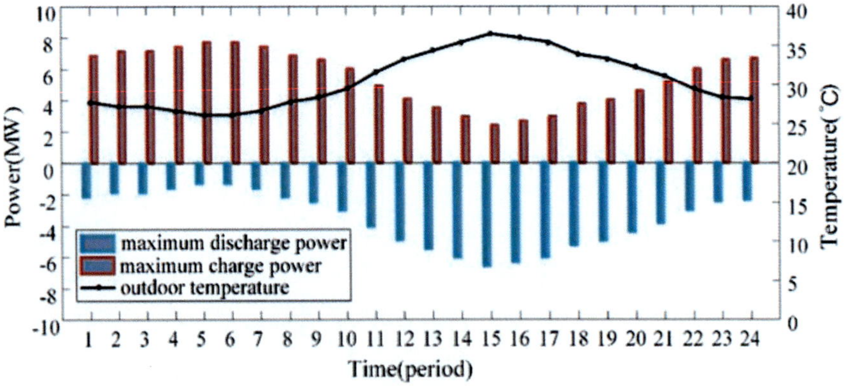

Fig. 7.8 The maximum aggregated charge and discharge power of PTBMs with varying outdoor temperatures

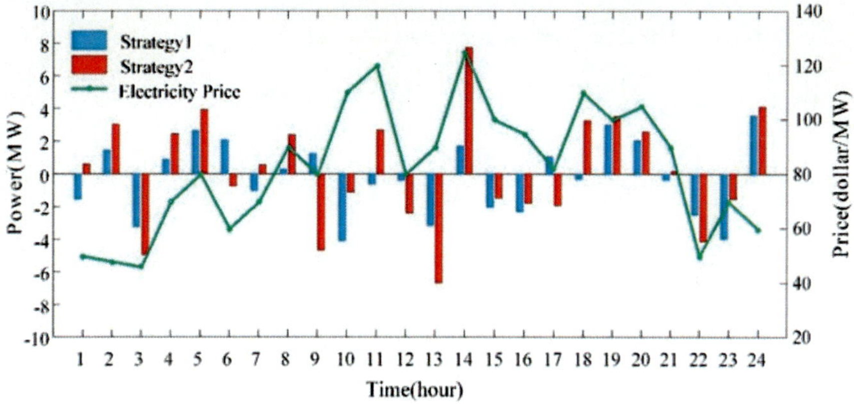

Fig. 7.9 The optimization results of the agssgregator with two strategies

charge power decreases when the outdoor temperature rises and increases when the outdoor temperature drops. This means that the maximum charge power of PTBMs has an opposite trend to the outdoor temperature. Likewise, the maximum discharge power of PTBMs follows the same trend as the outdoor temperature. This is because that the baseline power of PTBM [see (7.6)] has a positive linear relationship with the outdoor temperature. When the outdoor temperature rises, the baseline power of PTBM will increase, which will lead to a smaller charge power limit and a larger discharge limit according to the definitions of PTBM in (7.24) and (7.25).

Figuer 7.9 shows the behaviors of the aggregator with two different strategies in the real-time market. In strategy1, the aggregator without any energy storage devices purchases/sells electricity in the real-time market according to the power mismatches between the actual wind generation and load. In strategy 2, the aggregator optimizes the charge/discharge power of PTBMs in the real-time market for maximal economic benefit according to the dispatch model described in Sect. 7.6. The income of the aggregator with the two strategies are −399 dollars and 1106 dollars, respectively. It means that the economic loss is 399 dollars when there are no energy storage devices while the income is 1.106 dollars when PTBMs are dispatched to optimize the behaviors of the aggregator. PTBMs are beneficial to reduce the economic loss as a result of the fluctuations of wind generation. The positive values of the aggregator mean that the aggregator sells electricity and the negative values mean that the aggregator purchases electricity in the real-time market. It is obvious that the aggregator with PTBMs purchases more power at a lower price (e.g. the 3rd, 9th, 13th, 17th h) and sells more power at a higher price (e.g. the 4th, 8th, 14th, 18th h), which is similar to the operating way of the CBs. In other words, PTBMs can replace the CBs in some dispatch frameworks. PTBMs can make the aggregator operate more flexibly according to the electricity price and provide the scheduled power for the SO.

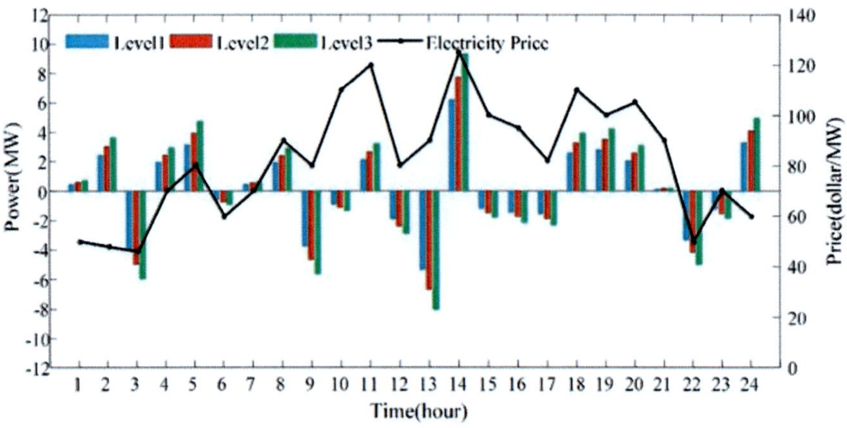

Fig. 7.10 The behaviors of the aggregator under different end-users' comfort

7.7.2 Impact of the End-users' Comfort Setting

This section focuses on the impact of temperature comfort levels on the real-time scheduling results of the aggregator. The comparisons of real-time dispatch results of the aggregator under different temperature comfort levels are shown in Fig. 7.10. The indoor temperature range of the level1, level2 and level3 in Fig. 7.10 are 1 °C ([20 °C, 21 °C]), 2 °C([20 °C, 22 °C]) and 4 °C ([20 °C, 24 °C]), respectively. The income of the aggregator under the three comfort levels is 889, 1106 and 1356 dollars, respectively. As shown in Fig. 7.10, the larger indoor temperature range can make the aggregator operate with more power potential. In other words, the aggregator with the wider temperature range will sell more power at the higher electricity price periods and purchases more power at the lower price periods. Although the selling power and purchase power are both increased, the income of the aggregator is increased because of the price difference. In a conclusion, the larger indoor temperature range allows PTBMs to be better at optimizing their behaviors in real-time scheduling for optimizing the fluctuations of the wind generation. However, this will also reduce users' comfort and cause more complaints, which will affect the enthusiasm of end-users to participate in demand response programs. Therefore, an appropriate temperature comfort level needs to be considered simultaneously in the signing of the agreement with end-users.

7.8 Conclusions

Inverter TCL is constructed as an energy storage device model to optimize the behaviors of wind power in the real-time market. First, the circuit and mathematical model of PTBM are established to show the energy storage characteristics of the inverter

TCL system. Secondly, for the comparison, CTBM is also presented and the differences between the two models are analyzed. Thirdly, PTBM is chosen to take part in the real-time market to solve the problem of the fluctuations of wind generation. The results are described as follows.

(1) Inverter TCL can be built as an energy storage model and have identical expressions with the CB because of its thermal energy storage characteristics, which helps the TCL be compatible with the current dispatch models;
(2) Compared with CTBM, PTBM has a faster response speed, larger power capacity and greater cost. Therefore, PTBM is better suited to provide ancillary services (e.g., load tracking and frequency regulation) in situations where the dispatch period is short and the response speed is fast, while CTBM is better suited to participate in demand response (e.g., load shaving) in situations where the dispatch period is long and the response speed is not required.
(3) PTBMs enable the aggregators to optimize their behaviors in the real-time market for reducing the economic losses caused bythe uncertainties of the wind generation in the way of the CBs. The wider comfort level can make the aggregator work more flexibly.

References

1. Yin, M., Han, F., Li, J., & Zhang, Y. (2009, September). Large-scale wind energy development in China and the relevant issues. In *2009 IEEE PES/IAS Conference on Sustainable Alternative Energy (SAE)* (pp. 1–6). Valencia, Spain.
2. Vartanian, C. K., Bentley, N., & Foster, R. (2012, July). Application of advanced battery energy storage systems for wind integration. In *2011 IEEE Power and Energy Society General Meeting* (pp. 1–4), San Diego, CA.
3. Song, M., Gao, C., Yang, J., & Yan, H. (2018). Energy storage modeling of inverter air conditioning for output optimizing of wind generation in the electricity market. *CSEE Journal of Power and Energy Systems, 4*(3), 305–315.
4. Ummels, B. C., Pelgrum, E., & Kling, W. L. (2008). Integration of large-scale wind power and use of energy storage in The Netherlands' electricity supply. *IET Renewable Power Generation, 2*(1), 34–46.
5. Song, M., Gao, C., Shahidehpour, M., Li, Z., Yang, J., & Yan, H. (2018). State space modeling and control of aggregated TCLs for regulation services in power grids. *IEEE Transactions on Smart Grid, 10*(4), 4095–4106.
6. Zhang, Y., & Lu, N. (2013). Parameter selection for a centralized thermostatically controlled appliances load controller used for intra-hour load balancing. *IEEE Transactions on Smart Grid, 4*(4), 2100–2108.
7. Lu, N., & Zhang, Y. (2012). Design considerations of a centralized load controller using thermostatically controlled appliances for continuous regulation reserves. *IEEE Transactions on Smart Grid, 4*(2), 914–921.
8. Lu, N. (2012). An evaluation of the HVAC load potential for providing load balancing service. *IEEE Transactions on Smart Grid, 3*(3), 1263–1270.
9. D&R International, Ltd. (2012). 2011 buildings energy data book. https://ieer.org/wp/wp-content/uploads/2012/03/DOE-2011-Buildings-Energy-DataBook-BEDB.pdf. Accessed March, 2012.

10. Mathieu, J. L., Koch, S., & Callaway, D. S. (2012). State estimation and control of electric loads to manage real-time energy imbalance. *IEEE Transactions on Power Systems, 28*(1), 430–440.
11. Song, M., Ciwei, G., Yang, J., Liu, Y., & Cui, G. (2017). Novel aggregate control model of air conditioning loads for fast regulation service. *IET Generation Transmission & Distribution, 11*(17), 4391–4401.
12. Molina-Garcia, A., Kessler, M., Fuentes, J. A., & Gómez-Lázaro, E. (2010). Probabilistic characterization of thermostatically controlled loads to model the impact of demand response programs. *IEEE Transactions on Power Systems, 26*(1), 241–251.
13. Lu, N., & Chassin, D. P. (2004). A state queuing model of thermostatically controlled appliances. *IEEE Transactions on Power Systems, 19*(3), 1666–1673.
14. Peppanen, J., Reno, M. J., & Grijalva, S. (2014, September). Thermal energy storage for air conditioning as an enabler of residential demand response. In *2014 North American Power Symposium* (pp. 1–6), Pullman, WA, USA.
15. Kamgarpour, M., Ellen, C., Soudjani, S. E. Z., Gerwinn, S., Mathieu, J. L., Müllner, N., Abate, A., Callaway, D.S., Fränzle, M., & Lygeros, J. (2013, August). Modeling options for demand side participation of thermostatically controlled loads. In *2013 IREP Symposium Bulk Power System Dynamics and Control-IX Optimization, Security and Control of the Emerging Power Grid* (pp. 1–15).
16. Mathieu, J. L., Kamgarpour, M., Lygeros, J., Andersson, G., & Callaway, D. S. (2014). Arbitraging intraday wholesale energy market prices with aggregations of thermostatic loads. *IEEE Transactions on Power Systems, 30*(2), 763–772.
17. Hao, H., Sanandaji, B. M., Poolla, K., & Vincent, T. L. (2014). Aggregate flexibility of thermostatically controlled loads. *IEEE Transactions on Power Systems, 30*(1), 189–198.
18. Sanandaji, B. M., Hao, H., Poolla, K., & Vincent, T. L. (2014, June). Improved battery models of an aggregation of thermostatically controlled loads for frequency regulation. In *2014 American Control Conference* (pp. 38–45), Portland, OR, USA.
19. Hao, H., Sanandaji, B. M., Poolla, K., & Vincent, T. L. (2013, October). A generalized battery model of a collection of thermostatically controlled loads for providing ancillary service. In *2013 51st Annual Allerton Conference on Communication, Control, and Computing* (pp. 551–558), Monticello, IL, USA.
20. Kim, Y. J., Norford, L. K., & Kirtley, J. L. (2014). Modeling and analysis of a variable speed heat pump for frequency regulation through direct load control. *IEEE Transactions on Power Systems, 30*(1), 397–408.
21. Kim, Y. J., Fuentes, E., & Norford, L. K. (2015). Experimental study of grid frequency regulation ancillary service of a variable speed heat pump. *IEEE Transactions on Power Systems, 31*(4), 3090–3099.
22. Wai, C. H., Beaudin, M., Zareipour, H., Schellenberg, A., & Lu, N. (2014). Cooling devices in demand response: A comparison of control methods. *IEEE Transactions on Smart Grid, 6*(1), 249–260.
23. Chiguchi, M., Yamaguchi, H., Higashino, T., & Shimoda, Y. (2016, November). Human thermal comfort estimation in indoor space by crowd sensing. In *2016 IEEE Smart Grid Comm* (pp. 45–50), Sydney, NSW, Australia.
24. Xing, H., Fu, M., Lin, Z., & Mou, Y. (2015). Decentralized optimal scheduling for charging and discharging of plug-in electric vehicles in smart grids. *IEEE Transactions on Power Systems, 31*(5), 4118–4127.
25. Shao, C., Wang, X., Wang, X., Du, C., & Wang, B. (2015). Hierarchical charge control of large populations of EVs. *IEEE Transactions on Smart Grid, 7*(2), 1147–1155.
26. Wang, D., Guan, X., Wu, J., Li, P., Zan, P., & Xu, H. (2015). Integrated energy exchange Scheduling for multi micro grid system with electric vehicles. *IEEE Transactions on Smart Grid, 7*(4), 1762–1774.
27. Wen, Y., Li, W., Huang, G., & Liu, X. (2016). Frequency dynamics constrained unit commitment with battery energy storage. *IEEE Transactions on Power Systems, 31*(6), 5115–5125.
28. Zareifard, M. T. (2015, September). Model predictive control of electricity market oriented wind farm dispatch with a battery energy storage system. In *2016 IEEE Conference on Control Applications* (pp. 727–732), Sydney, NSW, Australia.

29. Piphitpattanaprapt, N., & Bangerdpongchai, D. (2015, November). Optimal dispatch strategy of hybrid power generation with battery energy storage system in islanding mode. In *2015 IEEE Innovative Smart Grid Technologies Asian Conference* (pp. 1–6), Bangkok, Thailand.
30. Xu, Y., & Singh, C. (2012). Adequacy and economy analysis of distribution systems integrated with electric energy storage and renewable energy resources. *IEEE Transactions on Power Systems, 27*(4), 2332–2341.
31. Xu, Y., & Singh, C. (2012). Operation strategies of the load aggregator with electric energy storage. In *IEEE POWERCON 2012* (pp. 1–6), Auckland, New Zealand.
32. Xu, Y., Xie, L., & Singh, C. (2010, September). Optimal scheduling and operation of load aggregator with electric energy storage in power markets. In *2010 North American Power Symposium* (pp.1–7), Arlington, TX, USA.
33. Kang, M. S., Chen, C. S., Ke, Y. L., Lin, C. H., & Huang, C. W. (2007). Load profile synthesis and wind-power-generation prediction for an isolated power system. *IEEE Transactions on Industry Applications, 43*(6), 1459–1464.
34. E. Anastasio. (2017). Load Forecasting and Analysis. https://www.pjm.com/-/media/committees-groups/committees/pc/20170914/20170914-item-06d-m19-clean.ashx, Accessed January 12, 2017.

Chapter 8
Multi-time Scale Models and Parameter Identification Method of TCLs

Abstract The potential of TCL for DR has been demonstrated. One of the key challenges for TCLs to be integrated into the system-level operation is to build a compact aggregated model, in which the TCL primary behaviors are accurately captured. In this chapter, TCLs are aggregated as a virtual generator and two batteries according to their different compressor types and control methods for smoothing out multi-time-scale variability of wind power generation. This will greatly facilitate the system operator in managing TCLs and conventional components when making system-level decisions. Accordingly, accurate parameters of virtual generators and batteries are critical to effectively coordinate TCLs with other resources in the system operation. However, such integrated parameters are often difficult to obtain due to insufficient data for each TCL. To address this problem, HDMR is introduced to estimate the aggregated parameters of virtual generators and batteries using the probability distribution of TCL parameters. A numerical simulation study demonstrates that aggregated parameters of virtual generators and batteries can be accurately estimated by HDMR. Virtual generators and batteries can follow actual behaviors of TCL populations in power system operations.

Keywords Wind power generation · Virtual generator · Virtual battery · Parameter estimation

8.1 Introduction

TCLs which are able to store thermal energy in objects represent promising DR resources and can provide ancillary services to the power grid [1–5]. TCLs are divided into two categories of fixed-frequency TCLs (FTCLs) and inverter TCLs (ITCLs) [6], according to the compressor type. TCLs tend to have various operating characteristics (e.g., response time) under different control methods. Specifically, FTCLs whose power status is either 0 or rated power are mainly controlled by on-off switches [4, 7, 8] and temperature set-point changes [9–12]. The two FTCL control methods are compared in [13] which concludes that on-off switches can instantaneously change FTCL power consumptions but cannot maintain the designated adjustment for a long period. On the contrary, temperature set-point control can regulate power for

M. Song and C. Gao, *Integration of Distributed Resources in Smart Grids for Demand Response and Transactive Energy*, https://doi.org/10.1007/978-981-16-7170-8_8

longer periods, but the control impact may not be immediate. Therefore, FTCLs are suitable for taking part in short time-scale DR programs (e.g., frequency regulation) by on-off control and long time-scale DR programs (e.g., peak load shaving) by temperature set-point changes. Different from FTCL, the ITCL power can be adjusted continuously [14, 15], which allows a more flexible operation of ITCL particularly in the minute-based DR program which is utilized in this chapter.

The large number and wide distribution of TCLs makes it difficult to manipulate them individually. In addition, the adjustable capacity of TCL is too small to have significant impacts on power systems. Thus, aggregation is the key for TCL integration into power system operations. Currently, most studies focus on aggregated FTCL models managed by on-off controls [1, 7, 8, 16–18]. References [8, 18] proposed an accurate aggregated model of air conditioning loads for regulation services, in which the compressor time delay and communication delay were considered. The model predictive control was employed based on the aggregated TCL model in [1, 16] to regulate heterogeneous TCLs optimally. To reduce communication requirements between the centralized controller and end-users, the parameters of the aggregated TCL model were identified by state estimation using a model estimator controller system given in [17]. A finite-space stochastic dynamic model was introduced to address the aggregation of TCLs by formal abstractions which were extended to the controlled population [7].

Although there have been numerous studies on the aggregation of TCLs, these aggregated models are mainly utilized to describe the detailed evolutions of TCL indoor temperatures with external control signals at the level of aggregators and end-users. A concise system-level aggregated model that can accurately capture TCL primary performances is required to coordinate TCLs with system-level resources in power system operations. In particular, TCLs that are aggregated as conventional system components (e.g. generator and battery) can offer additional compatibility which is imperative for system operation [19]. References [20–22] presented battery models to describe the flexibility of aggregated TCLs, in which a TCL population served as a collection of power signals. However, such battery models cannot express the flexibility of conventional batteries because TCLs cannot be manipulated like a battery at the power system level.

Few studies have introduced aggregated FTCL models at the system level controlled by temperature set point changes, and the available studies rarely consider the effect of cold load pickup in aggregation. Thus, more efforts are required to present aggregated FTCL models at the system level. When aggregated TCL models are applied at the system level, accurate calculations of aggregated model parameters are critical for managing TCLs effectively. However, the corresponding parameter estimation methods for aggregated TCL models are deemed insufficient. Obviously, the parameters of aggregated TCL models are closely related to the parameters of each TCL. However, the limited data available for each TCL which corresponds to customer privacy, data measurement errors and customer random consumption behaviors, could make it difficult to calculate the pertinent parameters for aggregated TCL models. To address this problem, HDMR [23, 24] was utilized to estimate the parameters of virtual generators and batteries. HDMR stands out by establishing relationships between multiple inputs and outputs of a system and has been widely used in biology [25], communication technology [26, 27] and information storage [28].Compared to the data for each TCL, the probability distribution of TCL parameters is easier to ascertain using available technologies [29].

In this chapter, HDMR is applied to estimate the parameters of virtual generators and batteries using the TCL parameter distribution. HDMR can identify important TCL parameters, which allows aggregators to estimate critical parameters more efficiently. Accurate estimations of critical TCL parameters will reduce parameter uncertainties corresponding to insufficient TCL data for representing virtual generators and batteries. In practice, the forecasting accuracy of wind power generation will increase as the forecasting time scale becomes shorter. Accordingly, uncertainties in power system operations caused by variabilities in wind power generation are characterized by multiple time scales. Similarly, the operating behaviors of two types of TCLs, including the response speed and duration, depending on the control method that is considered. Therefore, an aggregated TCL model is needed to smooth the variation of wind power generation over multiple time scales, which not only increases the wind power generation but also improves the system efficiency.

The main work of this chapter includes (1) proposing a virtual generator model of FTCLs controlled by temperature set-point changes, in which the cold load pickup caused by temperature set-point adjustments is limited to reduce random fluctuations during power dispatch; (2) proposing a virtual battery model of FTCLs controlled by on-off switches to provide second time-scale ancillary services for the grid; (3) using HDMR to estimate parameters of the virtual generator and battery models of aggregated TCLS with the parameter distribution of TCL population.

8.2 Multi Time-Scale Dispatch Framework for Smoothing Out Wind Power Generation Variability

TCL aggregators serving as a bridge to system operators play an important role in DR programs. It takes responsibility to characterize the aggregated behaviors of heterogeneous TCLs for the optimal schedule of different resources (e.g. wind power generation and generators) at the power system level. Thus, TCL aggregators demonstrate comprehensive external characteristics of TCLs to the power system operators. In addition, TCL aggregators generate control signals to effectively manipulate the TCL population according to the data provided by system operators.

In this section, we focus on TCL aggregations provided by aggregators and the system operators' optimization strategies will be presented in our future work. To smooth out multi-time-scale variability of wind power generation, aggregated TCL

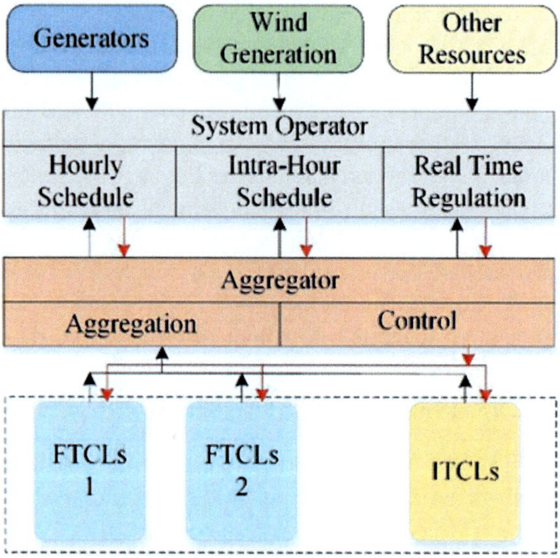

Fig. 8.1 Multi time-scale dispatch framework

models are considered on hourly (1 h), intra-hourly (15 min) and real-time (4 s) scales by aggregators. Figure 8.1 shows the dispatch framework.

8.3 Modeling of Virtual Generator of FTCLs on Hourly Time-Scale

8.3.1 Control Method

The indoor temperature evolution of TCLs is described by an ETP model [5] as:

$$C_a \cdot dT_i(t)/dt = (T_{out} - T_i(t))/R - Q(t) \tag{8.1}$$

There are heating and cooling operating modes for TCLs. However, only the cooling mode is considered in this chapter. The relationship between Q and P is closely tied to compressor types. For FTCLs, Q is formulated as

$$Q(t) = S(t) \cdot \eta \cdot P \tag{8.2}$$

FTCLs can only work at the rated frequency (i.e., rated power) before it is turned off so that indoor temperatures can change periodically within $[T_{\min}, T_{\max}]$ where, $T_{\min} = T_{\text{set}} - \delta$, $T_{\max} = T_{\text{set}} + \delta$. $S(t)$ is given as

$$S(t+\Delta t') = \begin{cases} 1 & T_i(t+\Delta t') > T_{\max} \\ 0 & T_i(t+\Delta t') < T_{\min} \\ S(t) & T_{\min} \le T_i(t+\Delta t') \le T_{\max} \end{cases} \tag{8.3}$$

Temperature setpoint control has been shown to regulate the aggregated FTCL power for a longer period [13].Thus, FTCLs manipulated by temperature set-point changes are aggregated as a virtual generator on an hourly time-scale to provide ancillary services to the grid. When FTCLs are controlled by temperature set-point changes, the aggregated power will be change gradually. Accordingly, the heterogeneity of FTCL operating states will disappear and the aggregated power for cold load pickup will look oscillatory, resulting in more variability in power system operations [30]. To address this problem, it is necessary to manipulate FTCLs by certain control strategies embedded in DR programs.

The control strategy for FTCLs is shown in Fig. 8.2. The temperature rise set-point is used as an example to illustrate the control mechanism. In order to limit cold load pickup, the heterogeneity of FTCL operating states should be restored quickly [30]. The control procedure is divided into four stages. At the first stage, FTCLs operate at T_{set} and their indoor temperatures change periodically within [$T_{\min}$, $T_{\max}$], which is given in Fig. 8.2a. At the second stage in Fig. 8.2b, FTCLs receive control signals and

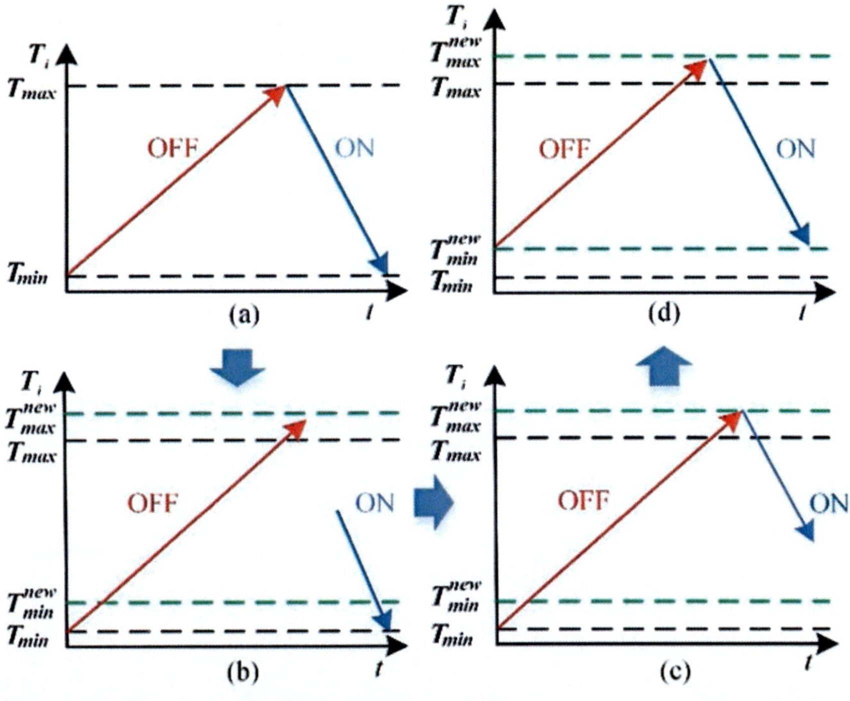

Fig. 8.2 Control strategy of FTCLs for limiting cold load pickup

their temperature set-points are increased from T_{set} to T_{set}^{new} (i.e. indoor temperature range changes from $[T_{min}, T_{max}]$ to $[T_{min}^{new}, T_{max}^{new}]$, where $T_{min}^{new} = T_{set}^{new} - \delta$, $T_{max}^{new} = T_{set}^{new} + \delta$). FTCLs in off state maintain their status until indoor temperatures reach T_{max}^{new}. FTCLs in on state will be turned off when their indoor temperatures are at T_{min}. At the third stage, FTCLs continue to operate similar to that at the second stage. As shown in Fig. 8.2c, FTCLs which are originally at on state are gradually turned off and some FTCLs in off state are turned on when their indoor temperatures reach T_{max}^{new}. At the fourth stage, FTCLs in on state are turned off at T_{min}^{new} and indoor temperatures of FTCLs begin to vary within in $[T_{min}^{new}, T_{max}^{new}]$, as shown in Fig. 8.2d.

In sum, FTCLs which are at off state with indoor temperatures within $[T_{min}^{new}, T_{max}^{new}]$, are governed by the new temperature range $[T_{min}^{new}, T_{max}^{new}]$. The other FTCLs operate as intended originally and are then governed by $[T_{min}^{new}, T_{max}^{new}]$ once they are at off state and indoor temperatures are within $[T_{min}^{new}, T_{max}^{new}]$. Accordingly, the heterogeneity of FTCL operating states is recovered gradually as FTCLs work uniformly within the new temperature range.

FTCL Parameters are characterized by normal distribution [31]. The corresponding mean and standard deviation of FTCL parameters are given in Table 8.1. The initial indoor temperatures of 10,000 FTCLs are uniformly distributed within $[T_{min}, T_{max}]$ and the outdoor temperature is 36 °C. The comparison of aggregated FTCL power with/without a control strategy is shown in Fig. 8.3. Before 0.2 h, FTCLs work at T_{set} and the aggregated power appears stable. At 0.2 h, the temperature set-points of FTCLs are increased by 0.5 °C. When the control strategy is applied, the corresponding aggregated FTCL power will drop as it reaches the valley at 0.35 h and then reaches a steady state fairly smoothly. By contrast, the aggregated FTCL

Table 8.1 FTCL Parameters

Parameter	R	C_a	P	T_{set}	δ	η
Mean	2.5	2.5	8	20	0.25	3
Standard deviation	0.2	0.2	0.5	0.5	0.02	0.2

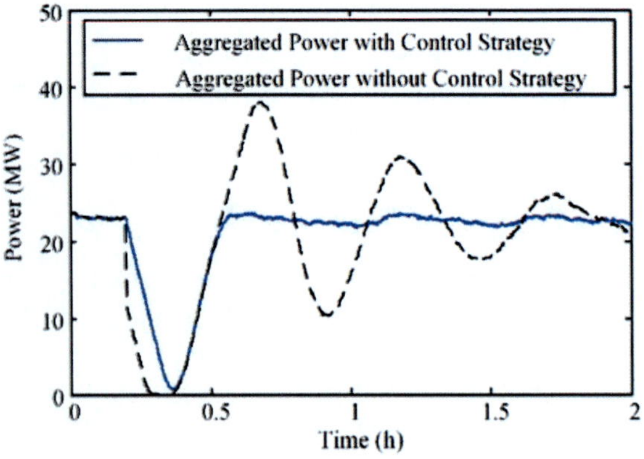

Fig. 8.3 Comparison of aggregated FTCL power

power is largely oscillatory when the control strategy is not utilized. Accordingly, the cold load pickup is successfully restricted using the proposed control strategy and only one hourly power valley is available to meet the dispatch requirements.

8.3.2 Virtual Generator Modeling of FTCLs

When FTCLs are regulated by temperature set-point changes, the aggregated power will vary within a certain range, which is similar to the operating characteristics of conventional generators. The difference between the power valley/peak and the power baseline of FTCLs is defined as the power output of a virtual generator. Hence, the power output of a virtual generator is only related with $\Delta T(k)$ for a given TCL population and known T_{out}, stated as

$$P_h(k) = f(\Delta T(k)) \tag{8.4a}$$

$$f(\Delta T(k)) = P_a(k) - P_{baseline} \tag{8.4b}$$

Using (8.1), $P_{baseline}$ is expressed as

$$P_{baseline} = \sum_{i=1}^{n} \left(P \frac{t_{on}}{t_{on} + t_{off}} \right)_i \tag{8.5a}$$

$$t_{on} = -RC_a \ln \frac{T_{out} - \eta PR - T_{\min}}{T_{out} - \eta PR - T_{\max}} \tag{8.5b}$$

$$t_{off} = -RC_a \ln \frac{T_{out} - T_{\max}}{T_{out} - T_{\min}} \tag{8.5c}$$

The specific expression of $f(\Delta T(k))$ will be obtained by the method proposed in Sect. 8.6. $\Delta T(k)$ is related to end users' comfort and determined by end-users. When $\Delta T(k)$ is constrained, $P_h^{\min}$ and $P_h^{\max}$ will be calculated according to the $f(\Delta T(k))$. Thus, the power output of a virtual generator is constrained as

$$P_h^{\min} \le P_h(k) \le P_h^{\max} \tag{8.6a}$$

$$\Delta T_{\min} \le \Delta T(k) \le \Delta T_{\max} \tag{8.6b}$$

As a result of the physical constraints, there are ramp up and down limits for conventional generators, which are utilized to describe the power adjustment capability of generators during a dispatch period [32]. For the virtual generator model of FTCLs, there are no specific physical constraints. Therefore, the virtual generator model is expressed using (8.6a). When the aggregator gets the desired $P_h(k)$ signal from a system operator, the required temperature set-point change, $\Delta T(k)$, will be calculated according to the relationship between $f(\Delta T(k))$ and $\Delta T(k)$. And FTCLs will be manipulated to change their temperature set-points as described in Sect. 8.3.A for the desired power.

8.4 Modeling of Virtual Battery of ITCLs on the Minute Time-Scale

8.4.1 Control Method

The indoor temperature evolution of ITCLs can be described by (8.1). Different from FTCLs whose electric power has only two states, the electric power of ITCLs can be continuously adjusted by frequency control. The relationship between Q and P is approximated as [15]

$$Q(t) = k_2/k_1 \cdot P(t) + (k_1 l_2 - l_1 k_2)/k_1 \tag{8.7a}$$

$$P_{\min} \leq P(t) \leq P_{\max} \tag{8.7b}$$

Accordingly, ITCLs with a considerable capacity can quickly respond to dispatch signals and take part in minute time-scale (e.g., 15 min) ancillary services. The detailed control strategy is provided in our previous work [19].

8.4.2 Virtual Battery Modeling of ITCLs

In [19], a virtual battery model of aggregated ITCLs with an expression identical to conventional batteries was expressed as

$$P_m^{\min} \leq P_m(k) \leq P_m^{\max} \tag{8.8a}$$

$$SOC_m(k+1) = \xi \cdot SOC_m(k) + P_m(k) + \gamma \tag{8.8b}$$

$$0 \leq SOC_m(k+1) \leq \tau \tag{8.8c}$$

The virtual battery parameters are calculated as

$$P_m^{\min} = -\sum_{i=1}^{n} \left(\frac{k_1(T_{out} - T_{set})}{k_2 R} + \frac{k_2 l_1 - k_1 l_2}{k_2} - P_{\min} \right)_i \tag{8.9a}$$

$$P_m^{\max} = \sum_{i=1}^{n} \left(P_{\max} - \frac{k_1(T_{out} - T_{set})}{k_2 R} + \frac{k_2 l_1 - k_1 l_2}{k_2} \right)_i \tag{8.9b}$$

$$\xi = \frac{1}{n} \sum_{i=1}^{n} \left[\exp(-\Delta t/(RC_a)) \right]_i \tag{8.9c}$$

$$\gamma = \sum_{i=1}^{n} \left[\frac{2k_1 \Delta T_c}{k_2 R} soc(0) \right]_i \tag{8.9d}$$

$$\tau = n \cdot \min\left\{ \left(\frac{2k_1 \Delta T_c}{k_2 R(1 - \exp(-\Delta t/(RC_a)))} \right)_i, 1 \le i \le n \right\} \tag{8.9e}$$

8.5 Modeling of Virtual Battery of FTCLs on the Second-Time Scale

8.5.1 Control Method

The response time for regulation services is limited to several seconds [33]. When FTCLs are controlled by on-off switches, their power can be changed instantaneously, which is consistent with the response time for regulation services. Therefore, FTCLs controlled by on-off switches are employed to provide regulation services for the grid.

The thermodynamic model of FTCLs was given in (8.1)–(8.3) in which FTCLs only operate at rated and 0 power states. The aggregator receives the desired power signal from a system operator and applies a certain control strategy to regulate the FTCL population. In this section, a control strategy, labeled as the priority stack method [5], is applied to reduce the long-term adverse effect of control signals on FTCL compressors. When additional FTCLs are to be turned on, FTCLs in off state are sorted in descending order based on their normalized indoor temperatures. The FTCLs whose indoor temperatures are closer to $T_{\max}$ will have higher priorities to be turned on. Likewise, when additional FTCLs are to be turned off, FTCLs in on state are sorted in ascending order based on their normalized indoor temperatures. The FTCLs whose indoor temperatures are closer to $T_{\min}$ will have higher priorities to be turned off.

8.5.2 Modeling of Virtual Battery of FTCLs

FTCLs are also modeled as virtual batteries to effectively operate and coordinate FTCLs at the level of power systems. Instead of the dead-band model (8.1)–(8.3), the continuous power model [20] is used to describe the operating performances of FTCLs. The continuous power model is expressed as

$$C_a \cdot \frac{dT_i(t)}{dt} = \frac{T_{out} - T_i(t)}{R} - \eta P^k(t) \tag{8.10a}$$

$$0 \le P^k(t) \le P \tag{8.10b}$$

Accordingly, P_o is formulated as

$$P_o = \frac{T_{out} - T_{set}}{\eta R} \tag{8.11}$$

Obviously, the stored energy of an FTCL is maximum when the indoor temperature is $T_{\min}$ while the stored energy is minimum when the indoor temperature is $T_{\max}$. Thus, the state of charge (SOC) of the virtual battery is defined as:

$$SOC(t) = \frac{T_{set} + \delta - T_i(t)}{2\delta} \tag{8.12}$$

$P_{cd}(t)$ is the difference between the adjusted and baseline power of an FTCL, which is given as:

$$P_{cd}(t) = P^k(t) - P_o \tag{8.13}$$

According to the definitions of SOC and P_{cd} in (8.12) and (8.13), SOC and P_{cd} are linked as:

$$SOC(k+1) = \sigma SOC(k) + \upsilon P_{cd}(k) + \psi \tag{8.14a}$$

$$\sigma = \exp(-\frac{\Delta t}{RC_a}) \tag{8.14b}$$

$$\upsilon = \frac{1 - \exp(-\Delta t/(RC_a))}{2\delta} \tag{8.14c}$$

$$\psi = \frac{1 - \exp(-\Delta t/(RC_a))}{2} \tag{8.14d}$$

Equation (8.14a) shows the relationship between the stored energy level update and the power adjustment of the virtual battery, which is similar to that of a conventional battery. For a large virtual battery population, the heterogeneity of parameters and operating states could increase the computational requirements for scheduling them at the system level. To address this problem, heterogeneous virtual batteries are aggregated as a succinct model that can accurately capture the behavior of virtual batteries. SOC evolutions of a virtual battery population are formulated as:

$$\begin{cases} SOC_1(k+1) = \sigma_1 SOC_1(k) + \upsilon_1 P_{cd\_1}(k) + \psi_1 \\ SOC_2(k+1) = \sigma_2 SOC_2(k) + \upsilon_1 P_{cd\_2}(k) + \psi_2 \\ \vdots \\ SOC_n(k+1) = \sigma_n SOC_n(k) + \upsilon_n P_{cd\_n}(k) + \psi_n \end{cases} \tag{8.15}$$

Generally, the dispatch period of regulation services is 4 s [33]. According to (8.14b), the relationship between RC_a and σ is shown in Fig. 8.4 in which the range of RC_a is set as [0.5, 25]. It is obvious that σ which approaches 1 with larger RC_a is approximately equal to 1. Here, R and C_a reflect the building energy storage characteristics in which RC_a is usually larger than 1 and σ is very close to 1. Accordingly, (8.15) is rewritten as

$$\begin{cases} SOC_1(k+1)/\upsilon_1 = SOC_1(k)/\upsilon_1 + P_{cd\_1}(k) + \psi_1/\upsilon_1 \\ SOC_2(k+1)/\upsilon_2 = SOC_2(k)/\upsilon_2 + P_{cd\_2}(k) + \psi_2/\upsilon_2 \\ \vdots \\ SOC_n(k+1)/\upsilon_n = SOC_n(k)/\upsilon_n + P_{cd\_n}(k) + \psi_n/\upsilon_n \end{cases} \tag{8.16}$$

$SOC_s(k)$ which is used to describe the energy storage level of the aggregated virtual battery model, is defined as

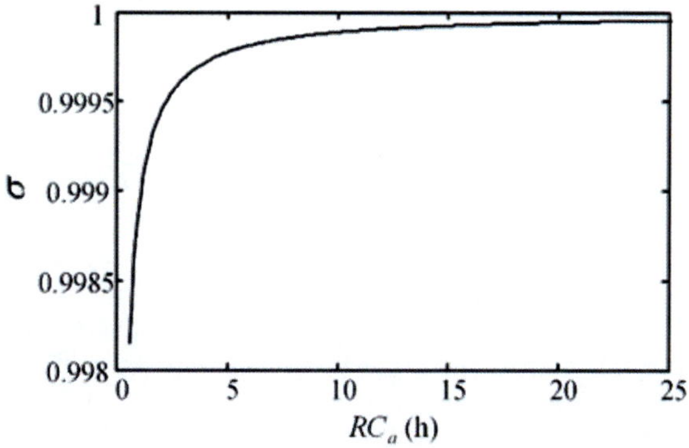

Fig. 8.4 Relationship between RC_a and σ

$$SOC_s(k) = \sum_{i=1}^{n} \frac{SOC_i(k)}{\upsilon_i} \tag{8.17}$$

$P_s(k)$ is the sum of charge/discharge power of the virtual battery population, which is expressed as

$$P_s(k) = \sum_{i=1}^{n} (P_{cd}(k))_i \tag{8.18}$$

From (8.16)–(8.18), the SOC for aggregated virtual batteries is formulated as

$$SOC_s(k+1) = SOC_s(k) + P_s(k) + \kappa \tag{8.19a}$$

$$\kappa = \sum_{i=1}^{n} \frac{\psi_i}{\upsilon_i} \tag{8.19b}$$

According to the definition of SOC given in (8.12), the constraint of $SOC_s(k+1)$ is given as

$$0 \leq SOC_s(k+1) \leq \nu \tag{8.19c}$$

$$\nu = \sum_{i=1}^{n} \frac{1}{\upsilon_i} \tag{8.19d}$$

From (8.18), $P_s(k)$ is constrained as

$$P_s^{\min} \leq P_s(k) \leq P_s^{\max} \tag{8.19e}$$

$$P_s^{\min} = -\sum_{i=1}^{n} (P_o)_i \tag{8.19f}$$

$$P_s^{\max} = \sum_{i=1}^{n} (P - P_o)_i \tag{8.19g}$$

Hence, the aggregated virtual battery on the second time scale is expressed by (8.19a)–(8.19e). The virtual battery model has similar expressions to those of conventional batteries. The aggregated model will offer system operators additional leverage for managing a multitude of batteries at normal and critical conditions.

8.6 Aggregated Parameter Estimation

To facilitate the multi time-scale TCL operations for smoothing out wind power generation variability at the power system level, two types of TCLs are modeled as a virtual generator (8.7a) and two virtual batteries (8.8a)–(8.8c), (8.19a), (8.19c) and (8.19e). In this section, HDMR is employed to estimate the parameters of virtual generator and battery models using the probability distribution of TCL parameters.

8.6.1 Heterogeneity and Uncertainty of TCL Parameters

Generally, parameters of a TCL population are heterogeneous because of different building structures, end users' preferences and device types [34]. In this chapter, TCL parameter heterogeneity is assumed to be characterized by a normal distribution $N(\mu, \chi^2)$ in which μ is mean and χ is standard deviation [31]. This means each TCL parameter is represented by a normal distribution with μ and χ.

For different populations of TCLs, μ and χ are usually characterized by a uniform distribution U(a, b) with a and b as the lower and upper limits [35]. The distributions of TCL parameters are shown in Table 8.2.

In addition to parameters representing the physical characteristics of buildings and devices, T_{out} presents a significant impact on TCL power consumption. T_{out} is assumed to be uniformly distributed within [30, 40] °C. Therefore, the parameters of an FTCL population considering uncertainties are described by $\{R\_\mu, C_{\text{a}}\_\mu, P\_\mu, T_{set}\_\mu, \delta\_\mu, \eta\_\mu, R\_\chi, C_{\text{a}}\_\chi, P\_\chi, T_{set}\_\chi, \delta\_\chi, \eta\_\chi, T_{\text{out}}\}$. Similarly, ITCL parameters are described by $\{R\_\mu, C_{\text{a}}\_\mu, T_{set}\_\mu, P_{\text{min}}\_\mu, P_{\text{max}}\_\mu, k_1\_\mu, l_1\_\mu, k_2\_\mu,$

Table 8.2 Distributions of TCL Parameters

Parameters	Distribution of μ	Distribution of χ
R	U(1.5, 4)	U(0.15, 0.4)
C_a	U(1.5, 4)	U(0.15, 0.4)
P	U(4, 10)	U(0.4, 1)
T_{set}	U(20, 24)	U(2, 2.4)
δ	U(0.25, 0.5)	U(0.025, 0.05)
η	U(2, 3)	U(0.2, 0.3)
P_{min}	U[0.4, 0.6]	[0.04, 0.06]
P_{max}	U[4, 6]	[0.4, 0.6]
k_1	[0.02, 0.04]	[0.002, 0.004]
l_1	[−0.5, −0.3]	[0.03, 0.05]
k_2	[0.05, 0.07]	[0.005, 0.007]
l_2	[−0.4, −0.2]	[0.02, 0.04]

l_2_μ, R_χ, C_a_χ, T_{set}_χ, P_{min}_χ, P_{max}_χ, k_1_χ, l_1_χ, k_2_χ, l_2_χ, $T_{out}\}$. These parameters are assumed to be independent [35, 36].

8.6.2 Parameter Estimation Via HDMR

The relationship between aggregated parameters of virtual models and the distributions of TCL parameters are constructed in order to estimate the parameters of virtual generators and batteries. When HDMR is employed to address this issue, a large number of inputs-output samples are required. Specifically, the TCL population controlled by aggregators is represented as a system in which μ and χ of TCL parameters are the inputs and the aggregated parameters are the outputs. In order to speed up the convergence, the Quasi Monte Carlo sampling method [35] is applied instead of the Monte Carlo method to generate input samples according to the ranges given in Table 8.2. Accordingly, the corresponding output will be calculated by individual simulations. The detailed estimation method is summarized below.

Algorithm 1: Parameter Estimation
1. Apply Quasi Monte Carlo method to generate N_0 input samples according to Table 8.2;
2. Generate parameters of n TCLs considering the log-normally distribution for each input sample;
3. Obtain parameters of virtual models by individual simulations and there will be N_0 input–output pairs;
4. Normalize N_0 input samples: $x^* = (x - x_{min})/(x_{max} - x_{min})$;
5. Utilize N_0 pairs of normalized input–output samples to build HDMR using GUI-HDMR [36] and calculate the associated coefficients α and β;
6. Validate HDMR modeling results and store the associated coefficients for estimating parameters of virtual models

8.7 Case Studies

A population of 10,000 heterogeneous air conditioners is sampled for virtual generators and batteries in this section. The parameter distributions of TCLs are assumed to be known and listed in Table 8.3. μ of TCLs is chosen using [19, 20], χ is 10% of μ, and T_{out} is 36 °C. Before TCLs are employed to take part in DR programs, they are assumed to operate at a steady state, which means indoor temperatures and switch states of FTCLs are uniformly distributed within $[T_{min}, T_{max}]$. Also, ITCLs work at their temperature set-points. The control strategies given in this chapter are utilized to regulate TCLs.

Table 8.3 TCL Parameter distributions

Parameter	M	χ
R	2	0.2
C_{a}	2	0.2
P	5.6	0.56
T_{set}	22.5	2.25
δ	0.3	0.03
η	2.5	0.25
P_{min}	0.5	0.05
P_{max}	5	0.5
k_1	0.03	0.003
l_1	−0.4	0.04
k_2	0.06	0.006
l_2	−0.3	0.03

8.7.1 HDMR Modeling Results

The following two parameter estimation methods are compared in order to show the HDMR merits for parameter estimation:

Method 1: HDMR is utilized to estimate the parameters of virtual generators and batteries. The parameter distributions of TCLs are presented in Table 8.2.

Method 2: Multiple linear regression (MLR) [35] as a common method to build the relationship between multiple inputs and the output is employed to obtain the expressions of aggregated parameters of TCLs. The input-output samples are the same as those of HDMR.

In addition to a TCL parameter, $\Delta T(k) \in [-1, 1]$ °C is also an input variable for the virtual generator. The expressions of $f(\Delta T(k))$ obtained by HDMR and MLR are given in Appendix terms (8.A1) and (8.A2), respectively. The input variables whose coefficients are less than 1 are ignored in MLR. In order to validate the modeling results of HDMR and MLR, 100 input samples $\{R\_\mu, C_{\mathrm{a}}\_\mu, P\_\mu, T_{set}\_\mu, \delta\_\mu, \eta\_\mu, R\_\chi, C_{\mathrm{a}}\_\chi, P\_\chi, T_{set}\_\chi, \delta\_\chi, \eta\_\chi, T_{\mathrm{out}}, \Delta T(k)\}$ as inputs to (8.A1) and (8.A2) are randomly chosen from their stated ranges in Table 8.2. The TCL aggregated power obtained by individual simulations of 10,000 TCLs is regarded as the actual value of $f(\Delta T(k))$. 100 outputs of HDMR, MLR and individual simulations are sorted in ascending order and demonstrated in Fig. 8.5.

It is clear that HDMR can follow individual simulations closely while the errors between MLR and individual simulations are much larger in Fig. 8.5. This means that HDMR offers a higher modeling accuracy as compared to MLR which is because multiple linear regressions, with a linear relationship between inputs and the output, can only present independent impacts of inputs on the system output. In (8.A2), only δ_μ and $\Delta T(k)$ play a linear role on $f(\Delta T(k))$. Thus, the cooperative effects of inputs and the nonlinear relationship between inputs and the output cannot be accurately

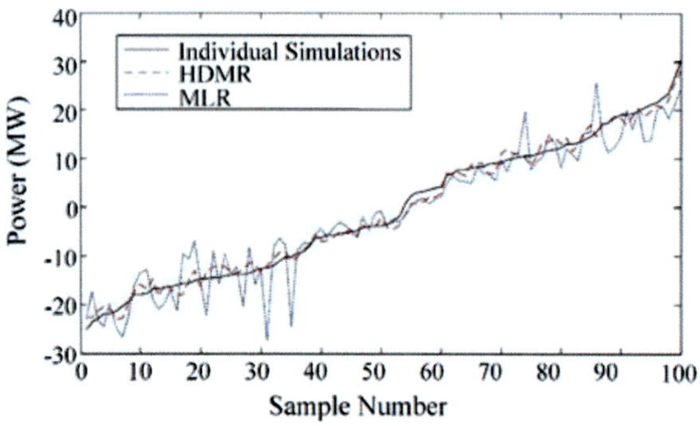

Fig. 8.5 Modeling comparisons of $f(\Delta T)$ using individual simulations, HDMR and MLR

modeled via MLR. In contrast, HDMR can model not only the cooperative effect of inputs acting collectively but also the nonlinear effect of a single input acting independently on the system output, which provides HDMR with a higher modeling accuracy. As shown in (8.A1), not only the nonlinear impacts of $\Delta T(k)$ on $f(\Delta T(k))$ but also the cooperative effects of $\Delta T(k)$ and other input variables (e.g., T_{out}) are given.

End users' comfort levels play an important role in the aggregated TCL models. Figure 8.6 shows $P_h^{\min}$ and $P_h^{\max}$ of the virtual generator for various end users' comfort levels that are characterized by $\Delta T_{\min}$ and $\Delta T_{\max}$ and a given population of TCLs shown in Table 8.3. It is assumed that T_{set} is the most comfortable temperature for end-users and indoor temperatures that can be adjusted within a range around T_{set} (i.e. $\Delta T_{\min} < 0$ and $|\Delta T_{\min}|$ is equal to $\Delta T_{\max}$). In Fig. 8.6, $P_h^{\max}$ becomes larger as $|\Delta T_{\min}|$ increases. However, $P_h^{\max}$ remains unchanged when $|\Delta T_{\min}|$ is larger than 0.8 °C. This is because the aggregated power of the TCL population reaches its maximum value (i.e., all TCLs work at the rated power state) when $|\Delta T_{\min}|$ is large. In addition, even though $|\Delta T_{\min}|$ continues to increase, the power consumption of TCLs cannot be higher than the aggregated rated power. Similarly, $|P_h^{\min}|$ grows as

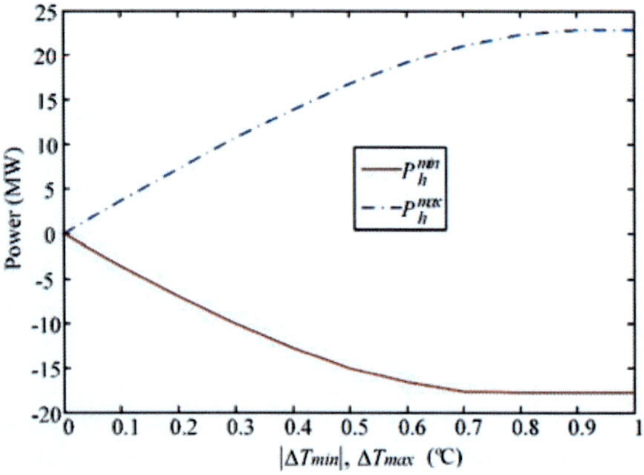

Fig. 8.6 $P_h^{\min}$ and $P_h^{\max}$ of the virtual generator for various end users' comfort levels

$\Delta T_{\max}$ (i.e., $|\Delta T_{\min}|$) is increased and then stays constant when $\Delta T_{\max}$ is higher than about 0.72 °C. Therefore, the capacity of the virtual generator can increase by setting wider comfort ranges, which provides a larger capability for tracking the desired power. However, end users' comfort level will have no impact on the capacity of the virtual generator when the comfort ranges are wide. Accordingly, HDMR outweighs MLR on parameter estimation, and only HDMR is employed here to estimate the parameters of the virtual battery model on a minute time-scale with the corresponding results given in Appendix (8.A3)–(8.A6). $\Delta T_{\mathrm{c}} \in [-1, 1]$ °C which characterizes end users' comfort level is also considered as one input variable.

Figure 8.7 shows the results for 100 samples that are randomly chosen to validate the four parameter expressions in (8.A3)–(8.A6). The results demonstrate good tracking behaviors compared with individual simulations, which means $P_{\mathrm{m}}^{\min}$, $P_{\mathrm{m}}^{\max}$, γ and τ can be accurately estimated by (8.A3)–(8.A6).

The modeling results via HDMR show that there are no input variables that have an impact on ξ. The probability distribution of ξ is given in Fig. 8.8. Here, ξ is limited to [0.90, 0.985] and the cumulative probability in [0.94, 0.985] is 90.3%. For simplicity, we assume ξ is 0.97. As shown in (8.A3)–(8.A6), only γ and τ are affected by ΔT_{c}. For a given TCL population stated in Table 8.3, γ and τ for different comfort levels are given in Fig. 8.9. Obviously, the relationship between γ/τ and ΔT_{c} is linear. γ and τ become larger with increasing ΔT_{c}. The comfort level impact on virtual battery is characterized by τ because τ as compared to γ is more influenced by ΔT_{c}. According to (8.8a)–(8.8b), the virtual battery capacity is determined by τ. The stored energy capacity will be larger when τ increases. Thus, a wider comfort

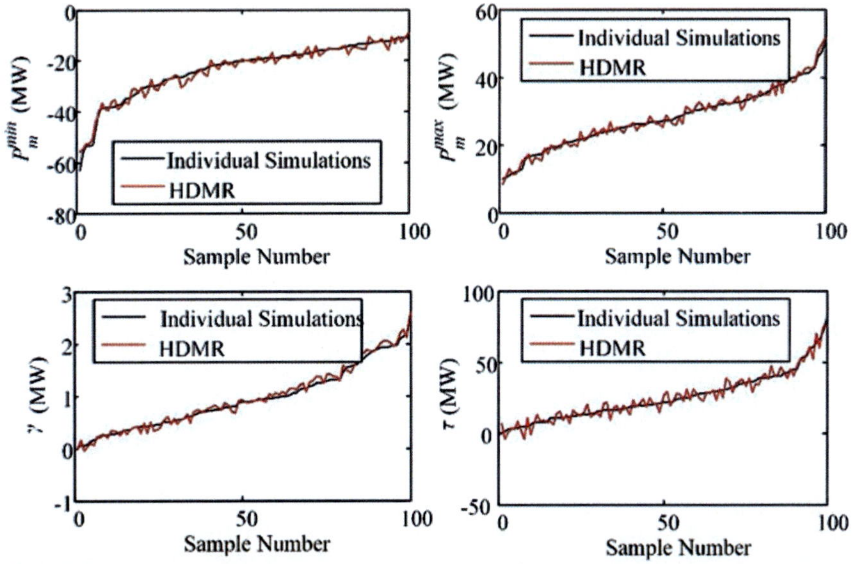

Fig. 8.7 Modeling result of the parameters of virtual battery model on a minute time-scale using HDMR and individual simulations

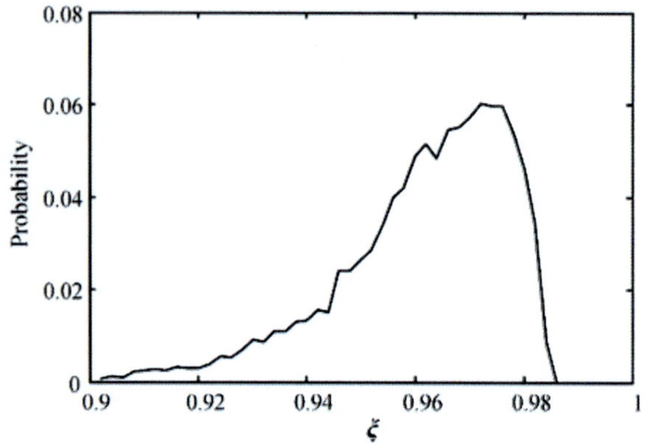

Fig. 8.8 Probability distribution of ξ

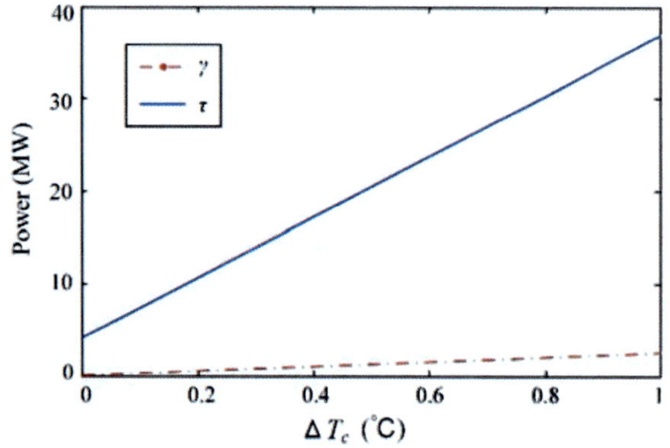

Fig. 8.9 γ and τ of the virtual battery model on a minute time-scale for various comfort levels

range for end-users is beneficial for the virtual battery to provide ancillary services for the grid.

Similarly, the parameters of virtual battery on a second time scale are given in Appendix (8.A7)–(8.A10). The results for 100 samples using HDMR and individual simulations are shown in Fig. 8.10. It is obvious that HDMR can be utilized to accurately estimate the parameters of a second time-scale virtual battery. As stated in Sect. 8.5.1, TCLs are controlled by on-off switches when a second time-scale battery model is considered in which indoor temperatures of end-users are kept at T_{set}. Hence, comfort level adjustments have no impact on the parameters of the virtual battery.

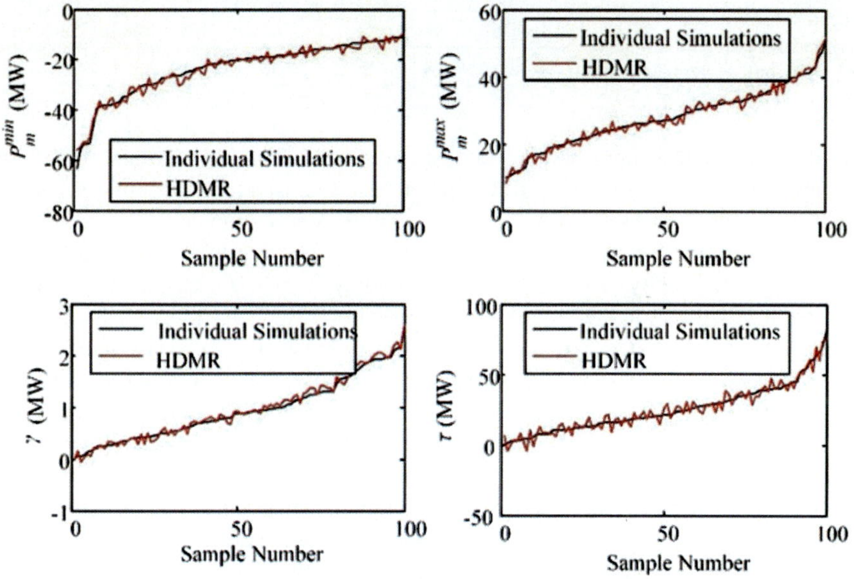

Fig. 8.10 Modeling results of the parameters of virtual battery model on a second time-scale using HDMR and individual simulations

8.7.2 *Impacts of the Number of Samples and TCL Parameter Distribution*

To show the impact of the number of samples on HDMR modeling, we apply different numbers of input-output samples to calculate the parameters of virtual models. The aggregated parameter of a second time-scale virtual battery, $P_s^{\max}$, is treated as an output example to examine the impact of sample numbers. The modeling results of HDMR using 100, 1000, 10,000, 50,000 input-output samples are expressed in Appendix (8.A11), (8.A12), (8.A8) and (8.A13), respectively. In addition, another 1000 samples are randomly chosen from Table 8.2 to test (8.A11), (8.A12), (8.A8) and (8.A13). The corresponding average relative errors are listed in Table 8.4. When the number of input-output samples is very small (e.g., 100), the average relative errors are large and the modeling results obtained by HDMR cannot accurately estimate the parameters of virtual models. As the number of input-output samples increases, the average relative errors become smaller, which means that the modeling accuracy of HDMR is improved by applying a higher number of input-output samples. However, it will take more effort to obtain the modeling results via HDMR when

Table 8.4 Average relative errors for various numbers of input–output samples

Number of samples	100	1000	10,000	50,000
Average relative error (%)	89.73	19.99	6.58	5.23

Table 8.5 Average relative errors with different numbers of input–output samples

Distribution type	Normal	Lognormal	Uniform
Average relative errors (%)	6.58	6.49	6.54

a higher number of input-output samples are utilized. Thus, it is critical to choose a proper number of input-output samples to get a satisfactory modeling accuracy with a reasonable workload when HDMR is employed to estimate the parameters of virtual generators and batteries.

There are generally three different distribution types of TCL parameters, including the normal distribution [31], lognormal distribution [12, 37] and uniform distribution [38]. In Sect. 8.6.1, TCL parameter heterogeneity is assumed to be characterized by the normal distribution. The other two distribution types, i.e., lognormal and uniform distributions are considered in this section to show the impacts of distribution types on the modeling results of HDMR. The $P_s^{\max}$ expressions are stated in Appendix (8.A14) and (8.A15) when TCL parameters are characterized by lognormal and uniform distributions. 1000 samples are randomly chosen from Table 8.2 to test (8.A8), (8.A14) and (8.A15). The corresponding relative errors are listed in Table 8.5 in which the average relative errors are almost the same, which demonstrates that the distribution types of TCL parameters have minute impacts on the modeling results of HDMR. Accordingly, any of the three distribution types can be utilized without influencing the results for estimating the parameters of virtual models via HDMR.

8.7.3 Multi Time-Scale Dispatch Results

In this section, the proposed three virtual models are employed to take part in power system operation for smoothing out the variability of multi time-scale wind power generation. For a TCL population shown in Table 8.3, the virtual generator and battery models can be obtained by an aggregator according to the HDMR modeling results. Given that the detailed optimization strategies are not the focus of this chapter, the virtual generator and the two battery models are assumed to be considered by the system operator in the optimization of the grid operation strategy. On an hourly time scale, the aggregator gets optimization results [39] from the system operator, calculates the temperature adjustment by (8.A1) and then distributes temperature control signals to FTCLs. Figure 8.11 shows the desired power, actual power by individual simulations and actual temperature adjustments. Here, the actual power of the FTCL population can closely track the desired power, indicating that (8.A1) could represent the actual relationship between the aggregated power of FTCLs and the temperature adjustment. The operating behavior of FTCLs can be accurately characterized by the virtual generator in (8.6a).

Similarly, Fig. 8.12 shows the actual aggregated performances of ITCLs according to the control signals generated by the aggregator for the desired power profile [40].

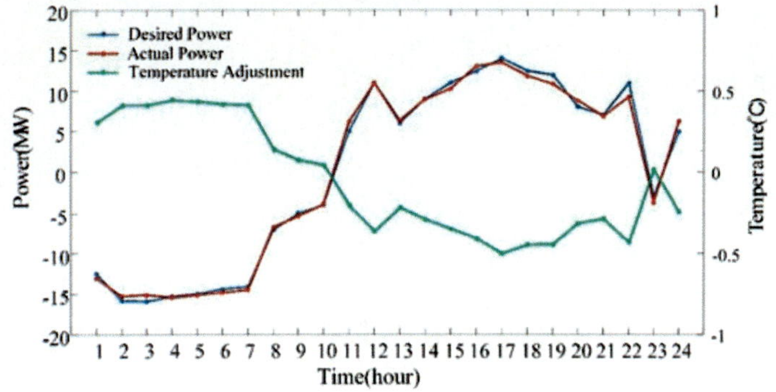

Fig. 8.11 Profiles of desired power, actual power and temperature adjustment

Fig. 8.12 Profiles of the desired power and the aggregated power of ITCLs by individual simulations

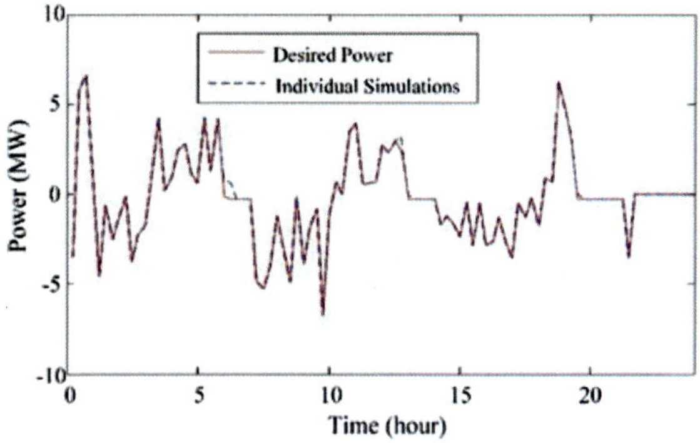

ΔT_c is 1 °C. Obviously, ITCLs can be accurately regulated to follow the desired power. The comparison of SOC for the virtual battery model and individual simulations are given in Fig. 8.13 to validate the virtual battery model on a minute time-scale. SOC obtained by individual simulations is the sum of SOCs of ITCLs [19].

Fig. 8.13 SOC comparisons between the virtual battery on the minute time-scale and individual simulations

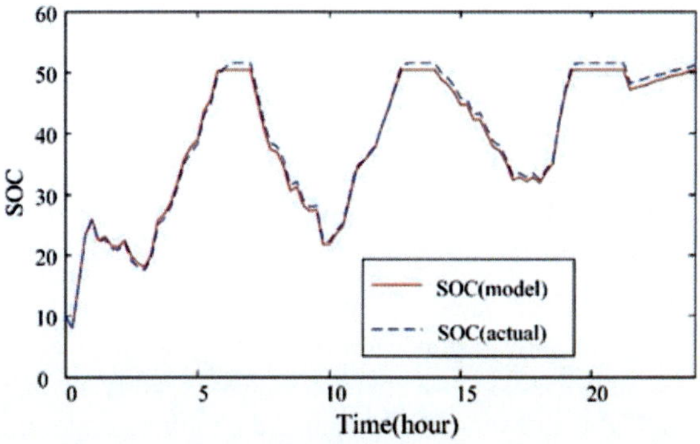

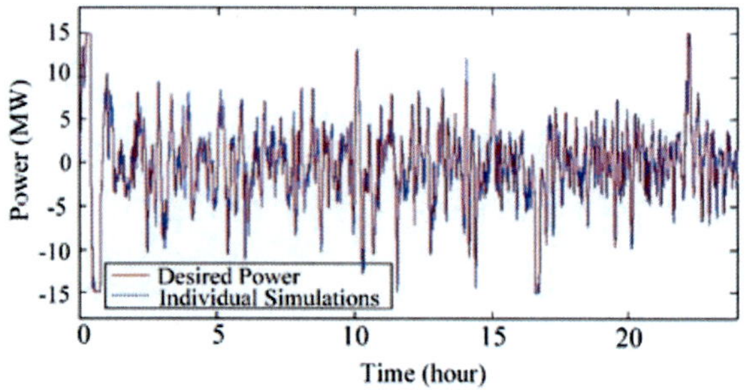

Fig. 8.14 Profiles of the desired power and the aggregated power of FTCLs by individual simulations

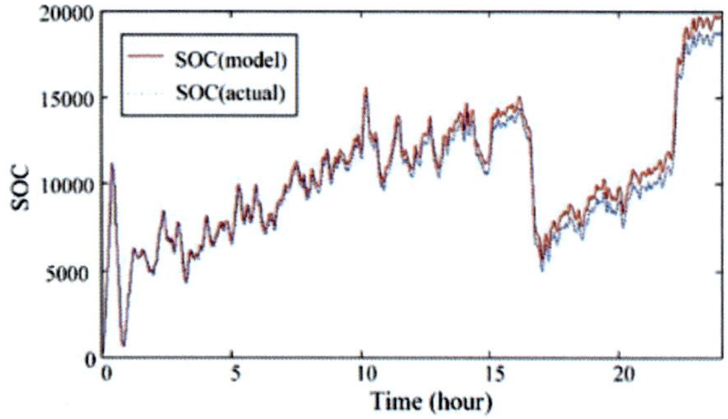

Fig. 8.15 SOC comparisons for the virtual battery on a second time-scale and individual simulations

As shown in Fig. 8.13, the model SOC has an almost identical evolution trajectory with that of the actual SOC, which means that the operating states of the ITCL population can be tracked by the virtual battery model. In other words, the virtual battery performs well on a minute time- scale.

Regulation signals are provided in [41] in which their magnitudes are scaled to match the power limits of the virtual battery on a second time-scale. The profiles of the desired power and individual simulations are given in Fig. 8.14. The desired power can be closely followed by the actual aggregated power of FTCLs. SOC comparisons for the virtual battery on a second time-scale and individual simulations are shown in Fig. 8.15. Here, SOC performances of the virtual battery are consistent with those of individual simulations, which validates the modeling of virtual battery on a second time-scale.

8.8 Conclusion

TCLs are aggregated as a virtual generator and two batteries to smooth out multi time-scale wind power generation variability at the level of power systems. In practice, it is inconceivable to assume that data are readily available for each TCL. Accordingly, HDMR is introduced to estimate the parameters of the virtual generator and battery models using the probability distribution of TCL parameters. Simulation results demonstrate that (1) parameters of virtual generator and batteries can be accurately estimated by HDMR and HDMR offers a better modeling performance compared to MLR; (2) capacity of the virtual generator and virtual battery on a minute time-scale can be enlarged to encompass wider comfort ranges; (3) modeling accuracy of HDMR can be improved by more input-output samples; (4) distribution types of TCL parameters have no impacts on the modeling results of HDMR; (5) proposed virtual generator and battery models of TCLs can closely track the actual aggregated performance of TCLs.

Appendix

The modeling results of the virtual generator via HDMR and MLR are given as follows.

$$\begin{aligned} f(\Delta T(k)) = & -0.55 - 14.99\phi_1(\Delta T(k)) \\ & + 1.06\phi_2(\Delta T(k)) + 2.58\phi_3(\Delta T(k)) - 2.57\phi_1(P\_\mu)\phi_1(\Delta T(k)) \\ & + 1.37\phi_1(\delta\_\mu)\phi_1(\Delta T(k)) - 1.56\phi_1(T_{out}(k))\phi_1(\Delta T(k)) \end{aligned} \tag{8.A1}$$

$$f(\Delta T(k)) = 26.17 - 2.15\delta\_\mu - 52.08\Delta T(k) \tag{8.A2}$$

The modeling results of the virtual battery on the minute time- scale via HDMR are shown as:

$$\begin{aligned} P_m^{\min} = & -21.46 + 7.51\phi_1(R\_\mu) - 1.95\phi_2(R\_\mu) + 2.33\phi_1(T_{set}\_\mu) \\ & - 5.82\phi_1(k1\_\mu) + 2.98\phi_1(k2\_\mu) \\ & - 5.86\phi_1(T_{out}) + 1.43\phi_1(R\_\mu)\phi_1(k1\_\mu) + 1.68\phi_1(R\_\mu)\phi_1(T_{out}) \end{aligned} \tag{8.A3}$$

$$\begin{aligned} P_m^{\max} = & 23.72 + 7.54\phi_1(R\_\mu) - 1.95\phi_2(R\_\mu) \\ & + 2.33\phi_1(T_{set}\_\mu) + 5.78\phi_1(P_{\max}\_\mu) \\ & - 4.25\phi_1(k1\_\mu) + 2.18\phi_1(k2\_\mu) - 5.85\phi_1(T_{out}) \\ & + 1.45\phi_1(R\_\mu)\phi_1(k1\_\mu) + 1.67\phi_1(R\_\mu)\phi_1(T_{out}) \end{aligned} \tag{8.A4}$$

$$\begin{aligned} \gamma = & 1.02 - 0.29\phi_1(R\_\mu) + 0.20\phi_1(k1\_\mu) - 0.10\phi_1(k2\_\mu) \\ & + 0.58\phi_1(\Delta T_c) - 0.17\phi_1(R\_\mu)\phi_1(\Delta T_c) + 0.11\phi_1(k1\_\mu)\phi_1(\Delta T_c) \end{aligned} \tag{8.A5}$$

$$\begin{aligned} \tau = & 26.32 + 9.53\phi_1(C_a\_\mu) - 2.63\phi_1(C_a\_\chi) \\ & + 7.21\phi_1(k1\_\mu) - 1.23\phi_2(k1\_\mu) \\ & + 2.59\phi_3(k1\_\mu) - 2.21\phi_2(k2\_\mu) \end{aligned}$$

$$+ 15.07\phi_1(\Delta T_c) + 2.87\phi_1(C_a\_\mu)\phi_1(k1\_\mu) + 5.41\phi_1(C_a\_\mu)\phi_1(\Delta T_c) + 4.15\phi_1(k1\_\mu)\phi_1(\Delta T_c) \quad (8.A6)$$

The modeling results of the virtual battery on the second time-scale via HDMR are expressed as:

$$P_s^{\min} = -21.39 + 6.18\phi_1(R\_\mu) - 1.66\phi_2(R\_\mu) + 1.86\phi_1(T_{set}\_\mu) + 2.57\phi_1(\eta\_\mu) - 4.73\phi_1(T_{out}) \quad (8.A7)$$

$$P_s^{\max} = 48.55 + 6.19\phi_1(R\_\mu) - 1.66\phi_2(R\_\mu) + 17.30\phi_1(P\_\mu) + 2.57\phi_1(\eta\_\mu) - 3.42\phi_2(\eta\_\mu) - 4.74\phi_1(T_{out}) \quad (8.A8)$$

$$\kappa = 3.74 + 0.72\phi_1(\delta\_\mu) \quad (8.A9)$$

$$\nu = 51098.2 + 13388.1\phi_1(R\_\mu) + 2363.4\phi_2(R\_\mu) + 13450.1\phi_1(C_a\_\mu) + 2157.2\phi_2(C_a\_\mu) + 9872.7\phi_1(\delta\_\mu) + 3506.6\phi_1(R\_\mu)\phi_1(C_a\_\mu) + 2561.8\phi_1(R\_\mu)\phi_1(\delta\_\mu) + 2566.0\phi_1(C_a\_\mu)\phi_1(\delta\_\mu) \quad (8.A10)$$

The expressions of $P_s^{\max}$ obtained by HDMR when the number of input–output samples is 100 (8.A11), 1000 (8.A12) and 50,000 (8.A13):

$$P_s^{\max} = 47.09 + 5.76\phi_1(R\_\mu) - 1.45\phi_2(R\_\mu) + 7.95\phi_3(R\_\mu) + 17.06\phi_1(P\_\mu) + 11.74\phi_2(P\_\mu) + 12.01\phi_3(P\_\mu) + 2.40\phi_1(\eta\_\mu) - 4.93\phi_1(T_{out}) + 2.49\phi_3(T_{out}) \quad (8.A11)$$

$$P_s^{\max} = 49.25 + 6.15\phi_1(R\_\mu) - 1.69\phi_2(R\_\mu) + 7.34\phi_3(R\_\mu) + 17.29\phi_1(P\_\mu) + 1.31\phi_2(P\_\mu) + 2.60\phi_1(\eta\_\mu) - 3.41\phi_2(\eta\_\mu) + 2.32\phi_3(\eta\_\mu) - 4.76\phi_1(T_{out}) \quad (8.A12)$$

$$P_s^{\max} = 48.67 + 6.12\phi_1(R\_\mu) - 1.59\phi_2(R\_\mu) + 17.32\phi_1(P\_\mu) + 2.54\phi_1(\eta\_\mu) - 4.72\phi_1(T_{out}) \quad (8.A13)$$

The expressions of $P_s^{\max}$ obtained by HDMR when the distribution types of TCL parameters are lognormal (8.A14) and uniform (8.A15):

$$P_s^{\max} = 48.34 + 6.20\phi_1(R\_\mu) - 1.59\phi_2(R\_\mu) + 17.42\phi_1(P\_\mu)$$

$$+ 2.54\phi_1(\eta\_\mu) - 3.38\phi_2(\eta\_\mu) - 4.78\phi_1(T_{out}) \quad (8.A14)$$

$$P_s^{\max} = 48.48 + 6.21\phi_1(R\_\mu) - 1.70\phi_2(R\_\mu) + 17.28\phi_1(P\_\mu) + 2.63\phi_1(\eta\_\mu) - 3.41\phi_2(\eta\_\mu) - 4.76\phi_1(T_{out}) \quad (8.A15)$$

References

1. Song, M., Gao, C., Shahidehpour, M., Li, Z., Lu, S., & Lin, G. (2018). Multi-time-scale modeling and parameter estimation of TCLs for smoothing out wind power generation variability. *IEEE Transactions on Sustainable Energy, 10*(1), 105–118.
2. Zhao, C., Wang, J., Watson, J. P., & Guan, Y. (2013). Multi-stage robust unit commitment considering wind and demand response uncertainties. *IEEE Transactions on Power Systems, 28*(3), 2708–2717.
3. Lu, N., Du, P., & Makarov, Y. V. (2012, July). The potential of thermostatically controlled appliances for intra-hour energy storage applications. In *2012 IEEE Power and Energy Society General Meeting* (pp. 1–6).
4. Lu, N., & Zhang, Y. (2012). Design considerations of a centralized load controller using thermostatically controlled appliances for continuous regulation reserves. *IEEE Transactions on Smart Grid, 4*(2), 914–921.
5. Lu, N. (2012). An evaluation of the HVAC load potential for providing load balancing service. *IEEE Transactions on Smart Grid, 3*(3), 1263–1270.
6. Zakula, T., Gayeski, N. T., Armstrong, P. R., & Norford, L. K. (2011). Variable-speed heat pump model for a wide range of cooling conditions and loads. *Hvac&R Research, 17*(5), 670–691.
7. Soudjani, S. E. Z., & Abate, A. (2014). Aggregation and control of populations of thermostatically controlled loads by formal abstractions. *IEEE Transactions on Control Systems Technology, 23*(3), 975–990.
8. Zhang, W., Lian, J., Chang, C. Y., & Kalsi, K. (2013). Aggregated modeling and control of air conditioning loads for demand response. *IEEE transactions on power systems, 28*(4), 4655–4664.
9. Hu, J., et al. (2017). Load following of multiple heterogeneous TCL aggregators by centralized control. *IEEE Transactions on Power Systems, 32*(4), 3157–3167.
10. Mahdavi, N., Braslavsky, J. H., & Perfumo, C. (2016). Mapping the effect of ambient temperature on the power demand of populations of air conditioners. *IEEE Transactions on Smart Grid, 9*(3), 1540–1550.
11. Erdinc, O., Taşcıkaraoğlu, A., Paterakis, N. G., Eren, Y., & Catalão, J. P. (2017). End-user comfort oriented day-ahead planning for responsive residential HVAC demand aggregation considering weather forecasts. *IEEE Transactions on Smart Grid, 8*(1), 362–372.
12. Mahdavi, N., Braslavsky, J. H., Seron, M. M., & West, S. R. (2017). Model predictive control of distributed air-conditioning loads to compensate fluctuations in solar power. *IEEE Transactions on Smart Grid, 8*(6), 3055–3065.
13. Wai, C. H., Beaudin, M., Zareipour, H., Schellenberg, A., & Lu, N. (2014). Cooling devices in demand response: A comparison of control methods. *IEEE Transactions on Smart Grid, 6*(1), 249–260.
14. Kim, Y. J., Norford, L. K., & Kirtley, J. L. (2014). Modeling and analysis of a variable speed heat pump for frequency regulation through direct load control. *IEEE Transactions on Power Systems, 30*(1), 397–408.
15. Kim, Y. J., Fuentes, E., & Norford, L. K. (2015). Experimental study of grid frequency regulation ancillary service of a variable speed heat pump. *IEEE Transactions on Power Systems, 31*(4), 3090–3099.

16. Liu, M., Shi, Y., & Liu, X. (2015). Distributed MPC of aggregated heterogeneous thermostatically controlled loads in smart grid. *IEEE Transactions on Industrial Electronics, 63*(2), 1120–1129.
17. Mathieu, J. L., Koch, S., & Callaway, D. S. (2012). State estimation and control of electric loads to manage real-time energy imbalance. *IEEE Transactions on Power Systems, 28*(1), 430–440.
18. Song, M., Ciwei, G., Yang, J., Liu, Y., & Cui, G. (2017). Novel aggregate control model of air conditioning loads for fast regulation service. *IET Generation Transmission & Distribution, 11*(17), 4391–4401.
19. Song, M., Gao, C., Yan, H., & Yang, J. (2017). Thermal battery modeling of inverter air conditioning for demand response. *IEEE Transactions on Smart Grid, 9*(6), 5522–5534.
20. Hao, H., Sanandaji, B. M., Poolla, K., & Vincent, T. L. (2014). Aggregate flexibility of thermostatically controlled loads. *IEEE Transactions on Power Systems, 30*(1), 189–198.
21. Sanandaji, B. M., Vincent, T. L., & Poolla, K. (2015). Ramping rate flexibility of residential HVAC loads. *IEEE Transactions on Sustainable Energy, 7*(2), 865–874.
22. Zhao, L., Zhang, W., Hao, H., & Kalsi, K. (2017). A geometric approach to aggregate flexibility modeling of thermostatically controlled loads. *IEEE Transactions on Power Systems, 32*(6), 4721–4371.
23. Rabitz, H., & Aliş, Ö. F. (1999). General foundations of high-dimensional model representations. *Journal of Mathematical Chemistry, 25*(2–3), 197–233.
24. Li, G., Wang, S. W., & Rabitz, H. (2002). Practical approaches to construct RS-HDMR component functions. *The Journal of Physical Chemistry A, 106*(37), 8721–8733.
25. Miller, M., Feng, X., Li, G., & Rabitz, H. (2009, September). Nonlinear bionetwork structure inference using the random sampling-high dimensional model representation (RS-HDMR) algorithm. In *2009 Annual International Conference of the IEEE Engineering in Medicine and Biology Society* (pp. 6412–6415).
26. Yücel, A. C., Bağcı, H., & Michielssen, E. (2015). An ME-PC enhanced HDMR method for efficient statistical analysis of multiconductor transmission line networks. *IEEE Transactions on Components, Packaging and Manufacturing Technology, 5*(5), 685–696.
27. Khalid, F., & Speidel, J. (2014). Robust hybrid precoding for multiuser MIMO wireless communication systems. *IEEE Transactions on Wireless Communications, 13*(6), 3353–3363.
28. Tipcharoen, W., Warisarn, C., & Kovintavewat, P. (2016). Effects of Island volume and hotspot position fluctuation for heated-dot magnetic recording. *IEEE Magnetics Letters, 7*, 1–4.
29. Mahdavi, N., Perfumo, C., & Braslavsky, J. (2014). Bayesian parameter estimation for direct load control of populations of air conditioners. *IFAC Proceedings Volumes, 47*, 9924–9929.
30. Sinitsyn, N. A., Kundu, S., & Backhaus, S. (2013). Safe protocols for generating power pulses with heterogeneous populations of thermostatically controlled loads. *Energy Conversion and Management, 67*, 297–308.
31. Bashash, S., & Fathy, H. K. (2011, June). Modeling and control insights into demand-side energy management through setpoint control of thermostatic loads. In *Proceedings of the 2011 American Control Conference* (pp. 4546–4553).
32. Shao, C., Wang, X., Shahidehpour, M., Wang, X., & Wang, B. (2017). Security-constrained unit commitment with flexible uncertainty set for variable wind power. *IEEE Transactions on Sustainable Energy, 8*(3), 1237–1246.
33. Sanandaji, B. M., Hao, H., & Poolla, K. (2014, January). Fast regulation service provision via aggregation of thermostatically controlled loads. In *2014 47th Hawaii International Conference on System Sciences* (pp. 2388–2397).
34. Song, M., Gao, C., Shahidehpour, M., Li, Z., Yang, J., & Yan, H. (2018). Impact of uncertain parameters on TCL power capacity calculation via HDMR for generating power pulses. *IEEE Transactions on Smart Grid, 10*(3), 3112–3124.
35. Sun, Y., Elizondo, M., Lu, S., & Fuller, J. C. (2014). The impact of uncertain physical parameters on HVAC demand response. *IEEE Transactions on Smart Grid, 5*(2), 916–923.
36. Ziehn, T., & Tomlin, A. S. (2009). GUI–HDMR–A software tool for global sensitivity analysis of complex models. *Environmental Modelling & Software, 24*(7), 775–785.

37. Callaway, D. S. (2009). Tapping the energy storage potential in electric loads to deliver load following and regulation, with application to wind energy. *Energy Conversion and Management, 50*(5), 1389–1400.
38. Mathieu, J. L., & Callaway, D. S. (2012, January). State estimation and control of heterogeneous thermostatically controlled loads for load following. In *2012 45th Hawaii International Conference on System Sciences* (pp. 2002–2011).
39. Zhang, X., Che, L., Shahidehpour, M., Alabdulwahab, A., & Abusorrah, A. (2016). Electricity-natural gas operation planning with hourly demand response for deployment of flexible ramp. *IEEE Transactions on Sustainable Energy, 7*(3), 996–1004.
40. Bao, Z., Zhou, Q., Yang, Z., Yang, Q., Xu, L., & Wu, T. (2014). A multi time-scale and multi energy-type coordinated microgrid scheduling solution—Part I: Model and methodology. *IEEE Transactions on Power Systems, 30*(5), 2257–2266.
41. "PJM regulation data", [online] Available: http://www.pjm.com/markets-and-operations/ancillary-services/mkt-based-regulation/fast-response-regulation-signal.aspx.

Chapter 9
Hierarchical Scheduling of TCL Flexibility for Transactive Energy

Abstract This chapter investigates a hierarchical approach to the optimal scheduling of flexibility offered as transactive energy by TCLs. The two-stage scheduling framework includes the lower stage in which TCLs are aggregated as a virtual battery. The aggregated TCL power can offer the required flexibility for the upper stage with significant impacts on power system scheduling as transactive energy. Comparisons are also made between the virtual battery model of TCLs and a conventional battery model. At the lower stage, a transactive control strategy is also employed to regulate TCLs for preserving the end-users' information privacy. At the upper stage, a transactive energy market is developed in which peer-to-peer trading of the available TCL flexibility is considered among aggregators. Accordingly, TCL scheduling at power system and device levels are coordinated to regulate TCLs in a distributed fashion. The simulation results demonstrate that the scalability concerns of traditionally centralized operations are addressed by the proposed distributed alternative solution. The upper stage transactive energy market allows aggregators to trade energy effectively without any significant concerns for maintaining information privacy. The results also point out that the lower stage virtual battery model can accurately characterize the TCL flexibility where TCLs can be effectively regulated in the proposed energy trading model.

Keywords Transactive energy · Flexibility · Electricity markets

9.1 Introduction

The variability of distributed energy resources (DER) can increase the power supply deficiency and culminate in a critical condition as DER proliferation continues to escalate in power systems [1–3]. TCLs can serve as indispensable DR resources. They are progressively utilized to address DER variability quickly, economically and effectively [4]. TCLs possess thermal inertia and have attracted additional attention to participate in power system operations. However, TCL's specific properties pose a key challenge to effectively participate in the power system scheduling and control.

First, it would be difficult to schedule TCLs individually as their adjustable capacities are too small to have significant impacts on power systems [5]. Also, the number

M. Song and C. Gao, *Integration of Distributed Resources in Smart Grids for Demand Response and Transactive Energy*, https://doi.org/10.1007/978-981-16-7170-8_9

of TCLs could traditionally be very large, which would require significant computation resources when TCLs are individually scheduled. Accordingly, TCL aggregation is required for their participation in power system operations. Meanwhile, TCLs with thermal inertia act like batteries that provide various grid services owing to their flexible charging and discharging capabilities [6]. References [7, 8] proposed a stochastic battery model for representing the aggregated TCL flexibility with a concise collection of power signals. Reference [9] developed a rough time-varying thermal energy storage model to denote the aggregated TCL flexibility in which suboptimal control trajectories were calculated. In [10], a leaky storage unit was proposed to embody the aggregated flexibility of TCLs in the optimization of multi-service portfolios. A generalized battery model that described flexibilities of TCLs and conventional batteries for delivering power grid and end-use services was presented in [11]. A virtual battery model of HVAC was also studied in [12]. TCLs show great potentials in terms of virtual energy storage which can replace conventional batteries to serve the power grid. Nevertheless, differences between the battery model of aggregated TCLs and conventional batteries are seldom highlighted, which need to be investigated further for the provision of TCLs and conventional batteries in power system operations.

Second, widespread TCLs throughout a power network are often regulated in a centralized fashion. For example, [10] conducted a linear optimization model of TCLs for frequency and energy arbitrage. Hierarchical centralized load control with TCLs was proposed in [13] to provide load-following services. In [14], a sequential setpoint control of the TCL population was provided in which aggregated TCLs were characterized by an extended Markov model. However, a centralized TCL operation is more complicated which could expose TCLs' information privacy [15, 16]. In practice, a distributed scheduling and control is more practical for regulating largely distributed TCLs which can protect the corresponding information privacy effectively.

Transactive control that employs market mechanisms to enable self-interested entities to serve the power grid, is a promising distributed way to preserve the proprietary data and respect specific performances of participating entities [17]. Reference [18] proposed a real-time electric vehicle (EV) charging management scheme in a transactive energy market. EV owners provide charging requirements and price signals, while end-users' information privacy is preserved. A day-ahead transactive market framework was presented in [19] in which transactive energy in transmission and distribution networks was cleared simultaneously for optimal operations of distributed resources. The clearing process for trading in a transactive energy market was studied in [16, 20]. Here, Lagrangian relaxation was introduced to describe the trading process. Reference [16] proposed an optimal strategy for multi microgrids using dynamic programming. Reference [20] developed a distributed optimization framework by Lagrangian relaxation for energy trading among microgrids. Transactive control provided an innovative solution to regulate TCLs in a distributed fashion; however, additional investigations are required to fully explore the stated potentials.

Additional studies focused on TCL regulation by transactive control. Reference [21] provided a framework for transaction-based building controls. Reference [22] proposed a single-auction market for commercial buildings to respond to external price signals. Each building provides a price curve for adjusting its temperature setpoint. In [23], a transactive control strategy was developed for regulating commercial buildings to shave power peaks and shift loads. In this case, building bidding strategies and market clearing process were both provided. The device bidding and market clearing strategies were investigated in [24, 25] to motivate self-interested TCLs to pursue efficient energy allocations at peak hours, in which TCLs were regulated by a market-based coordination framework. The transactive control of residential buildings was analyzed in [26]. Even though the transactive TCL control was analyzed earlier at the level of devices/buildings, the optimal scheduling of aggregated TCLs was rarely analyzed at the power system level. Moreover, specific TCL performances in transactive control (i.e., lockout time effects) were often disregarded.

In this chapter, we focus on hierarchical optimal scheduling and control of TCL populations in power systems using market-based distributed modes. Accordingly, specific operating characteristics of TCLs will be fully considered and participants' information privacy will be preserved. The main contributions of this chapter are summarized as follows. First, a hierarchical framework is proposed for TCLs, in which the TCL scheduling at power system and device levels are coordinated to regulate TCLs in a distributed fashion. At the lower stage, a virtual energy market is designed to control TCLs at the device level in which specific TCL characteristics are fully considered. At the upper stage, a transactive energy market is adopted for aggregated TCLs in which the proposed peer-to-peer optimal trading strategy among aggregators would minimize the cost of power system operation. Each aggregator makes proper decisions according to other aggregators' trading signals without revealing any private information. Second, based on the virtual battery model for the provision of aggregated TCL flexibility, the chapter investigates the differences between virtual and conventional battery models in terms of operating behaviors.

9.2 TCL Scheduling Framework

The hierarchical framework for distributed TCL scheduling is shown in Fig. 9.1. In this chapter, TCLs connected to a power network node are regulated by an aggregator [27]. The implementation procedure is summarized as follows. As TCLs are widely distributed in the grid, they are aggregated as a virtual battery model at the lower stage (aggregating/de-aggregating) to qualify their system scheduling potential (see Sect. 9.3.1). When aggregators receive the virtual battery model of TCLs, each aggregator (AGG) uses the aggregated data of TCL flexibility to participate in a transactive energy market at the upper stage (aggregator's transactive trading, see Sect. 9.4). System-level peer-to-peer trading among aggregators with aggregated TCLs is to minimize the power system cost.

TCLs are controlled in a distributed fashion by hosting a virtual market at the lower stage as aggregators receive transactive results from the upper stage (see Sect. 9.3.2). Each TCL determines its operating state in terms of its bid and market-clearing

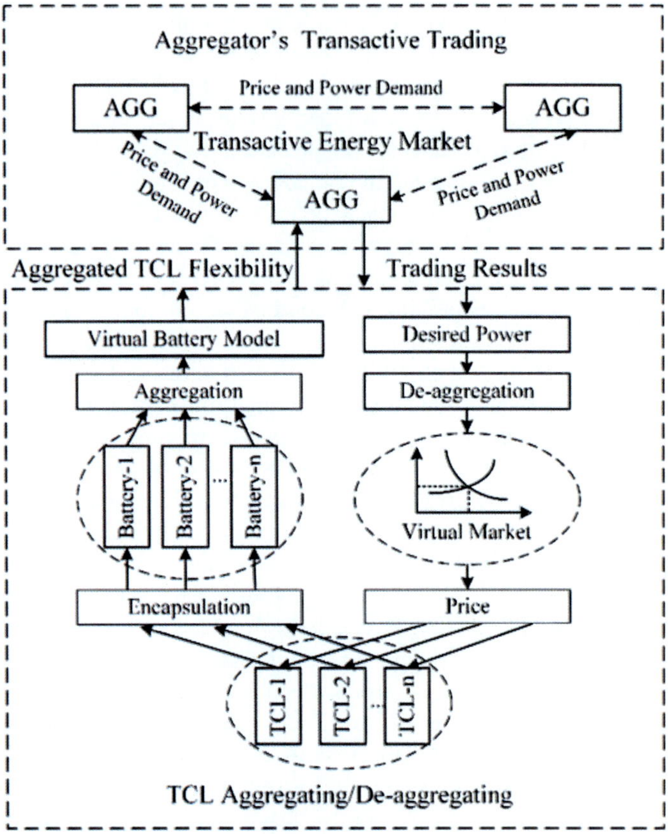

Fig. 9.1 Hierarchical TCL scheduling framework

price. Accordingly, TCLs are optimally regulated in a distributed fashion by the coordination of system and device-level schedules.

The virtual battery model is employed to characterize the aggregated TCL flexibility, which guarantees that the virtual battery is scheduled at the upper stage to reside within TCL capabilities stated at the lower stage. Also, eligible TCLs are regulated by aggregators to respond to the desired power at the lower stage (see Sect. 9.3.2). The proposed hierarchical framework allows TCLs at the lower stage to follow the aggregated TCLs' power scheduled at the upper stage.

9.3 Aggregation and De-aggregation of TCLs

9.3.1 Aggregation of TCLs at the Lower Stage

There are two operating modes (i.e., heating and cooling) for certain TCLs (e.g., air conditioning) and the cooling mode is presented as an example in this chapter. The thermodynamic behavior of TCLs is described by a discrete equivalent model [9, 13, 28], which is given as:

$$C_a \frac{dT_{in}(t)}{dt} = \frac{T_{out} - T_{in}(t)}{R} - S(t) \cdot COP \cdot P_{rated} \tag{9.1}$$

TCLs working at the rated power (on state) or zero power (off state) have their indoor temperatures adjusted periodically within $[T_{\min}, T_{\max}]$ ($T_{\min} = T_{\text{set}} - \delta$, $T_{\max} = T_{\text{set}} + \delta$) [29]. A continuous TCL model for closely following the aggregated TCL power is proposed in [7]. The continuous model, utilized as a more convenient choice than the discrete equivalent model, describes the TCL performance as:

$$C_a \cdot \frac{dT_{in}(t)}{dt} = \frac{T_{out} - T_{in}(t)}{R} - COP \cdot P_{cont}(t) \tag{9.2a}$$

$$0 \leq P_{cont}(t) \leq P_{rated} \tag{9.2b}$$

When there are no external control signals, TCL operates at T_{set} with the corresponding baseline power, $P_{baseline}$, expressed as:

$$P_{baseline} = \frac{T_{out} - T_{set}}{COP \cdot R} \tag{9.3}$$

TCLs have thermal inertia and share similar operating behaviors with batteries. Accordingly, TCL operation is presented as a virtual battery model in this chapter. When there are external control signals, the TCL power $P_{cont}(t)$ will be changed and the internal temperature will deviate from T_{set}. If $P_{cont}(t)$ is larger than $P_{baseline}$, TCL absorbs power from the grid to store energy in the virtual battery. Otherwise, TCL releases power to the grid by discharging the energy stored in the virtual battery. Thus, the difference between $P_{cont}(t)$ and $P_{baseline}$ is defined as the charging/discharging power of the virtual battery, which is stated as:

$$P_{cd}(k) = P_{cont}(k) - P_{baseline} \tag{9.4a}$$

$$-P_{baseline} \leq P_{cd}(k) \leq P_{rated} - P_{baseline} \tag{9.4b}$$

When $T_{\text{in}}(t)$ is at $T_{\max}$ or $T_{\min}$, the stored energy is either the lowest or the highest, respectively. Considering that T_{set} is the user's preferred temperature setpoint, its corresponding energy storage is set at 0 in this chapter. Thus, energy storage is defined as:

$$x(k) = \frac{T_{set} - T_{in}(k)}{\delta} \tag{9.5}$$

where, $T_{in}(k)$ is within $[T_{\min}, T_{\max}]$ to guarantee end-users' comfort. Thus, $x(k)$ is constrained as:

$$-1 \le x(k) \le 1 \tag{9.6}$$

Using (9.2a), (9.3), (9.4a) and (9.5), the updated TCL energy storage level is derived as:

$$x(k+1) = \kappa x(k) + \gamma P_{cd}(k) \tag{9.7a}$$

where

$$\kappa = \exp\left(-\frac{\Delta t}{RC_a}\right) \tag{9.7b}$$

$$\gamma = \frac{COP \cdot R \cdot (1-\kappa)}{\delta} \tag{9.7c}$$

Accordingly, TCL is encapsulated in a battery model using (9.4b), (9.6) and (9.7a). TCL energy storage levels are presented as

$$\begin{cases} x_1(k+1) = \kappa_1 x_1(k) + \gamma_1 P_{cd,1}(k) \\ x_2(k+1) = \kappa_2 x_2(k) + \gamma_2 P_{cd,2}(k) \\ \vdots \\ x_n(k+1) = \kappa_n x_n(k) + \gamma_n P_{cd,n}(k) \end{cases} \tag{9.8}$$

To represent the aggregated TCL's virtual battery model, (9.8) is rewritten as

$$\sum_{j=1}^{n} \frac{x_j(k+1)}{\gamma_j} = \sum_{j=1}^{n} \kappa_j \frac{x_j(k)}{\gamma_j} + \sum_{j=1}^{n} P_{cd,j}(k) \tag{9.9}$$

n

Here, $SOC(k)$ and $P(k)$ for aggregated TCLs' virtual battery are defined as

$$SOC(k) = \sum_{j=1}^{n} \frac{x_j(k)}{\gamma_j} \tag{9.10a}$$

$$P(k) = \sum_{j=1}^{n} P_{cd,j}(k) \tag{9.10b}$$

Considering (9.8)–(9.10b), the aggregated TCL flexibility is approximately expressed as

$$P_{\min} \leq P(k) \leq P_{\max} \tag{9.11a}$$

$$SOC(k+1) = \alpha SOC(k) + P(k) \tag{9.11b}$$

$$SOC_{\min} \leq SOC(k+1) \leq SOC_{\max} \tag{9.11c}$$

where

$$P_{\max} = \sum_{j=1}^{n} \left(P_{rated,j} - \frac{T_{out} - T_{set,j}}{COP_j \cdot R_j} \right) \tag{9.11d}$$

$$P_{\min} = -\sum_{j=1}^{n} \frac{T_{out} - T_{set,j}}{COP_j \cdot R_j} \tag{9.11e}$$

$$SOC_{\min} = -\sum_{j=1}^{n} \frac{\delta_j}{COP_j \cdot R_j \cdot \left(1 - \exp\left(-\frac{\Delta t}{R_j \cdot C_{a,j}}\right)\right)} \tag{9.11h}$$

$$SOC_{\max} = \sum_{j=1}^{n} \frac{\delta_j}{COP_j \cdot R_j \cdot \left(1 - \exp\left(-\frac{\Delta t}{R_j \cdot C_{a,j}}\right)\right)} \tag{9.11f}$$

$$\alpha = \frac{1}{n} \sum_{j=1}^{n} \exp\left(-\frac{\Delta t}{R_j \cdot C_{a,j}}\right) \tag{9.11g}$$

It is noted that κ in (9.9) varies for different TCLs' virtual batteries, which makes it difficult to aggregate a virtual battery population. For simplicity, α in (9.11b) is the average κ for all virtual batteries as described in (9.11g). Hence, a TCL population's aggregated flexibility managed by an aggregator is built as a virtual battery model stated in (9.11a)–(9.11c). The aggregator will access virtual battery parameters using the TCL's probability distribution derived from the total TCL power [5]. Accordingly, TCLs will be scheduled like conventional batteries.

9.3.2 De-aggregation of TCLs at the Lower Stage

When aggregators receive trading results from the upper-stage, they decomposed the desired power into control signals to regulate TCLs in the de-aggregation process. To preserve TCL information privacy (i.e., TCL parameters, including C_a, R, δ, COP) and customer preferences (i.e., T_{set}), the aggregator organizes a virtual market, which is a market-based control mechanism to drive self-interested TCLs to follow the desired power. Price signals are applied as control information to guide TCLs' performances; although, economic behavior is not a consideration at this stage. Specifically, TCLs as sellers submit bids for power and price according to their operating status. It is assumed that there is only one virtual buyer for the desired power which pays the market-clearing price. The aggregator collects bids, clears the virtual market, and sends the market-clearing price to TCLs. TCLs will respond accordingly by submitting revised bids. The virtual market mechanism is portrayed as follows.

As charging/discharging schedules of virtual batteries are calculated at the upper stage, the desired TCL power is stated as:

$$P_{desired}(k) = \sum_{j=1}^{n} P_{baseline,j} + P(k) \tag{9.12}$$

Accordingly, the power adjustment (i.e., virtual buyer's purchased power) is represented as:

$$\Delta P(k) = P_{desired}(k) - P_{actual}(k) \tag{9.13}$$

If $\Delta P(k) > 0$, certain TCLs will be turned on to be charged (charging market) and others will be turned off to be discharged (discharging market), which allows TCLs to submit bids strategically and per their operating states. Each TCL is installed with a smart controller for monitoring its operating status, calculate the related parameters, and communicate with the aggregator. Also, TCL bids are determined by their operating states. The market-based TCL control process is described as follows.

Step 1: TCLs decide if they are eligible to partake in a virtual market. In the charging market, only TCLs in off state can submit bids to charge their virtual batteries. Other TCLs in on state will participate in the discharging market to discharge their virtual batteries.

Step 2: When eligible TCLs submit bids, the clearing price will be determined. There are two main components to be considered in the bid price.

(1) TCL compressor wear-and-tear: Excessive switching can increase the stress on TCL compressors and potentially shorten their operating lives [28]. To relieve the mechanical stress, ρ_1 determined by the location of TCL internal temperature is developed. In the charging market, ρ_1 is calculated as:

$$\rho_1 = \frac{T_{\max} - T_{in}(k)}{T_{\max} - T_{\min}} \tag{9.14a}$$

indicating that the TCL with an internal temperature closer to $T_{\max}$ has a lower price and is easier to be scheduled in the charging market, which reduces switching cycles. Similarly, ρ_1 in the discharging market is given as:

$$\rho_1 = \frac{T_{in}(k) - T_{\min}}{T_{\max} - T_{\min}} \tag{9.14b}$$

(2) Fairness in regulation: Some TCLs are turned off/on excessively, while others are rarely manipulated during the regulation process. This unfair regulation process can irritate certain end-users. Thus, ρ_2 is proposed to influence the fairness in bid price:

$$\rho_2 = \frac{N_r}{N_t} \tag{9.15}$$

Higher numbers of TCL switching will result in a larger ρ_2, indicating that the TCL is more difficult to be scheduled. Here, ρ_{bid} consists of two price components which are stated as

$$\rho_{bid} = \omega_1 \cdot \rho_1 + \omega_2 \cdot \rho_2 \tag{9.16}$$

$$\omega_1 + \omega_2 = 1 \tag{9.17}$$

Each customer can decide ω_1 and ω_2 according to its preferences. For simplicity, ω_1 and ω_2 are assumed to be 1/2 in this chapter. In addition, as TCL power can only be changed from P_{rated} to 0 or 0 to P_{rated}, P_{rated} is used as bid power.

Step 3: Aggregators collect TCL bids, sort out prices in ascending order and clear the market based on virtual buyer's required power, $\Delta P(k)$, as depicted in Fig. 9.2. As scheduling results are obtained from the upper stage using the aggregated TCLs' virtual battery model, there is always a market-clearing price between the lowest and the highest TCL bid prices as depicted in Fig. 9.2.

The aggregator sends the corresponding market clearing price (ρ_{bid}^{cl}) to participating TCLs for comparison with respective bid prices. If ρ_{bid}^{cl} is higher, TCL will

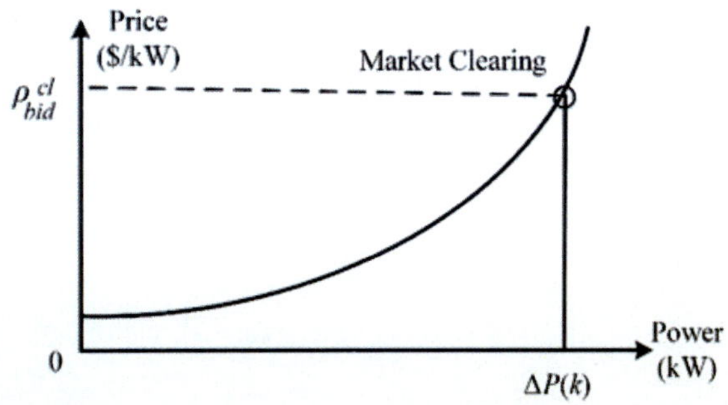

Fig. 9.2 Market clearing in TCL de-aggregation process

switch its state; otherwise, TCL will keep its state. In this regard, TCLs will not provide their parameters to aggregators to protect end-users' information privacy.

9.4 Transactive Energy Market Operation with Aggregated TCL Flexibility

It is assumed that distributed generators and loads connected to a power network node are also managed by TCL aggregators. Accordingly, distributed generators, virtual batteries and inelastic loads are regulated by an aggregator to take part in the power system operation. To preserve information privacy, aggregators at the upper stage coordinate peer-to-peer energy trading in power systems. The information transacted among aggregators includes electricity price and demand. However, distributed resource parameters managed by aggregators will not be communicated. The detailed scheduling strategy is described as follows.

The primal objective is to minimize the total operation cost of N_{agg} aggregators, which is formulated as

$$L = \min \sum_{k=1}^{N_k} \sum_{i=1}^{N_{agg}} \begin{pmatrix} C_g(P_{G,i}(k)) + \rho_i(k) \cdot P_{pur,iSO}(k) \\ + \sum_{v=1}^{N_{agg}} d_{iv} \cdot (P_{pur,iv}(k))^2 \end{pmatrix} \tag{9.18}$$

where the first term is the generator cost expressed as $C_g(x) = ax^2 + bx + c$. The second term is the aggregator's cost for purchasing power from the system operator (SO). The aggregator which purchases power from other aggregators also pays for power losses in the trading process. Thus, the third term is the delivery cost when aggregators purchase power from other aggregators in which the quadratic power loss is taken into account [30].

The primal constraints are given as

$$\begin{aligned} P_{G,i}(k) + P_{pur,iSO}(k) + \sum_{v=1}^{N_{agg}} P_{pur,iv}(k) - P_{sel,i}(k) \\ - P_{battery,i}(k) - P_{load,i}(k) = 0, \;\; 1 \le i \le N_{agg} \end{aligned} \tag{9.19a}$$

$$0 \le P_{G,i}(k) \le P_{G,\max}, \;\; 1 \le i \le N_{agg} \tag{9.19b}$$

$$0 \le P_{pur,iv} \le P_{pur,\max}, \; P_{pur,ii} = 0, 1 \le i \le N_{agg}, 1 \le v \le N_{agg} \tag{9.19c}$$

$$P_{sel,i}(k) \ge 0, 1 \le i \le N_{agg} \tag{9.19d}$$

$$P_{sel,i}(k) = \sum_{v=1}^{N_{agg}} P_{pur,vi}(k),\ 1 \le i \le N_{agg} \tag{9.19e}$$

$$P_{pur,iSO}(k) \ge 0,\ 1 \le i \le N_{agg} \tag{9.19f}$$

$$\text{Constraints of virtual battery model (9.2a)–(9.2c)} \tag{9.19g}$$

where (9.19a) is the power balance of AGG i ($1 \le i \le N_{\text{agg}}$). (9.19b) is the generator I power constraint ($1 \le i \le N_{\text{agg}}$). (9.19c)–(9.19e) are aggregators' transacted power constraints. (9.19f) is the purchased power constraint of the aggregator from SO. (9.19g) is the set of power constraints of aggregated TCLs provided in Sect. 9.3.

Lagrangian relaxation is utilized here to address peer-to-peer transactions that would protect each aggregator's information privacy. The primal problem is decoupled into aggregator subproblems. Each aggregator will transact with other aggregators using trading signals to achieve optimal operations. The Lagrangian function is stated as

$$F = \min \sum_{k=1}^{N_k} \sum_{i=1}^{N_{agg}} \begin{pmatrix} C_g(P_{G,i}(k)) + \rho_i(k) \cdot P_{pur,iSO}(k) \\ + \sum_{v=1}^{N_{agg}} d_{iv} \cdot (P_{pur,iv}(k))^2 + \lambda_i(k)(P_{pur,1i}(k) \\ + P_{pur,2i}(k) + \cdots + P_{pur,N_{agg}i}(k) - P_{sel,i}(k)) \end{pmatrix} \tag{9.20}$$

Here, (9.20) will be decomposed into aggregator subproblems in which the optimization model of AGG i for peer-to-peer transactions are expressed in (9.21). Each aggregator solves its cost minimization problem expressed as (9.21). However, N_{agg} aggregator solutions cannot satisfy the power balance constraint in the primal problem (9.19e). Accordingly, λ_i and $P_{pur,i1}$, $P_{pur,i2}$, ..., $P_{pur,iNagg}$ are exchanged between AGG i and other aggregators. The subgradient method is employed to update the multipliers.

$$F_i = \min \sum_{k=1}^{N_k} \begin{pmatrix} C_g(P_{G,i}(k)) + \rho_i(k) \cdot P_{pur,iSO}(k) \\ + \sum_{v=1}^{N_{agg}} d_{iv} \cdot (P_{pur,iv}(k))^2 + \lambda_1(k) \cdot P_{pur,i1}(k) \\ + \lambda_2(k) \cdot P_{pur,i2}(k) + \cdots + \lambda_{N_{agg}}(k) \cdot P_{pur,iN_{agg}}(k) \\ - \lambda_i(k) \cdot P_{sel,i}(k) \end{pmatrix} \tag{9.21}$$

s.t. (9.19a)–(9.19d) and (9.19f)–(9.19g).

$$\lambda_i(k)[m+1] = \lambda_i(k)[m] + \frac{1}{h \cdot m + l} \left(\sum_{v=1}^{N_{agg}} P_{pur,vi}(k)[m] - P_{sel,i}(k)[m] \right) \tag{9.22}$$

In (9.22), the updated λ_i reflects the AGG *i's* supply and demand relation in the market. When demand is less than supply (i.e., $\sum_{v=1}^{N_{agg}} P_{pur,vi}(k) < P_{sel,i}(k)$), λ_i will be lowered. Otherwise, λ_i will become larger. Thus, λ_i is treated as the AGG *i's* biding price and the final λ_i is the AGG *i*'s market-clearing price. Moreover, $\lambda_1(k)P_{pur,i1}(k) + \lambda_2(k)P_{pur,i2}(k) + \ldots + \lambda_{Nagg}(k)P_{pur,iNagg}(k)$ in (9.21) is the AGG *i's* cost when it purchases power from other aggregators. $\lambda_i(k)P_{sel,i}(k)$ is the AGG *i's* income when it sells power. The difference between the two is the AGG *i's* cost for energy transactions, which reflects AGG *i's* market cost and revenue. Such a peer-to-peer market mechanism that balances supply and demand using electricity price as a key parameter is presented as the transactive energy market [31].

The flowchart of aggregators' trading process is given in Fig. 9.3, and the aggregators' transactive process is described as follows.

(1) Form the Lagrangian relaxation problem (9.20) and initialize parameters, including λ_i $(1 \leq i \leq N_{\text{agg}})$, the power grid structure, AGG contract prices, generator cost curve, load curve, and virtual battery model;
(2) Solve each aggregator's subproblem (9.21) with the latest λ_i $(1 \leq i \leq N_{\text{agg}})$. Then each aggregator *i* gets its $P_{pur,iv}$ and $P_{sel,I}$ $(1 \leq v \leq N_{\text{agg}}, v \neq i)$ according

Fig. 9.3 Flowchart of aggregators' trading process

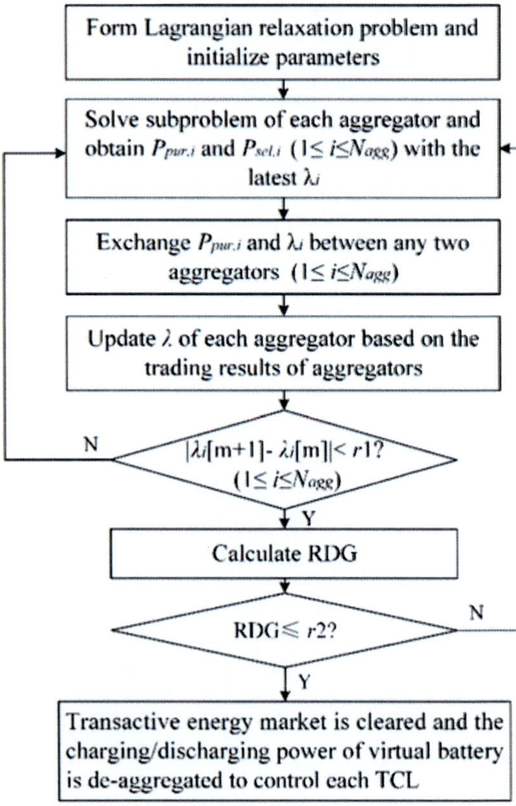

to other aggregators' prices. Repeat Step (2), and calculate P_{pur} and P_{sel} for each aggregator;

(3) It is assumed that aggregators are able to communicate with each other by a local communication network. Exchange P_{pur} and λ between any two aggregators. Each aggregator gets access to other aggregators' trading results and prices;

(4) Update each aggregator's λ using the trading results in terms of (9.22). If each aggregator's λ does not change much in two consecutive iterations (i.e., $|\lambda[m+1]-\lambda[m]|\leq r1$, with r1 representing the termination threshold), λ is a feasible solution and go to Step (5); otherwise, go to Step (2);

(5) Calculate the primal objective value L in (9.18) and F in (9.20) according to each aggregator's latest trading results. The upper bound of the optimal value L^* of (9.18) is given by L and the lower bound is provided by F. Analyze the relative duality gap (RDG) (see (9.23)). Check whether RDG $\leq$ r2, where r2 is the termination threshold. If satisfied, the trading results in the transactive market are the same as the primal optimization problem described in (9.19a)–(9.19g); go to Step (6). Otherwise, more trading iterations are required; go back to Step (2).

$$RDG = \left|\frac{L-F}{F}\right| \tag{9.23}$$

(6) Transactive energy market is cleared and each aggregator's trading is determined. Accordingly, each virtual battery's charging/discharging power regulated by aggregators is also decided, which is sent to the lower stage for de-aggregation to control individual TCLs.

Hence, TCLs are regulated by virtual market signals at the lower stage and scheduled in a transactive energy market at the upper stage. In this way, the TCL information will not be leaked to aggregators and aggregators will not expose the information to their peers.

9.5 Case Studies

We assume there are three aggregators represented as AGG1-AGG3. TCL parameters of each aggregator are shown in Table 9.1. The number of TCLs in each AGG is 1000. The ambient temperature is 36 °C. The time slots at the upper and lower stages are 5 min and 1 min, respectively. Generation cost function coefficients for AGG1-AGG3 are shown in Table 9.2. The generator capacities for AGG1-AGG3 are 7.5 MW, 10 MW, and 12 MW, respectively. Loads are scaled [32]; r1 and r2 are 0.05; d_{12}, d_{13} and d_{23} (representing actual distances) are 4×10^{-4} \$/MW^2, 1.4×10^{-3} \$/MW^2 and 1.8×10^{-3} \$/MW^2.

To validate the proposed methods and highlight our main contributions, case studies are structured as follows. The simulations in Sects. 9.5.1 and 9.5.2 are

Table 9.1 TCL parameter ranges of aggregators

Parameters	AGG1	AGG2	AGG3
T_{set}	15–25 °C	20–25 °C	20–22.5 °C
δ	0.25–1 °C	0.25–0.35 °C	0.3–0.5 °C
R	1.5–2.5 C/kW	1.8–2.2 °C/kW	2 °C/kW
C_a	1.5–2.5 kWh/°C	1.8–2.2 kWh/°C	2 kWh/°C
P_{rated}	10–18 kW	4.5–7 kW	6–14 kW
COP	2.5	2.5	2.5

Table 9.2 Coefficients of generator cost function

Aggregator	a ($/kW^2)	b ($/kW)	c ($)
AGG1	1.22e (−6)	0.04	0.35
AGG2	1.12e (−6)	0.02	0.25
AGG3	8.65e (−7)	0.03	0.40

employed to analyze aggregators' trading behaviors at the upper stage and the control performance of TCLs at the lower stage, respectively. The scheduling of TCLs at two stages is coordinated to regulate TCLs. Section 9.5.3 compares virtual and conventional battery models based on the virtual battery's scheduling results provided in Sect. 9.5.1.

9.5.1 *Aggregators' Transactive Trading Results at the Upper Stage*

Using our proposed model, virtual battery parameters are calculated and shown in Table 9.3. The contract price of AGG1–AGG3 is previously determined by SO and depicted in Fig. 9.4 [33]. Aggregators with flexible TCLs participate in the proposed peer-to-peer trading. Each aggregator makes trading decisions based on other aggregators' prices, SO contract prices and local generation costs, without leaking any proprietary information. Simulation results are presented as follows. The behaviors of entities (including aggregators, generators, and virtual batteries) and prices (including SO contract prices, generators' marginal costs and aggregators' trading

Table 9.3 Parameters of virtual batteries

Aggregator	P_{min} (MW)	P_{max} (MW)	SOC_{min} (MW)	SOC_{max} (MW)	α
AGG1	−3.24	10.76	−6.15	6.15	0.98
AGG2	−2.71	3.02	−2.89	2.89	0.98
AGG3	−2.94	6.93	−3.86	3.86	0.98

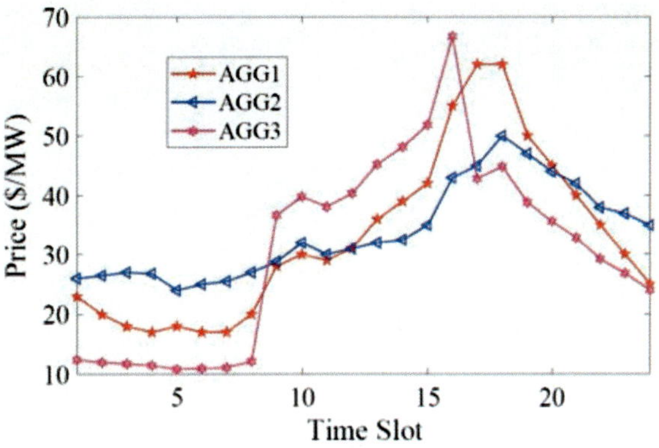

Fig. 9.4 SO's contract prices of AGG1–AGG3

prices) are all analyzed to show the operating mechanism of the proposed peer-to-peer trading. Two Cases with and without peer-to-peer trading are also compared to show that peer-to-peer trading can reduce the operation cost in the power system.

(1) *Analyses of power provided by SO and local generation*: Figs. 9.5 and 9.6 show the power purchased from SO and generators' power and marginal costs, respectively. Here, any purchases depend on the SO's price and the generator marginal cost. Considering the results in Figs. 9.5, 9.6 and 9.7, AGG3 purchases more power from SO at time slots 1–8, when SO the price of SO is lower than that of AGG1-AGG2

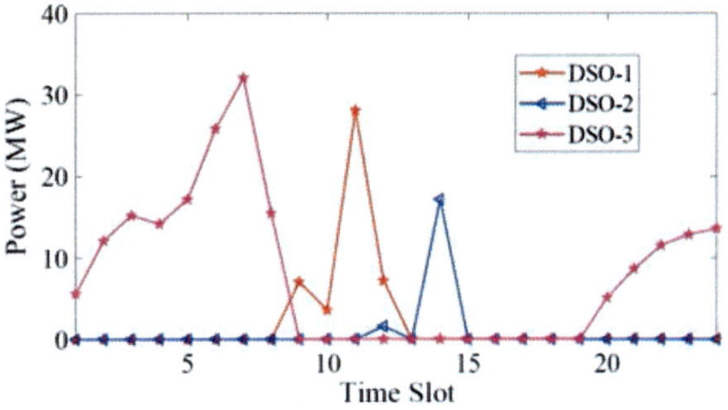

Fig. 9.5 Power purchased by AGG1–AGG3 from SO

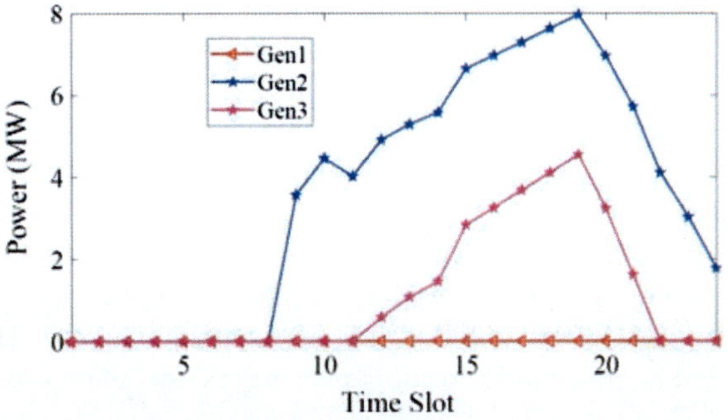

Fig. 9.6 Generator power of AGG1–AGG3

Fig. 9.7 Generator marginal cost of AGG1–AGG3

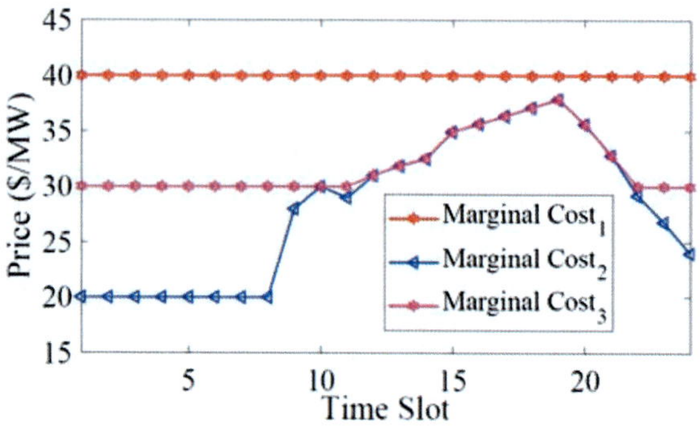

Fig. 9.8 Transactive trading price of AGG1–AGG3

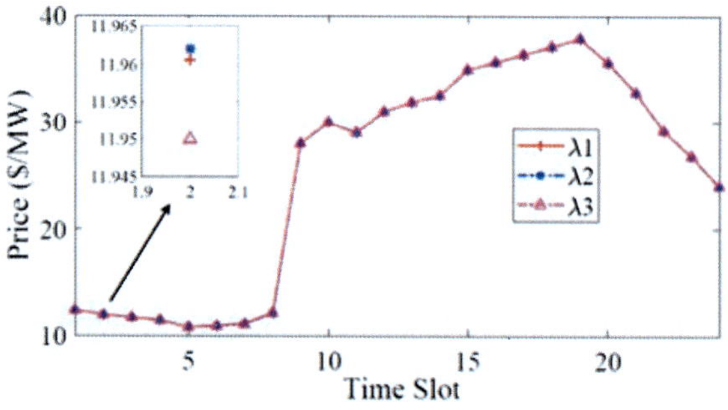

and the generator marginal cost of AGG1-AGG3. At this time, generators as more expensive resources will not be dispatched. When the SO's price is higher (e.g., time slots 15–19), generators of AGG2 and AGG3 with lower marginal costs, as compared with those of AGG1, are dispatched.

(2) *Analyses of aggregators' trading prices*: Fig. 9.8 shows transactive trading prices (λ_1, λ_2 and λ_3) of AGG1-AGG3 which have overlapped due to small differences of λ_1, λ_2 and λ_3. It is clear that the trends in aggregator trading prices are consistent with the SO's price for AGG3 at time slots 1–8, when the corresponding SO's price is the lowest as compared to other SO prices and generator marginal costs. When any aggregators' generator marginal costs are lower than the SO's price, the trading price trends are consistent with generator marginal costs in AGG1 and AGG2 at time slots 15–19. In this example, aggregators' trading prices reflect those of the cheapest dispatched resources (SO or generators).

Take λ_1–λ_3 at time slot 2 to analyze the transactive trading among AGG1-AGG3. The convergence of λ_1–λ_3 at time slot 2 is depicted in Fig. 9.9. The corresponding trading power is shown in Fig. 9.10. Here, $\lambda_3 = 11.9501$ is the lowest compared with $\lambda_1 = 11.9608$ and $\lambda_2 = 11.9623$, which means the cheapest resource is scheduled by AGG3. Thus, both AGG1 and AGG2 purchase power from AGG3. Moreover, purchasing aggregators also pay for transmission losses. So, AGG2 also buys power from AGG1 to reflect a shorter electrical distance. The differences among λ1, λ2 and λ3 describe the trading power direction among AGG1-AGG3. Aggregators will

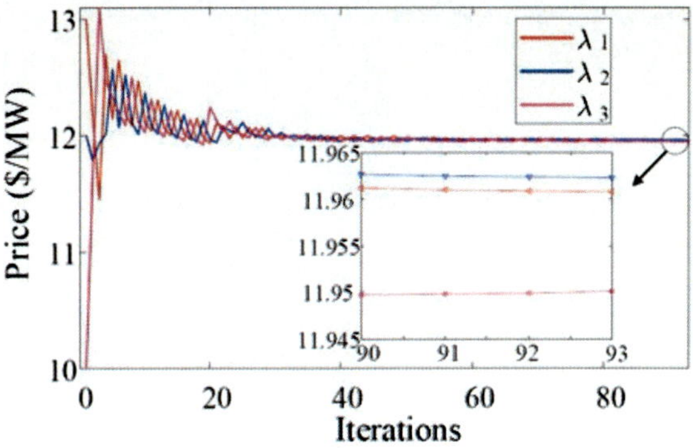

Fig. 9.9 Transactive trading price convergence in time slot 2

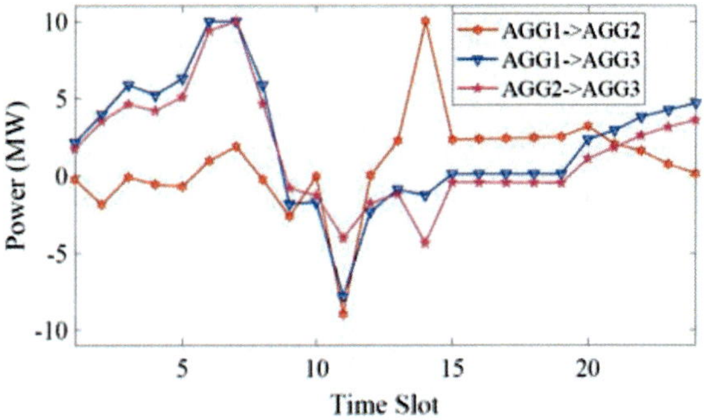

Fig. 9.10 Transactive trading power between AGG1–AGG3

operate economically by adopting the trading price λ in the peer-to-peer transaction process.

(3) *Analysis of virtual battery*: The virtual batteries' operations are shown in Fig. 9.11 in which virtual batteries in AGG1-AGG3 have similar operating behaviors. Virtual batteries are either charging or discharging when transactive trading prices in Fig. 9.8 are lower (e.g., at time slots 5–7) or higher (e.g., at time slots 15–19),

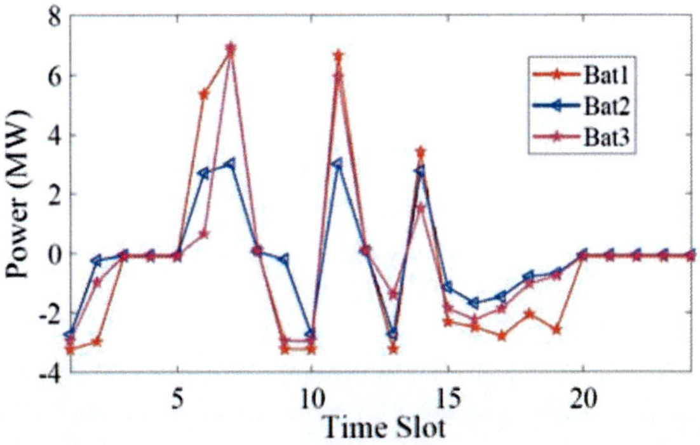

Fig. 9.11 Virtual battery charging/discharging power of AGG1–AGG3

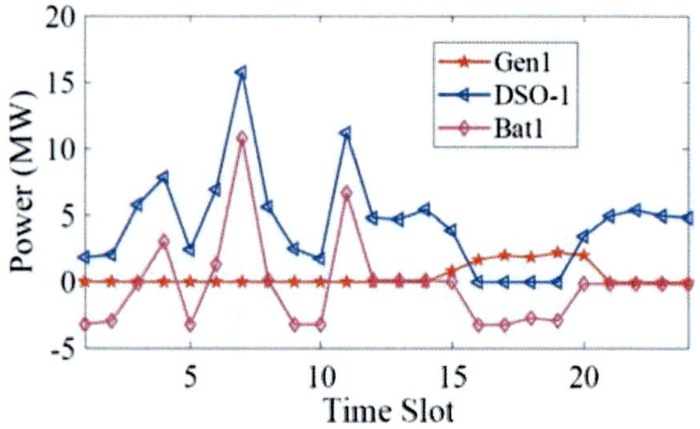

Fig. 9.12 AGG1 operation in Case 2

respectively. This Case demonstrates that virtual batteries' operating behaviors are determined by the trend in aggregator trading prices.

(4) *Comparison of aggregator's behaviors with and without peer-to-peer transactive trading:* Two Cases are compared to show the effectiveness of the proposed peer-to-peer transactive trading. Aggregators with flexible TCLs participate in the proposed peer-to-peer trading in Case 1. Such aggregators do not participate in the proposed transactive energy market and only purchase electricity from SO in Case 2 (i.e., in Case 2, there are only transactions between SO and aggregators). The dispatched power of generators and SO are determined by generator marginal costs and SO prices. The operating behaviors of AGG1 in Case 2 are presented in Fig. 9.12. If the SO's price is lower than generator marginal costs, aggregators will purchase power from SO and virtual batteries are charged (e.g., time slot 4 for AGG1 in Fig. 9.12). If the SO's price is higher than generator marginal prices, aggregators will generate power and virtual batteries are discharged (e.g., time slots 17–18 for AGG1). The total costs in the two cases are \$8151.8 and \$9486.1, respectively. In Case 2, aggregators cannot trade any cheaper power with other aggregators, which increases the operation cost as compared to that in Case 1.

9.5.2 Control Performances of TCLs at the Lower Stage

The scheduled power of aggregated TCLs (i.e., charging/ discharging power of virtual battery) is determined when the transactive market is cleared at the upper stage. Then the aggregator will de-aggregate the scheduled power to control each TCL in terms of the proposed method in Sect. 8.3.2.

Take AGG1 as an example to validate the proposed virtual battery model and the control method of TCLs. The TCL's actual power and errors (representing differences between the desired and the actual TCL power) are shown in Fig. 9.13, and the corresponding internal temperature profiles are given in Fig. 9.14. Different colors in Fig. 9.14 represent TCLs' profiles for respective control signals [28, 29]. TCLs' diverse internal temperatures are normalized in Fig. 9.14. In Fig. 9.13, power errors

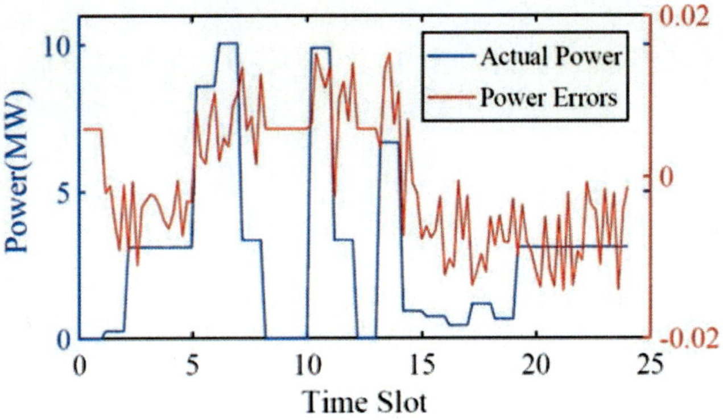

Fig. 9.13 Actual power and power errors of TCLs in AGG1

Fig. 9.14 TCL internal temperature profiles of AGG1

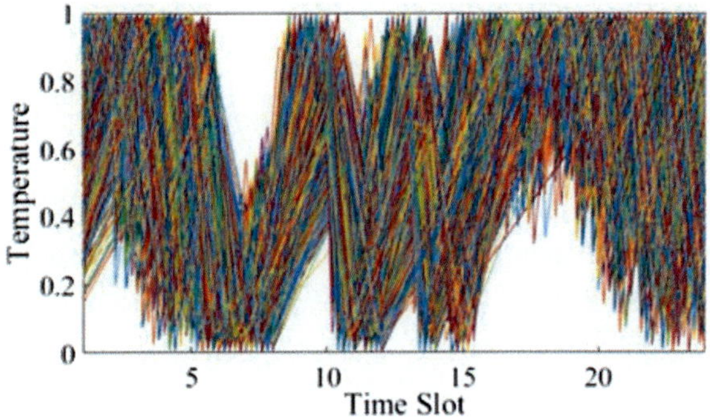

are very small, which means TCLs can closely track the virtual battery's scheduled power that is determined at the upper stage. In Fig. 9.14, TCLs' internal temperatures are limited within [0, 1] for satisfying end users' comfort requirements. The outcome demonstrates that virtual batteries accurately describe the TCLs' aggregated flexibility as TCLs are effectively controlled by the proposed method.

When TCLs are regulated to follow the desired power, it is imperative to relieve the switching's mechanical stress as TCL's internal temperature is considered for implementing ascending/descending orders in centralized control [7, 28]. The switching's mechanical stress is also relieved in the proposed transactive control when TCL's internal temperature is considered as bid. Two Cases are compared to show the other merits of the proposed transactive control at the lower stage. In Case 1, the TCL bid is composed of TCL's internal temperature and regulation fairness (i.e., $\omega_1 = 1/2$, $\omega_2 = 1/2$). In Case 2, only TCL's internal temperature is considered in the TCL bid (i.e., $\omega_1 = 1$, $\omega_2 = 0$). The numbers of TCL switching in Cases 1 and 2 are provided in Fig. 9.15. The corresponding variances of the number of TCL switching in the two Cases are 23.34 and 55.15, respectively. It is clear that the proposed method will effectively avoid unfair regulations where some TCLs are turned off/on excessively.

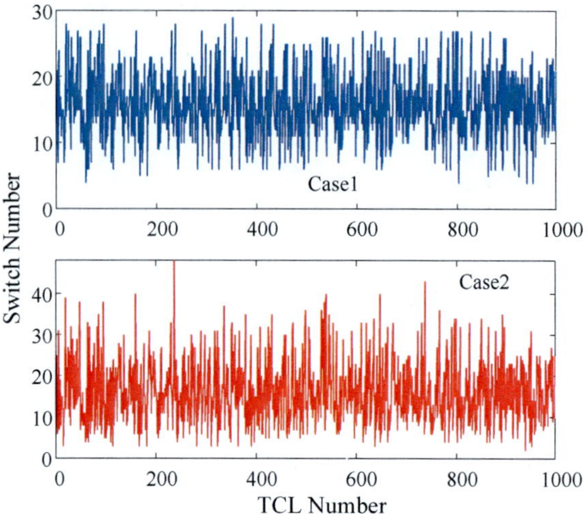

Fig. 9.15 Number of TCL switching in Cases 1 and 2

9.5.3 Comparison of Conventional and Virtual Batteries

Although the virtual battery shares the same models as the conventional lithium-ion battery, there are differences between the two models, as presented below.

(1) *Virtual battery is sensitive to external temperatures.* External temperatures affect virtual batteries' operating behaviors. Generally, the external temperature would change continuously, and we assume external temperatures remain fixed in half-hour intervals. The P_{min} and P_{max} in AGG1's virtual battery for different external temperatures are calculated in Table 9.4. The conventional battery is required to work within a certain range of ambient temperature. Even though, a conventional battery's operating behavior cannot change as significantly as that of a virtual battery when external temperature changes within a normal range.

(2) *Virtual battery shows an infinite capacity in some Cases.* The virtual battery's operating behavior in AGG1 is shown in Fig. 9.16, where the aggregators' total cost is \$8151.8. To compare virtual and conventional batteries in terms of operating behaviors, a lithium-ion battery with parameters stated in Table 9.5 is scheduled in the peer-to-peer transactive trading with a similar total cost of \$8131.8. In Fig. 9.16, SOC is fixed even though the virtual battery is slightly charged/discharged in certain time slots (e.g., hours 20–24) with an infinite capacity that will never be fully charged/discharged. By contrast, SOC can only stay fixed when the lithium-ion

Table 9.4 P_{min} and P_{max} of virtual battery for different external temperatures

T_{out} (°C)	36	37.5	38	37
P_{min} (MW)	−3.24	−3.58	−3.66	−3.44
P_{max}(MW)	10.76	10.40	10.37	10.64

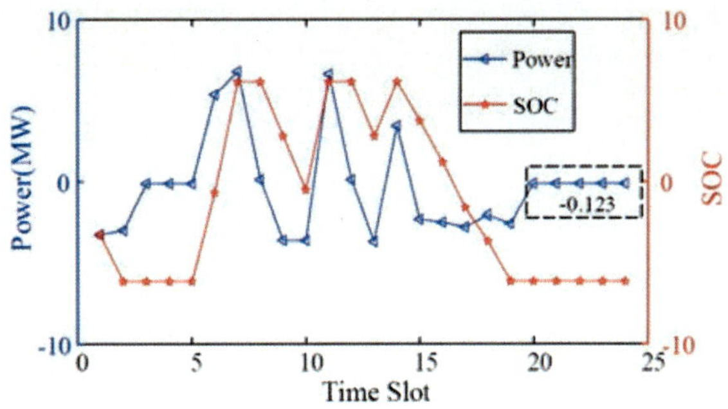

Fig. 9.16 Operating behavior of virtual battery in AGG1

Table 9.5 Parameters of lithium-ion battery

Parameter	P^d	P^c	η^c	η^d
Value	5/MW	5/MW	0.95	0.95
Parameter	o	$soc_{\min}$	$soc_{\max}$	soc_0
Value	9/MWh	0	1	0.5

battery's charging/discharging power is 0 (e.g., time slot 20–24 h in Fig. 9.17). The corresponding explanation is given as follows.

For a virtual battery, $SOC(k+1)$ is equal to $SOC(k)$ when $P(k)$ is given as

$$P(k) = (1 - \alpha) \cdot SOC(k) \tag{9.24}$$

In (9.24), $P(k)$ will not be 0 as $SOC(k)$ is not within $(0 < \alpha < 1)$. This means that the energy storage level of a virtual battery does not change even though it is charging/discharging. In this case, we assume the TCL's virtual battery has an infinite capacity that cannot be fully charged/discharged. A TCL will consume a certain amount of power, which will be adjusted with temperature, to maintain a fixed internal temperature. $P(k)$ is defined as the difference between the actual power and baseline power of TCLs according to (9.4a) and (9.10b). When the internal temperature is kept at a point that is different from T_{set}, the TCL's actual power

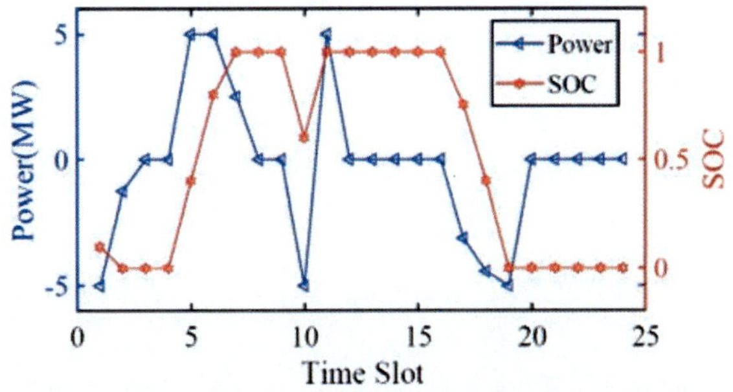

Fig. 9.17 Operating behavior of the lithium-ion battery

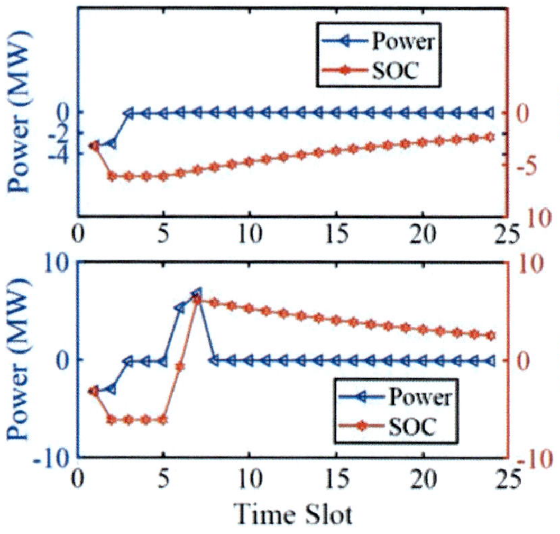

Fig. 9.18 Operating behavior of the virtual battery

is always different from the baseline power. In other words, a virtual battery would need to absorb/release some power to maintain its energy storage levels (i.e., $SOC(k) \neq 0$). Thus, a virtual battery possesses an unlimited capacity when $P(k)$ is equal to the power difference corresponding to TCL's internal temperatures at T^* and T_{set} ($T^* \neq T_{\text{set}}$), respectively. At this time, the virtual battery's energy storage level is fixed at $SOC(k)$ and the corresponding TCL internal temperature is T^*. By contrast, a conventional battery's energy storage level will change when charging/discharging power is not 0. This means a conventional battery has a limited capacity that can be fully charged/discharged.

(3) The *virtual battery has an automatic restoration capability*. The charging/discharging power is set at 0 between time slots 6 and 24. The corresponding relationship between $P(k)$ and $SOC(k)$ is simulated in the upper Fig. 9.18, where the SOC gradually approaches 0. Similarly, the relationship between $P(k)$ and $SOC(k)$ is shown in the lower Fig. 9.18, when $P(k)$ is 0 beyond the time slot 7 when $SOC(k)$ also approaches 0. The corresponding explanation is given as follows.

$SOC(k+1)$ changes according to $SOC(k)$ and the absolute value of $SOC(k+1)$ will be less than that of $SOC(k)$ when $P(k)$ is 0 ($0 < \alpha < 1$) in (9.11b). If $SOC(k) > 0$, $SOC(k+1)$ is smaller than $SOC(k)$, which means the energy storage level is reduced. If $SOC(k) < 0$, $SOC(k+1)$ is larger than $SOC(k)$, which means the energy storage level is increased. Hence, $SOC(k+1)$ tends to approach 0 (its original state) when $P(k)$ is 0. This phenomenon is related to the self-dissipation power of a virtual battery. When $SOC(k) > 0$, a virtual battery will self-dissipate power and $P(k)$ is too small to maintain the current energy storage level, which will result in a lower SOC. When $SOC(k) < 0$, the self-dissipated power of a virtual battery is smaller than 0 and $P(k)$ is too large to maintain the current energy storage level, which will result in a higher SOC. However, for a conventional battery, $SOC(k+1)$ is equal to $SOC(k)$ when $P(k)$ is 0. This means a conventional battery's energy storage level remains constant when it is not deployed.

9.6 Conclusions

A hierarchical coordination mechanism is proposed to manage the TCL flexibility in a distributed fashion. TCLs are regulated by the transactive control and TCL flexibilities are aggregated using a virtual battery model at the lower stage. Compared with a conventional battery, the proposed virtual battery is more sensitive to external temperatures when the temperature is changing within a normal range, offers an infinite capacity in some cases, and has an automatic restoration capability. A transactive energy market enables aggregators with TCL flexibility to trade with each other for minimizing the cost of the power system at the upper stage. The effectiveness of the proposed methods is validated by several case studies. Our future work will focus on enhancing the resilience of power systems via transactive energy markets by utilizing distributed resources, including TCLs, electric vehicles and conventional batteries.

References

1. Song, M., Sun, W., Wang, Y., Shahidehpour, M., Li, Z., & Gao, C. (2019). Hierarchical scheduling of aggregated TCL flexibility for transactive energy in power systems. *IEEE Transactions on Smart Grid, 11*(3), 2452–2463.
2. Sajjad, I. A., Chicco, G., & Napoli, R. (2016). Definitions of demand flexibility for aggregate residential loads. *IEEE Transactions on Smart Grid, 7*(6), 2633–2643.
3. Kim, Y. J., Norford, L. K., & Kirtley, J. L. (2014). Modeling and analysis of a variable speed heat pump for frequency regulation through direct load control. *IEEE Transactions on Power Systems, 30*(1), 397–408.
4. Lu, N., & Zhang, Y. (2012). Design considerations of a centralized load controller using thermostatically controlled appliances for continuous regulation reserves. *IEEE Transactions on Smart Grid, 4*(2), 914–921.
5. Song, M., Gao, C., Shahidehpour, M., Li, Z., Lu, S., & Lin, G. (2018). Multi-time-scale modeling and parameter estimation of TCLs for smoothing out wind power generation variability. *IEEE Transactions on Sustainable Energy, 10*(1), 105–118.
6. Hughes, J. T., Domínguez-García, A. D., & Poolla, K. (2016). Identification of virtual battery models for flexible loads. *IEEE Transactions on Power Systems, 31*(6), 4660–4669.
7. Hao, H., Sanandaji, B. M., Poolla, K., & Vincent, T. L. (2014). Aggregate flexibility of thermostatically controlled loads. *IEEE Transactions on Power Systems, 30*(1), 189–198.
8. Zhao, L., Zhang, W., Hao, H., & Kalsi, K. (2017). A geometric approach to aggregate flexibility modeling of thermostatically controlled loads. *IEEE Transactions on Power Systems, 32*(6), 4721–4731.
9. Mathieu, J. L., Kamgarpour, M., Lygeros, J., Andersson, G., & Callaway, D. S. (2014). Arbitraging intraday wholesale energy market prices with aggregations of thermostatic loads. *IEEE Transactions on Power Systems, 30*(2), 763–772.

10. Trovato, V., Tindemans, S. H., & Strbac, G. (2016). Leaky storage model for optimal multi-service allocation of thermostatic loads. *IET Generation, Transmission & Distribution, 10*(3), 585–593.
11. Hao, H., Wu, D., Lian, J., & Yang, T. (2017). Optimal coordination of building loads and energy storage for power grid and end user services. *IEEE Transactions on Smart Grid, 9*(5), 4335–4345.
12. Raman, N. S., & Barooah, P. (2019). On the round-trip efficiency of an HVAC-based virtual battery. *IEEE Transactions on Smart Grid, 11*(1), 403–410.
13. Hu, J., Cao, J., Chen, M. Z., Yu, J., Yao, J., Yang, S., & Yong, T. (2016). Load following of multiple heterogeneous TCL aggregators by centralized control. *IEEE Transactions on Power Systems, 32*(4), 3157–3167.
14. Radaideh, A., Vaidya, U., & Ajjarapu, V. (2017). Sequential set-point control for heterogeneous thermostatically controlled loads through an extended Markov chain abstraction. *IEEE Transactions on Smart Grid, 10*(1), 116–127.
15. Hussain, A., Bui, V. H., & Kim, H. M. (2016). A resilient and privacy-preserving energy management strategy for networked microgrids. *IEEE Transactions on Smart Grid, 9*(3), 2127–2139.
16. Wu, J., & Guan, X. (2013). Coordinated multi-microgrids optimal control algorithm for smart distribution management system. *IEEE Transactions on Smart Grid, 4*(4), 2174–2181.
17. Rahimi, F., & Ipakchi, A. (2016). Using a transactive energy framework: Providing grid services from smart buildings. *IEEE Electrification Magazine, 4*(4), 23–29.
18. Wu, Q., Shahidehpour, M., Li, C., Huang, S., & Wei, W. (2018). Transactive real-time electric vehicle charging management for commercial buildings with PV on-site generation. *IEEE Transactions on Smart Grid, 10*(5), 4939–4950.
19. Renani, Y. K., Ehsan, M., & Shahidehpour, M. (2017). Optimal transactive market operations with distribution system operators. *IEEE Transactions on Smart Grid, 9*(6), 6692–6701.
20. Gregoratti, D., & Matamoros, J. (2014). Distributed energy trading: The multiple-microgrid case. *IEEE Transactions on industrial Electronics, 62*(4), 2551–2559.
21. Somasundaram, S., Pratt, R. G., Akyol, B., Fernandez, N., Foster, N., Katipamula, S., ... & Taylor, Z. (2014). Reference guide for a transaction-based building controls framework. Pacific Northwest National Laboratory.
22. Katipamula, S., Chassin, D. P., Hatley, D. D., Pratt, R. G., & Hammerstrom, D. J. (2006). Transactive controls: A market-based GridWiseTM controls for building systems (No. PNNL-15921). Pacific Northwest National Lab.(PNNL), Richland, WA, USA.
23. Hao, H., Corbin, C. D., Kalsi, K., & Pratt, R. G. (2016). Transactive control of commercial buildings for demand response. *IEEE Transactions on Power Systems, 32*(1), 774–783.
24. Li, S., Zhang, W., Lian, J., & Kalsi, K. (2015). Market-based coordination of thermostatically controlled loads—Part I: A mechanism design formulation. *IEEE Transactions on Power Systems, 31*(2), 1170–1178.
25. Li, S., Zhang, W., Lian, J., & Kalsi, K. (2015). Market-based coordination of thermostatically controlled loads—Part II: Unknown parameters and case studies. *IEEE Transactions on Power Systems, 31*(2), 1179–1187.
26. Vidyamani, T., & Swarup, K. S. (2018, December). Analysis of active and transactive demand response strategies for smart residential buildings. 2018 20th National Power Systems Conference (NPSC) (pp. 1–5).
27. Ikäheimo, J., Evens, C., & Kärkkäinen, S. (2010). DER Aggregator business: the Finnish case. Technical Research Centre of Finland (VTT): Espoo, Finland.
28. Lu, N. (2012). An evaluation of the HVAC load potential for providing load balancing service. *IEEE Transactions on Smart Grid, 3*(3), 1263–1270.
29. Song, M., Gao, C., Shahidehpour, M., Li, Z., Yang, J., & Yan, H. (2018). State space modeling and control of aggregated TCLs for regulation services in power grids. *IEEE Transactions on Smart Grid, 10*(4), 4095–4106.
30. Wang, Y., Huang, Z., Shahidehpour, M., Lai, L. L., Wang, Z., & Zhu, Q. (2019). Reconfigurable distribution network for managing transactive energy in a multi-microgrid system. *IEEE Transactions on Smart Grid, 11*(2), 1286–1295.

31. Liu, Z., Wu, Q., Huang, S., & Zhao, H. (2017, June). Transactive energy: A review of state of the art and implementation. 2017 IEEE Manchester PowerTech (pp. 1–6).
32. Song, M., Gao, C., Yan, H., & Yang, J. (2017). Thermal battery modeling of inverter air conditioning for demand response. *IEEE Transactions on Smart Grid, 9*(6), 5522–5534.
33. Bai, L., Wang, J., Wang, C., Chen, C., & Li, F. (2017). Distribution locational marginal pricing (DLMP) for congestion management and voltage support. *IEEE Transactions on Power Systems, 33*(4), 4061–4073.

Chapter 10
Multi-time Scale Transactive Scheduling of TCLs for Smoothing Microgrid Tie Flow Fluctuations

Abstract To address the microgrid tie flow errors caused by wind generation variability, this chapter proposes and develops a multi-time scale coordinated control and scheduling strategy of inverter TCLs. First, in hour-time scale, inverter-based TCLs with adjusting temperature set-point are modeled as virtual generators to compensate for the tie flow deviations from the day-ahead plan. Next, in a minute-time scale, virtual batteries representing the operating behavior of inverter-based TCLs with frequency control are determined by the control of virtual generators in an hour-time scale. The virtual batteries are scheduled to smooth out tie flow errors corresponding to the day-ahead plan and hour-time scale scheduling. In this work, multi-time scale control methods are coordinated to employ the response potential of inverter-based TCLs. And transactive scheduling based on the response curve is proposed to regulate inverter TCLs considering customer privacy. The multi-time scale stochastic scheduling of TCLs is coordinated to accommodate wind generation variability. Simulation results demonstrate that the microgrid tie flow errors are effectively mitigated by the proposed multi-time scale coordinated control and scheduling of inverter TCLs.

Keywords Transactive scheduling · Multi-time scale · Response curve · Wind generation

10.1 Introduction

The proliferation of wind generation could increase the real-time fluctuations in power delivery and introduce tremendous challenges to the operation and planning of power systems [1, 2]. Accordingly, more flexibility ought to be placed in power systems to address the related uncertainties. As for typical DR resources, TCLs provide promising opportunities to compensate for power imbalances triggered by wind generation variabilities [3, 4].

When regulated by different control methods, TCLs present various operating behaviors in terms of response time, duration and capacity. For example, it is shown that TCL power consumption can be changed either instantaneously with a short response duration by on-off switches or regulated for a longer period but with

M. Song and C. Gao, *Integration of Distributed Resources in Smart Grids for Demand Response and Transactive Energy*, https://doi.org/10.1007/978-981-16-7170-8_10

delayed effects by temperature set-point adjustment [5]. Therefore, TCLs can provide multiple types of energy services for a reliable and secure power system operation. Reference [6] proposes a novel framework based on partial differential equations and a sliding mode controller to regulate air conditioning loads in real-time to compensate for variable wind power generation. A centralized and hierarchical control strategy is employed in [7] to manipulate TCLs to provide load-following services. In [8], a priority-based TCL algorithm with a large time constant is proposed for smoothing out the variability of renewable energy.

In the available literature on ancillary services provided by TCLs [9, 10], only a single time-scale TCL control has already been considered. There have been limited research efforts on the coordination of multi-time scale TCL control, which makes it difficult to fully employ TCL capabilities in DR programs. Moreover, wind power generation has multi-time scale characteristics where its forecast accuracy would increase as the forecast time-scale becomes small [11]. Therefore, multi-time scale TCL regulations have great potentials in addressing wind power generation variabilities more effectively.

TCLs are generally controlled in a centralized manner. Reference [12] proposed an aggregated model to capture the collective TCL behavior and a centralized control strategy to manage designated demand curves. A priority-stack-based control architecture is developed in [13] where TCLs are sorted and controlled in sequence to provide frequency regulation. In [7], TCLs are controlled by a hierarchical and centralized control method for load following. Although centralized control methods present good predictability and controllability, they often cause communication burdens and would have to deal with end users' privacy concerns [14].

Compared to centralized control, decentralized control offers more flexibility and preserves customer privacy more elegantly. For example, a decentralized TCL implementation is proposed in [15] which allows each TCL to consider its temperature limits. Based on stochastic TCL switching, a decentralized control scheme is developed for primary frequency control [16]. However, decentralized control also has the disadvantage of less predictability and difficulty in achieving global optimality [14].

This chapter proposes an improved centralized control strategy, which also considers the merits of decentralized control to regulate TCLs. In other words, we develop a TCL control method that protects end users' privacy and offers higher implementation efficiency with fewer communication burdens and better options for controllability. Price as an important parameter is embedded into the control and scheduling strategy of TCLs. Accordingly, the proposed method can also be seen as transactive control which the price is utilized to regulate the behaviors of devices.

The TCL compressor has evolved from a fixed to variable speed unit [17]. Inverter TCLs have obtained remarkable market share because of the higher energy efficiency [18]. However, fixed-speed TCLs still dominate the research, and inverter TCL applications in DR programs are very limited in the literature. Thus, a multi-time scale coordinated control and scheduling of inverter TCLs are developed to address the microgrid tie flow errors caused by wind generation variabilities. Load fluctuations would have a profound impact on microgrid tie flows. Here, we only consider wind generation variabilities as wind generation poses a higher variability in such cases.

The main work is summarized as follows.

1. Multi-time scale control methods are coordinated to fully utilize the response potential of inverter TCLs, which the model parameters of inverter TCLs in post-time scale are determined by the control in pre-time scale;
2. Transactive control of inverter TCLs based on response curve is proposed, which will not only provide the DR adjustment fee for inverter TCLs but also protect end users' privacy and improve the implementation efficiency on TCLs with less communication burden;
3. Multi-time scale stochastic scheduling strategies are coordinated according to response curves of inverter TCLs to address the tie flow errors caused by wind generation variabilities.

10.2 The Framework of Multi-time Scale Coordinated Control and Scheduling

Consider a microgrid with wind generators and loads, which is connected to the main grid by the tie. Loads are divided into inverter TCLs (i.e., elastic) and inelastic loads. In the day-ahead time scale, the microgrid submits its scheduling plan to the main grid according to the daily forecast of wind generation and load. However, the wind generation in real-time is different because of forecast errors stemming from the wind speed variability. The tie flow errors may cause voltage and frequency issues in the main grid. Thus, the tie flow is required by the microgrid operator (MO) to keep close track of the day-ahead plan for maintaining the grid reliability [19]. Considering the multi-time scale characteristics of wind power generation, inverter TCLs are coordinated in multiple time scales for control and scheduling. The configuration and scheduling framework are depicted in Fig. 10.1.

- A smart controller is installed at each end-users' site to measure, calculate and distribute control signals. In the hour-time scale (time step is 1 h), inverter TCLs with temperature set-point adjustments are modeled as virtual generators (Sect. 10.3.2). In minute-time scale (time step is 15 min), inverter TCLs with frequency control are modeled as virtual batteries (Sect. 10.3.3), whose parameters are determined by the control of virtual generators in an hour-time scale (Sect. 10.3.4). In hour/minute time interval, each smart controller submits the response curve of virtual generator/battery to MO according to end-users' preferences. And control signals are generated to regulate devices based on the adjustment price from the MO (Sect. 10.4). It is noted that the adjustment price in this chapter is defined as the ratio of the adjustment fee paid by MO for end-users in DR programs and the adjusted power of inverter TCLs.
- In the view of scheduling, multi-time scale stochastic scheduling strategies of inverter TCLs are coordinated to address the tie flow errors caused by wind generation variabilities. Prior to each hour, the MO organizes a local market to purchase the flexibility from the virtual generators to compensate for the tie flow

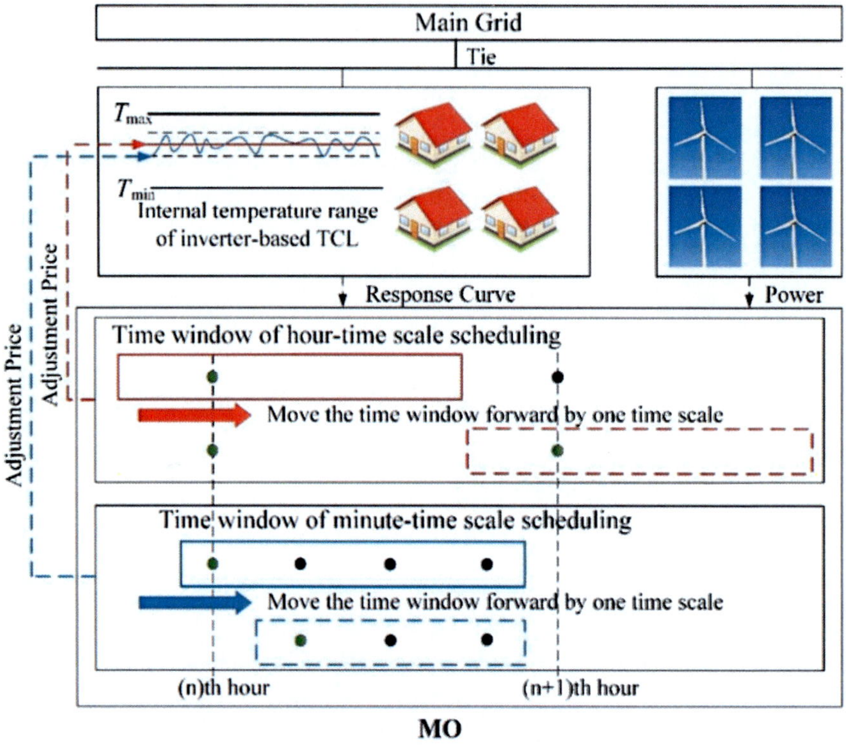

Fig. 10.1 The framework of multi-time scale control and scheduling of TCLs

errors compared to the day-ahead plan (Sect. 10.5.1). Then the MO submits its hour-time scale scheduling plan to the main grid and distributes the price signal to inverter TCLs to determine their internal temperature. In each minute-time scale interval, MO applies a model predictive control (MPC)-based transactive strategy [20] to smooth out the tie flow errors compared to the day-ahead plan and hour-time scheduling results (Sect. 10.5.2). The corresponding adjustment price derived from the cleared market is provided to inverter TCLs to determine their compressor frequency.

- The customer's response capability depends on MO's price, which is characterized by the customer's response curve (Sect. 10.4). The corresponding adjustment fee is calculated according to the adjustment price and the TCL response capability. The MO will control the corresponding TCLs using the prices. Moreover, the internal temperatures of customers are kept within a certain range when they are under the regulation of MO, which has little impact on customers' comfort. Thus, the number of customers in DR programs is seen as a fixed value.

10.3 Multi-time Scale Scheduling Models of TCLs

10.3.1 Inverter TCL Model

ACs are taken as a typical TCLs to be investigated in this chapter. The thermodynamic model of an inverter TCL [17, 21] is given as

$$C_a \cdot dT_{in}/dt = (T_{out} - T_{in})/R - Q \tag{10.1a}$$

$$Q = k_q \varphi + l_q \tag{10.1b}$$

$$P = k_p \varphi + l_p \tag{10.1c}$$

$$P_{\min} \le P \le P_{\max} \tag{10.1d}$$

According to (10.1b) and (10.1c), the relationship between Q and P is rewritten as

$$Q = s \cdot P + l \tag{10.1e}$$

where s is k_q/k_p and l is l_q-$k_q l_p/k_p$.

When there is no external control signal, the inverter TCL works at T_{set0} with a P_{baseline} power consumption where

$$P_{baseline} = (T_{out} - T_{set0})/(sR) - l/s \tag{10.2}$$

Inverter TCLs are mainly controlled by temperature set-point adjustment and compressor frequency control. Their response speed and duration depend on the two control methods [22] which can be applied in multi time-scale scheduling.

10.3.2 Hour-Time Scale Control Model

When controlled by temperature set-point, the adjusted TCL power can be maintained for longer periods but with a slow response rate [22]. Therefore, inverter TCLs with temperature adjustment are applied to hour-time scale scheduling. When T_{in} varies within a certain range (i.e. $[T_{\min}, T_{\max}]$, $T_{\min} = T_{\text{set0}} - \Delta T$, $T_{\max} = T_{\text{set0}} + \Delta T$), customers cannot feel obviously uncomfortable. Thus, the corresponding TCL operational behavior is characterized by a virtual generator [22], as

$$P_g^{\min} \le P_g(k) \le P_g^{\max} \tag{10.3a}$$

$$P_g^{\max} = \Delta T/(sR) \tag{10.3b}$$

$$P_g^{\min} = -\Delta T/(sR) \tag{10.3c}$$

Here, $P_g(k)$ is defined as the difference between the actual TCL power consumption and P_{baseline}. Thus, the virtual generator is generating power (i.e., $P_g(k) < 0$) when the actual power is smaller than P_{baseline}. Similarly, the virtual generator is consuming power (i.e., $P_g(k) > 0$) when the actual power is larger than P_{baseline}.

10.3.3 Minute-Time Scale Control Model

Inverter TCLs can respond rapidly by compressor frequency control. However, they cannot provide a considerable power adjustment capability in hour and even longer time-scale scheduling because of temperature set-point limits [17]. Thus, inverter TCLs with frequency control are modeled as virtual batteries [17, 23] in minute-time scale DR programs, as

$$P_b^{\min} \le P_b(k) \le P_b^{\max} \tag{10.4a}$$

$$SOC(k+1) = \alpha SOC(k) + \beta P_b(k) + \gamma \tag{10.4b}$$

$$SOC^{\min} \le SOC(k+1) \le SOC^{\max} \tag{10.4c}$$

Similar to the definition of $P_g(k)$, $P_b(k)$ is the difference between the actual power consumption of TCLs and the corresponding power consumption when the internal temperature is kept at the set-point determined in the hour-time scale. Thus, loads can be divided into virtual generator/battery and undispatchable loads. There will be an extra virtual generator and virtual battery to be regulated based on the original loads when inverter TCLs are scheduled by MO.

10.3.4 Coordination of Hour and Minute-Time Scale Control of Inverter TCLs

Wind generation forecast accuracy will increase as the forecast time scale becomes small [11]. Thus, the hour- and minute-time scale scheduling of inverter TCLs is employed to handle large and small tie flow errors (Sect. 10.5), respectively. Temperature set-points are re-set in the hour-time scale, and frequency control accordingly regulates inverter TCLs in a minute-time scale. It is noted that the compressor frequency is kept fixed at a given temperature set point [21] and the transient process of temperature adjustment is not considered in this chapter. When the TCL temperature set-point is determined in the hour-time scale scheduling, the TCL frequency can be regulated around a fixed value for power balancing in the minute-time scale scheduling. However, there will also be temperature deviations when TCLs are manipulated by frequency control in minute-time scale scheduling [17]. Considering that TCLs provide different services for microgrids with separate payments in terms of their control variables in hour and minute-time scale scheduling, internal temperatures of inverter TCLs are not allowed to be drastically changed for frequency control. Accordingly, the minute-time scale scheduling will have a minute impact on $T*$ in.

Therefore, TCL internal temperatures in minute-time scale are required to stay within a small range around T_{in}^*. In terms of the virtual battery model described in [17], the virtual battery parameters, $P_b^{\min}$, $P_b^{\max}$, β, and γ in (10.4a) and (10.4b) depend on T_{in}^*, $T_{\min}$ and $T_{\max}$, as

$$P_b^{\min} = P_{\min} - (T_{out} - T_{in}^*)/(sR) + l/s \tag{10.5a}$$

$$P_b^{\max} = P_{\max} - (T_{out} - T_{in}^*)/(sR) + l/s \tag{10.5b}$$

$$\beta = [1 - \exp(\Delta t_m/RC_a)] \cdot s \cdot R/(T_{\max} - T_{\min}) \tag{10.5c}$$

$$\gamma = [1 - \exp(-\Delta t_m/RC_a)] \cdot (T_{\max} - T_{in}^*)/(T_{\max} - T_{\min}) \tag{10.5d}$$

Moreover, the virtual battery's SOC is determined by the TCL internal temperature [17]. Considering that TCL internal temperatures in minute-time scale stay within a small range around T_{in}^*, the virtual battery's SOC will change within $[SOC_0 + \Delta SOC^{\min}, SOC_0 + \Delta SOC^{\max}]$. SOC_0 is determined by T_{in}^*, as stated below:

$$SOC_0 = (T_{\max} - T_{in}^*)/(T_{\max} - T_{\min}) \tag{10.6a}$$

$\Delta SOC^{\min}$ and $\Delta SOC^{\max}$ are defined as

$$\Delta SOC^{\max} = (T_{in}^{\max} - T_{in}^*)/(T_{\max} - T_{\min}) \tag{10.6b}$$

$$\Delta SOC^{\min} = (T_{in}^{\min} - T_{in}^*)/(T_{\max} - T_{\min}) \tag{10.6c}$$

where $T_{\text{in}}^{\max}$ and $T_{\text{in}}^{\min}$ are the permissible maximum and minimum internal temperatures of inverter TCLs in minute-time scale scheduling, which are close to T_{in}^* (shown in a black dashed line within $T_{\min}$ and $T_{\max}$ in Fig. 10.1).

Thus, the SOC limits of virtual battery are expressed as

$$SOC^{\min} = SOC_0 + \Delta SOC^{\min} \tag{10.6d}$$

$$SOC^{\max} = SOC_0 + \Delta SOC^{\max} \tag{10.6e}$$

To sum up, T_{in}^*, $T_{\max}$ and $T_{\min}$ are provided by the hour-time scale scheduling. Thus, the virtual battery in minute-time scale is set by the control of the virtual generator in hour-time scale.

10.4 Transactive Control of TCLs

10.4.1 Hour-Time Scale Response Curve

The predicted mean vote is an index that expresses the human thermal comfort degree and is mainly determined by indoor air temperature and vapor pressure in the air [24]. It shows that there are different uncomfortable levels for individuals when T_{in}^* deviates from the preferred temperature set T_{set0} (vapor pressure in air is assumed to keep constant in this section). The uncomfortable level will be higher when there is a larger difference between T_{in}^* and T_{set0}. Accordingly, MO organizes a local market to purchase the flexibility and the virtual generators submit their response curves to MO for the hour-time scale scheduling according to end users' preferences.

For simplicity, a linear response curve is employed here to show the TCL response to external price signals. The response curve consists of power limits and the sensitivity of regulated power to adjustment prices, as illustrated in Fig. 10.2a. For higher adjustment prices, end users might be willing to sacrifice their comfort level within the power flexibility limit. The virtual generator's response in hour-time scale is formulated as

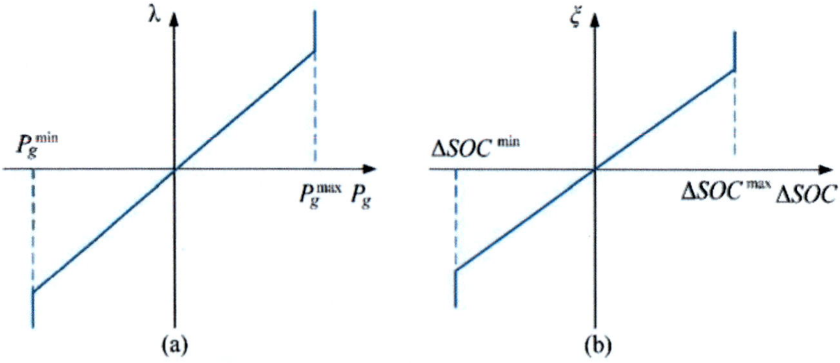

Fig. 10.2 The response curve of **a** virtual generators, **b** and batteries in local markets

$$P_g = \begin{cases} \mu\lambda,\ P_g^{\min}/\mu \le \lambda \le P_g^{\max}/\mu \\ P_g^{\max},\ \lambda > P_g^{\max}/\mu \\ P_g^{\min},\ \lambda < P_g^{\min}/\mu \end{cases} \tag{10.7}$$

Each virtual generator decides its internal temperature according to its response curve and the adjustment price, as modeled below

$$T_{in}^* = \begin{cases} T_{set0} - \mu\lambda s R,\ P_g^{\min}/\mu \le \lambda \le P_g^{\max}/\mu \\ T_{\min},\ \lambda > P_g^{\max}/\mu \\ T_{\max},\ \lambda < P_g^{\min}/\mu \end{cases} \tag{10.8}$$

The hour-time scale scheduling is made after MO collects participating virtual generators' response curves (Sect. 10.5.2). The adjustment price will be determined and distributed to all virtual generators. Virtual generators receive the adjustment fee for offering flexibility to the microgrid, as formulated below

$$R_h = P_g \cdot \lambda \tag{10.9}$$

10.4.2 Minute-Time Scale Response Curve

To ensure that the minute-time scale scheduling has a minute impact on T_{in}^*, ΔSOC^{m} is assumed small enough so that TCL internal temperature can be seen as a fixed value in the minute-time scale scheduling. Here, ΔSOC^{m} is determined by MO and kept the same for all virtual batteries.

The Inverter TCLs also submit their response curves to MO when participating in minute-time scale scheduling. To reduce the number of submitted parameters, the response curve of ΔSOC shown in Fig. 10.2b is submitted, in which, only SOC_0 instead of $\Delta SOC^{\min}$ and $\Delta SOC^{\max}$ is provided by each virtual battery. The MO gets access to $\Delta SOC^{\min}$ and $\Delta SOC^{\max}$ of virtual batteries, in terms of SOC_0 and ΔSOC_0, as given below

$$\Delta SOC^{\max} = \min\{\Delta SOC^m, 1 - SOC_0\} \tag{10.10a}$$

$$\Delta SOC^{\min} = -\min\{\Delta SOC^m, SOC_0\} \tag{10.10b}$$

On the other hand, compared to the P_b constraint in (10.4a), the SOC constraint in (10.4c) plays a more significant role on $P_b(k)$ because ΔSOC^{m} is very small. In most cases, (10.11a) and (10.11b) are satisfied, which means that the virtual battery's operating behavior is described by (10.4b) and (10.4c), and the P_b constraint in (10.4a) has no impacts on virtual battery. Thus, the ΔSOC response curve can mitigate some of the virtual battery's constraints, enhance the calculation efficiency in minute-time scale schedule, and relieve communication burdens.

$$[SOC_0 + \Delta SOC^{\max} - \alpha(SOC_0 + \Delta SOC^{\min}) - \gamma]/\beta \leq P_b^{\max} \tag{10.11a}$$

$$[SOC_0 + \Delta SOC^{\min} - \alpha(SOC_0 + \Delta SOC^{\max}) - \gamma]/\beta \geq P_b^{\min} \tag{10.11b}$$

ΔSOC represents the virtual battery's regulation degree, which an end-user is willing to take for the adjustment fee. In Fig. 10.2b, the virtual battery's response behavior is formulated as

$$\Delta SOC = \begin{cases} \upsilon\xi, \Delta SOC^{\min}/\upsilon \leq \xi \leq \Delta SOC^{\max}/\upsilon \\ \Delta SOC^{\max}, \xi > \Delta SOC^{\max}/\upsilon \\ \Delta SOC^{\min}, \xi < \Delta SOC^{\min}/\upsilon \end{cases} \tag{10.12}$$

The adjustment price in the minute-time scale is calculated and distributed to virtual batteries after MO collects all response curves of virtual batteries. Each virtual battery will control its compressor frequency [17] to adjust its power consumption in terms of the price and response curve, stated as

$$\varphi = \begin{cases} [SOC_0 + \upsilon\xi - \alpha SOC(k-1) - \gamma]/(k_p \cdot \beta) - l_p/k_p, \\ \Delta SOC^{\min}/\upsilon \leq \xi \leq \Delta SOC^{\max}/\upsilon \\ [SOC_0 + \Delta SOC^{\max} - \alpha SOC(k-1) - \gamma]/(k_p \cdot \beta) \\ - l_p/k_p, \xi > \Delta SOC^{\max}/\upsilon \\ [SOC_0 + \Delta SOC^{\min} - \alpha SOC(k-1) - \gamma]/(k_p \cdot \beta) \\ - l_p/k_p, \xi < \Delta SOC^{\min}/\upsilon \end{cases} \tag{10.13}$$

The corresponding adjustment fee for a virtual battery is given as

$$R_m = \varepsilon \cdot \Delta SOC \cdot \xi \tag{10.14}$$

In Fig. 10.2, the sign of λ and ξ will identify the status of the virtual generator and battery power. When virtual generator power is less than 0 and the virtual battery is discharging, λ and ξ are negative; otherwise, λ and ξ are positive.

For maintaining the customer privacy, inverter TCL parameters and preferences (e.g., R, C_a, $T_{\min}$, $T_{\max}$) are concealed from MO according to the modeling of virtual generator and battery. In this regard, virtual generator parameters ($P_g^{\min}$ and $P_g^{\max}$) and only a fraction of virtual battery parameters (SOC_0, α, β, and γ) are submitted to MO when participating in minute-time scale scheduling. The limited data exchange is also helpful to relieve communication burdens. Accordingly, customer freewill and preferences are fully respected. In the proposed control methods, the adjustment price is determined by MO, which is applied to inverter TCLs according to response curves. It will not only provide a reasonable adjustment fee for inverter TCLs according to their response behaviors but also improve the manipulation efficiency of the MO on TCLs.

10.5 Problem Formulation of Microgrid Scheduling

10.5.1 Hour-Time Scale Stochastic Control Strategy

Considering the wind generation uncertainty in the hour-time scale scheduling, the objective of the local market is to minimize MO' expected economic loss in the following hour:

$$\min \sum_{\omega \in \Psi} \pi_\omega [|P_{g,t} + P_{load,t} - P_{w,t,\omega} - P_{da,t}^{sch}| \cdot \rho \ + P_{g,t} \cdot \lambda_t] \cdot \Delta t_h \quad (10.15a)$$

Subject to

$$P_{g,t} = \sum_{i \in N} P_{g,i,t}, \forall i \in N, \forall t \in T \quad (10.15b)$$

$$(10.7) is\ satisfied, \forall i \in N, \forall t \in T \quad (10.15c)$$

The first term in (10.15a) is the MO's economic loss due to the difference between the actual and the day-ahead tie flow schedule. The second term in (10.15a) is the virtual generator's economic adjustment fee paid by MO. Constraint (10.15b) is the virtual generator's limit. Constraint (10.15c) is the virtual generator's response curve limit.

10.5.2 Minute-Time Scale Control Strategy

The minute-time scale market's objective is to minimize MO' sum of economic loss in the current time slot and the expected economic loss in the next time slots:

$$\begin{aligned} \min\Big\{ & |P_{b,t,v} + P_{g,t}^{sch} + P_{load,t,v}^{a} - P_{w,t,v}^{a} - P_{da,t}^{sch}| \cdot \rho \\ & + \varepsilon \cdot \Delta SOC_{t,v} \cdot \xi_{t,v} + \sum_{\omega \in \Pi} \pi_\omega \Bigg[\sum_{\kappa \in \mathrm{E}} \Big(|P_{b,t,\kappa}^{f} + P_{g,t}^{sch} + P_{load,t,\kappa}^{f} \\ & - P_{w,t,\kappa,\omega}^{f} - P_{da,t}^{sch}| \cdot \rho + \varepsilon \cdot \Delta SOC_{t,\kappa}^{f} \cdot \xi_{t,\kappa}^{f} \Big) \Bigg] \Big\} \cdot \Delta t_m \end{aligned} \quad (10.16a)$$

Subject to:

$$P_{b,t,v} = \sum_{i \in N} P_{b,i,t,v}, \forall i \in N, \forall t \in T, \forall v \in V \tag{10.16b}$$

$$P_{b,i,t}^{\min} \le P_{b,i,t,v} \le P_{b,i,t}^{\max}, \forall i \in N, \forall t \in T, \forall v \in V \tag{10.16c}$$

$$\begin{aligned} SOC_{i.t,v+1} &= \alpha_{i,t} SOC_{i,t,v} + \beta_{i,t} P_{b,i,t,v} + \gamma_{i,t} \\ &\forall i \in N, \forall t \in T, \forall v \in V \end{aligned} \tag{10.16d}$$

$$(10.12) is\ satisfied, \forall i \in N, \forall t \in T, \forall v \in V \tag{10.16e}$$

$$\Delta SOC_{t,v} = \sum_{i \in N} \Delta SOC_{i,t,v}, \forall i \in N, \forall t \in T, \forall v \in V \tag{10.16f}$$

$$\begin{aligned} &(10.16b)-(10.16f) \text{ are also satisfied for virtual batteries} \\ &\text{in the prediction-horizon time slots} \end{aligned} \tag{10.16g}$$

The first two terms are the MO's economic loss in the current time slot. The next two terms are the of MO's economic loss in the prediction-horizon time slots. Constraint (10.16b) is the virtual battery's total charging/discharging power. Constraints (10.16c) and (10.16d) are virtual battery limits. Constraint (10.16e) is the virtual battery's response curve, and (10.16f) limits the virtual battery's total SOC. Constraints (10.16b)–(10.16f) provide limits in the current time slot. Constraint (10.16g) provides limits in the predicted horizon's time slots.

10.5.3 Linearization

Both optimization models in hour- and minute-time scales are nonlinear and non-convex. Two linearization techniques are employed to covert the models into mixed-integer linear programming (MILP) problems.

First, the hour-time scale optimization model is re-formulated as:

$$\min \sum_{\omega \in \Psi} \pi_\omega [Y_{t,\omega} \cdot \rho + Z_t] \cdot \Delta t_h \tag{10.17a}$$

Subject to:

$$Y_{t,\omega} \ge P_{g,t} + P_{load,t} - P_{w,t,\omega} - P_{da,t}^{sch}, \forall t \in T, \forall \omega \in \psi \tag{10.17b}$$

$$Y_{t,\omega} \ge -(P_{g,t} + P_{load,t} - P_{w,t,\omega} - P_{da,t}^{sch}), \forall t \in T, \forall \omega \in \psi \tag{10.17c}$$

$$a_{g,i,t,ra} \in \{0, 1\}, \forall i \in N, \forall t \in T, \forall ra \in R_1 \tag{10.17d}$$

$$\sum_{ra \in R_1} a_{g,i,t,ra} = 1, \forall i \in N, \forall t \in T \tag{10.17e}$$

$$m_{g,i,t,rm} \geq 0, \forall i \in N, \forall t \in T, \forall rm \in R_2 \tag{10.17f}$$

$$\sum_{rm \in R2} m_{g,i,t,rm} = 1, \forall i \in N, \forall t \in T \tag{10.17g}$$

$$m_{g,i,t,1} 1 \leq a_{g,i,t,1}, \forall i \in N, \forall t \in T \tag{10.17h}$$

$$m_{g,i,t,rm} \leq a_{g,i,t,rm-1} + a_{g,i,t,rm}, \forall t \in T, \forall i \in N, 2 \leq rm \leq r_0 \tag{10.17i}$$

$$m_{g,i,t,r_0+1} \leq a_{g,i,t,r_0}, \forall i \in N, \forall t \in T \tag{10.17j}$$

$$\lambda_t = \sum_{rm \in R_2} m_{g,i,t,rm} \cdot \lambda_{0,i,t,rm}, \forall i \in N, \forall t \in T \tag{10.17k}$$

$$P_{g,i,t} = \sum_{rm \in R_2} m_{g,i,t,rm} \cdot P_{g0,i,t,rm}, \forall i \in N, \forall t \in T \tag{10.17l}$$

$$P_{g,t} = \sum_{i \in N} P_{g,i,t}, \forall t \in T \tag{10.17m}$$

$$Z_{i,t} = \sum_{rm \in R_2} m_{g,i,t,rm} \cdot Z_{0,i,t,rm}, \forall i \in N, \forall t \in T \tag{10.17n}$$

$$Z_t = \sum_{i \in N} Z_{i,t}, \forall t \in T \tag{10.17o}$$

There are two barriers to solve the optimization model in (10.15a)–(10.15c). The first one is that the tie flow errors compared to the day-ahead plan is described by $|P_{g,t} + P_{load,t} - P_{w,t,\omega} - P_{da,t}^{sch}|$ in (10.15a). Thus, $Y_{t,\omega}$ constrained by (10.17b) and (10.17c) is introduced to substitute for $|P_{g,t} + P_{load,t} - P_{w,t,\omega} - P_{da,t}^{sch}|$. The second barrier is that the virtual generator's adjustment fee described by $P_{g,t} \cdot \lambda_t$ in (10.15a) is non-convex and nonlinear. Then the piecewise linear technique is utilized to address this issue. $\lambda_{i,t}$ is divided into several segments. The relationship between auxiliary variables $Z_{i,t}$ representing the virtual generator's adjustment fee and $\lambda_{i,t}$ is considered linear in each segment. $\lambda_{i,t}$ in segment *ra* is calculated by $\lambda_{0,i,t,ra} \cdot \mathrm{m}_{g,i,t,ra} + \lambda_{0,i,t,ra+1} \cdot \mathrm{m}_{g,i,t,ra+1}$, where $\mathrm{m}_{g,i,t,ra} + \mathrm{m}_{g,i,t,ra+1} = 1$ as constrained by (10.17d)–(10.17k). Moreover, (10.17d)–(10.17j) guarantee that $\lambda_{i,t}$ can only be located in one of these segments. And $\lambda_{i,t}$ is the same for all virtual generators in a time slot, which is constrained by (10.17k). Constraints (10.17l)–(10.17o) are used to obtain $P_{g,t}$ and Z_t based on (10.17l) and (10.17n), which is similar to that of $\lambda_{i,t}$. Constraints (10.17m) and (10.17o) provide limits on total power and virtual generator's adjustment fee. As a result, the optimization model in (10.17a)–(10.17o) is a convex MILP problem.

Similarly, the minute-time scale optimization problem is re-formulated as:

$$\min[Y_{t,v} \cdot \rho + G_{t,v} + \sum_{\omega \in \prod} \pi_{\omega} \sum_{\kappa \in \mathrm{E}} (Y_{t,\kappa,\omega} \cdot \rho + G_{t,\kappa})] \cdot \Delta t_m \tag{10.18a}$$

Subject to

$$Y_{t,v} \geq -(P_{b,t,v} + P_{g,t}^{sch} + P_{load,t,v}^{a} - P_{w,t,v}^{a} - P_{da,t}^{sch}), \forall t \in T, \forall v \in V \tag{10.18b}$$

$$Y_{t,v} \geq P_{b,t,v} + P_{g,t}^{sch} + P_{load,t,v}^{a} - P_{w,t,v}^{a} - P_{da,t}^{sch}, \forall t \in T, \forall v \in V \tag{10.18c}$$

$$a_{b,i,t,v,ha} \in \{0, 1\}, \forall i \in N, \forall t \in T, \forall v \in V, \forall ha \in \mathrm{H}_1 \tag{10.18d}$$

$$\sum_{ha \in \mathrm{H}_1} a_{b,i,t,v,ha} = 1, \forall i \in N, \forall t \in T, \forall v \in V \tag{10.18e}$$

$$m_{b,i,t,v,hm} \geq 0, \forall i \in N, \forall t \in T, \forall v \in V, \forall hm \in \mathrm{H}_2 \tag{10.18f}$$

$$\sum_{hm \in H_2} m_{g,i,t,hm} = 1, \forall i \in N, \forall t \in T \tag{10.18g}$$

$$m_{b,i,t,v,1} \leq a_{b,i,t,v,1}, \forall i \in N, \forall t \in T, \forall v \in V \tag{10.18h}$$

$$\begin{aligned} &m_{b,i,t,v,hm} \leq a_{b,i,t,v,hm-1} + a_{b,i,t,v,hm} \\ &\forall i \in N, \forall t \in T, \forall v \in V, 2 \leq hm \leq h_0 \end{aligned} \tag{10.18i}$$

$$m_{g,i,t,h_0+1} \leq a_{g,i,t,h_0}, \forall i \in N, \forall t \in T \tag{10.18j}$$

$$\xi_{t,v} = \sum_{hm \in H_2} m_{b,i,t,v,hm} \cdot \xi_{0,i,t,v,hm}, \forall i \in N, \forall t \in T, \forall v \in V \tag{10.18k}$$

$$P_{b,t,v} = \sum_{i \in N} P_{b,i,t,v}, \forall t \in T, \forall v \in V \tag{10.18l}$$

$$P_{b,i,t}^{\min} \leq P_{b,i,t,v} \leq P_{b,i,t}^{\max}, \forall i \in N, \forall t \in T, \forall v \in V \tag{10.18m}$$

$$SOC_{i.t,v+1} = \alpha_{i,t} SOC_{i,t,v} + \beta_{i,t} P_{b,i,t,v}(k) + \gamma_{i,t}$$

$$\forall i \in N, \forall t \in T, \forall v \in V \tag{10.18n}$$

$$\Delta SOC_{t,v} = \sum_{hm \in H_2} m_{b,i,t,v,hm} \cdot \Delta SOC_{0,i,t,v,hm} \quad \forall i \in N, \forall t \in T, \forall v \in V \tag{10.18o}$$

$$SOC_{i,t,v+1} = SOC_{i,t,v} + \Delta SOC_{i,t,v}, \forall i \in N, \forall t \in T, \forall v \in V \tag{10.18p}$$

$$G_{i,t,v} = \sum_{hm \in H_2} m_{b,i,t,v,hm} \cdot G_{0,i,t,v,hm}, \forall i \in N, \forall t \in T, \forall v \in V \tag{10.18q}$$

$$G_{t,v} = \sum_{i \in N} G_{i,t,v}, \forall t \in T, \forall v \in V \tag{10.18r}$$

$$\text{(10.18b)} - \text{(10.18r) are also satisfied for virtual batteries in the prediction-horizon time slots} \tag{10.18s}$$

Constraints (10.18b)–(10.18c) make Y equivalent to the absolute value of tie flow errors. The piecewise linear technique is applied to obtain (10.18d)–(10.18j). Adjustment price and operating behaviors of virtual batteries are constrained by (10.18k)–(10.18p). Constraints (10.18p)–(10.18q) limit the virtual battery's total adjustment fee. Constraints (10.18b)–(10.18r) provide limits in the current time slot, and (10.18s) is for the predicted horizon's time slots. The minute-time scale optimization model (10.18a)–(10.18s) is successfully transformed to a convex MILP problem.

10.6 Case Studies

10.6.1 Parameter Setup

The number of inverter TCLs is 200 with parameters provided in [17]. μ is assumed to be randomly distributed within [0.4, 0.8], and υ is randomly distributed within [0.01, 0.1]. Outdoor temperature is for a typical summer profile in the PJM region [17]. It is clear that changing outdoor temperature has no impact on the operating behaviors of TCLs in hour-time scale in terms of the virtual generator model described in Sect. 10.3.2. The relationship between the operating behaviors of virtual batteries and changing outdoor temperature is provided in [17]. ρ is 1 and ε is 4. The profile of wind generation and load are provided in [11, 25], with scaled magnitudes to match the test case. The wind and load forecasts follow the method in [26], considering 10 scenarios with a uniform probability of 0.1. When TCLs are controlled by temperature set-point changes, ΔT is randomly distributed within [0.1 °C, 2 °C]; while, ΔSOC^{m} is set as 0.05 when regulated by frequency control. Matlab R2018b is utilized to simulate the microgrid scheduling and operating behaviors of inverter TCLs. The CPU model of the desktop is Intel(R) Core (TM) i7-7700 and the frequency is 3.60GHz.

10.6.2 Hour-Time Scale Control Results

(1) *Tie flow scheduling*: Actual errors between load and wind generation (i.e. $P_{load,t} - P_{w,t}$), and day-ahead plan (i.e. $P_{da,t}^{sch}$) are given in Fig. 10.3. The MO's hour-time scale scheduling results are shown in Fig. 10.4. The average simulation time for each time step is 6.97 seconds. Due to wind generation variability, there are considerable tie flow errors (i.e., $P_{load,t} - P_{w,t} - P_{da,t}^{sch}$). When the actual wind generation is less than its forecast (i.e., $P_{load,t} - P_{w,t} - P_{da,t}^{sch} > 0$), the virtual generator will operate at the power generation state (i.e., $P_g < 0$). Otherwise, virtual generator will operate at the power consumption state (i.e., $P_g > 0$) to consume the surplus wind generation

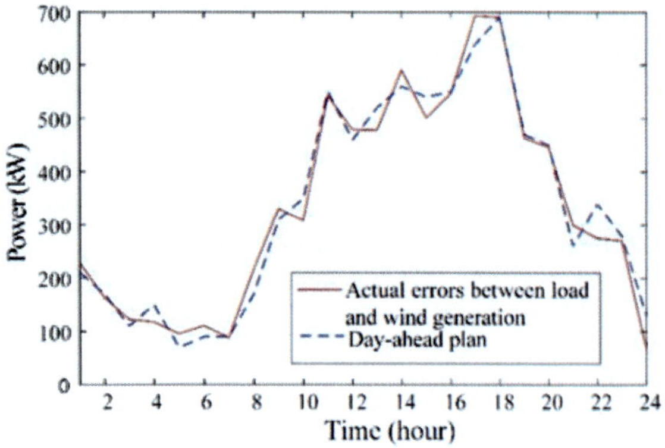

Fig. 10.3 Diagrams of actual errors between load and wind generation, and day-ahead plan

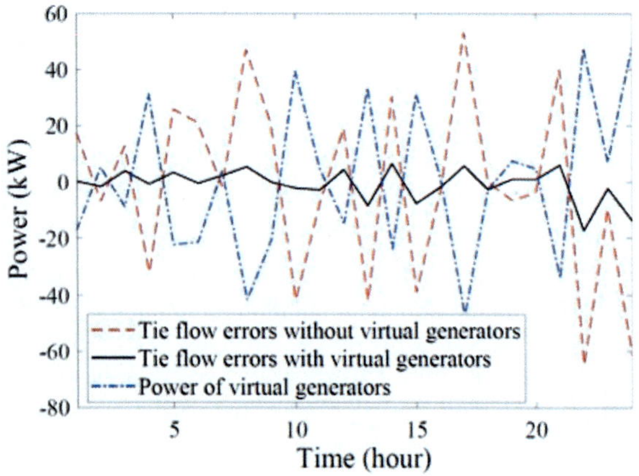

Fig. 10.4 Hour-time scale scheduling results

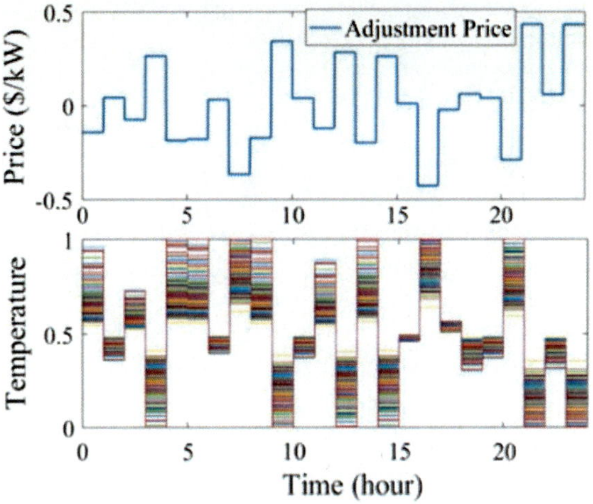

Fig. 10.5 Hour–time scale adjustment prices and response behaviors of TCLs

(i.e., $P_{load,t} - P_{w,t} - P^{sch}_{da,t} < 0$). Thus, virtual generator power is always presented as the opposite to tie flow errors in Fig. 10.4. Moreover, the economic loss of MO is $608.0 when there are no virtual generators. Tie flow errors are restrained and the economic loss is reduced by 58.7% to $251.4 when TCLs are scheduled as virtual generators.

(2) Response behaviors of inverter TCLs: The virtual generator's adjustment price (λ) and the corresponding TCL internal temperatures are shown in Fig. 10.5. Different colors in Fig. 10.5 represent the participating TCLs. The TCL temperature limits are heterogeneous which are normalized in terms of customer preferences. Since the transient process of temperature adjustment is not considered in this chapter, each TCL's hourly internal temperature is assumed to remain fixed in Fig. 10.5. All normalized internal temperatures are within [0, 1], which means that end users' preferences are respected. It is also shown that internal temperatures of inverter TCLs are changed in the opposite direction of λ. Specifically, when λ is larger than 0, the power consumption is increased by allowing TCL internal temperatures to be lower than the initial temperature set-point of 0.5, in which case the virtual generators absorb power. Otherwise, a negative λ will reduce the power consumption by allowing TCL internal temperatures to be higher than 0.5, in which case the virtual generators produce power. Moreover, although TCLs operate similarly in each time slot, their internal temperatures vary according to the submitted response curves, which result in different sensitivities to λ.

(3) Analysis of the marginal cost (MC) of virtual generators: The marginal costs of virtual generators as a function of λ are shown in Fig. 10.6. In Fig. 10.2a, ΔR_{h} is positive if $|\lambda|$ increases (in both first and third quadrants). Since MC is the ratio of ΔR_{h} to ΔP_g, ΔP_g and MC are positive in the first quadrant and negative in the third quadrant as $|\lambda|$ increases. The left-hand side of Fig. 10.6 shows the virtual generator's MC when λ varies between \$−1.1/kW and \$1.1/kW. The virtual generator's MC adjusts its direction according to λ. For example, MC changes slightly when λ is

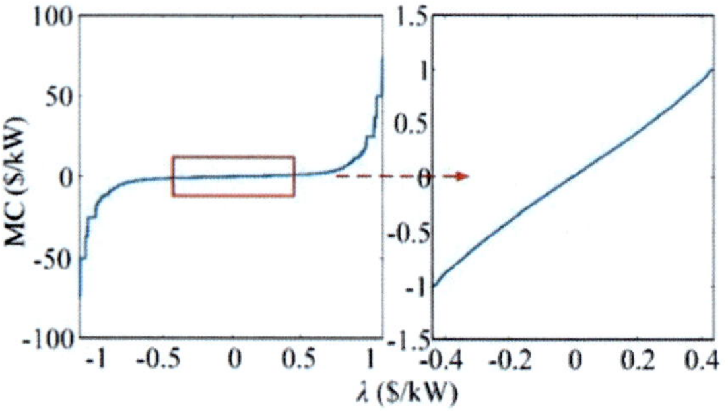

Fig. 10.6 Marginal cost of virtual generators for different adjustment prices

within [$−0.8/kW, $0.8/kW]. This is because MC has a linear relationship with λ, as shown in Fig. 10.2a when the virtual generator's power output is within its limits. On the other hand, several virtual generators reach their lower and upper power limits if λ continues to decrease or increase within [$−1.1/kW, $−0.8/kW] and [$0.8/kW, $1.1/kW], respectively, where MC changes significantly.

(4) Sensitivity analysis of penalty factor: When the absolute value of virtual generators' MC is less than ρ (i.e. $1/kW), λ changes between $−0.45/kW and $0.45/kW, as shown on the right-hand-side of Fig. 10.6. This means that virtual generators representing cheaper resources will be scheduled to reduce the economic penalty caused by wind generation variability. Otherwise, if λ is out of the range of [$−0.45/kW, $0.45/kW], virtual generators will not be scheduled due to their higher scheduling costs. Thus, λ should vary within this range, which is consistent with the upper figure in Fig. 10.5. The tie flow errors are shown in Fig. 10.7 for different values of ρ. Here, the tie flow errors will diminish as ρ increases. This is because λ can be increased and more power from virtual generators will be scheduled when ρ is larger. Thus, a larger ρ can help smooth out the tie flow errors caused by wind generation uncertainty in hour-time scale scheduling.

(5) Impact of linearization on the model accuracy: Two linearization methods are employed to solve the hour-time scale optimization model in (10.15a)–(10.15c).

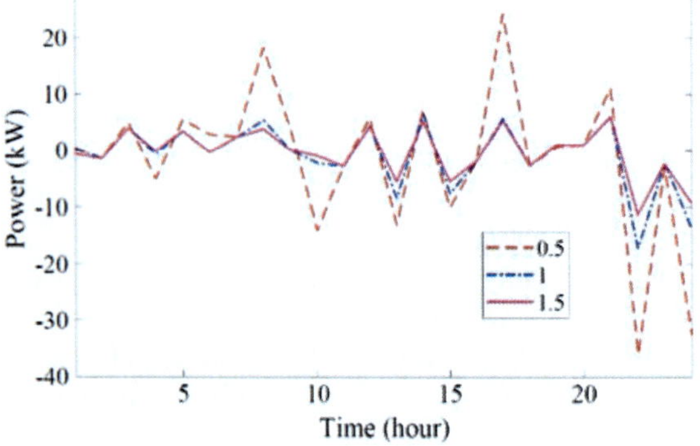

Fig. 10.7 Tie flow errors with different ρ

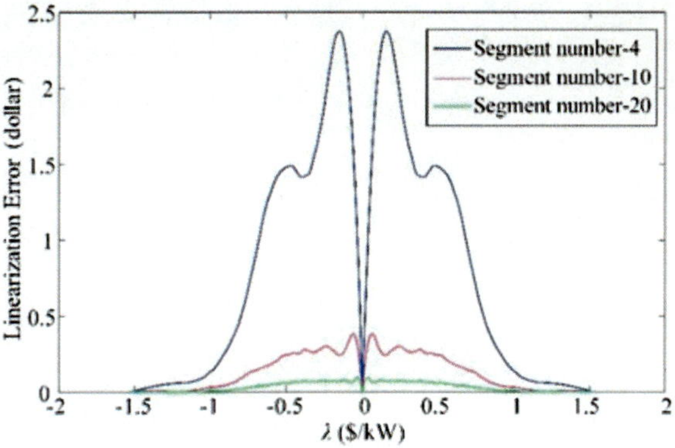

Fig. 10.8 Linearization errors for different segment numbers of λ

The first one is that $Y_{t,\omega}$ constrained in (10.17b) and (10.17c) are introduced to substitute $|P_{g,t} + P_{load,t} - P_{w,t,\omega} - P^{sch}_{da,t}|$, which has no impact on the model accuracy. The second one is the piecewise linear technique that converts the non-convex and nonlinear components of the virtual generator adjustment fee, stated as $P_{g,t}{\cdot}\lambda_t$ in (10.15a), where λ is divided into several segments. The linearization errors are presented in Fig. 10.8 when segment numbers of λ are designated as 4, 10 and 20, respectively. Here, the linearization error will increase as the segment number decreases. When the segment number is relatively large, the linearization error is small and the piecewise linear technique has a minute effect on the original model accuracy. Thus, the linearized model will have a high accuracy as long as the segment number of λ (e.g., 20 in this section) is properly selected.

10.6.3 Minute-Time Scale Control Results

(1) Tie flow scheduling: The comparison of the tie flow errors without and with virtual batteries in minute-time scale scheduling is given in the upper of Fig. 10.9. The corresponding economic loss of MO is \$251.5 and \$70.2, respectively, which shows that virtual batteries can smooth out tie flow errors using TCL flexibility. All virtual battery SOCs are normalized in the lower Fig. 10.9 which demonstrates that all SOCs work within [0, 1] and customer preferences are preserved. Different colors in the lower Fig. 10.9 represent participating virtual batteries.

(2) Response of inverter TCLs: The tie flow errors ($P^{\mathrm{sch}}_{\mathrm{g,t}} + P^{a}_{load,t} - P^{a}_{w,t} - P^{sch}_{da,t}$) in the 13th hour is given in Fig. 10.10. The adjustment price (ξ) and virtual battery charging/discharging power in the 13th hour are shown in Fig. 10.11, with the corresponding SOC profiles presented in Fig. 10.12. In Fig. 10.11, the charging/discharging power of virtual batteries is closely related to ξ. For example, at second and fourth-time slots, ξ increases as compared to the previous time slot and the virtual battery's charging/discharging power is positive. This is because the relationship between ξ and ΔSOC of virtual batteries is mostly linear. When ξ increases,

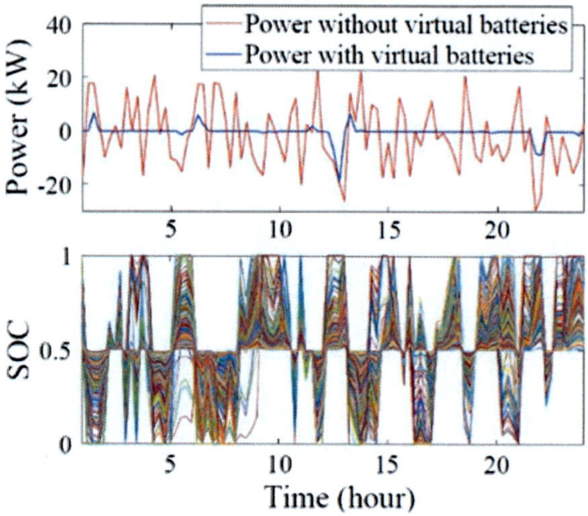

Fig. 10.9 Tie flow errors and virtual battery SOCs

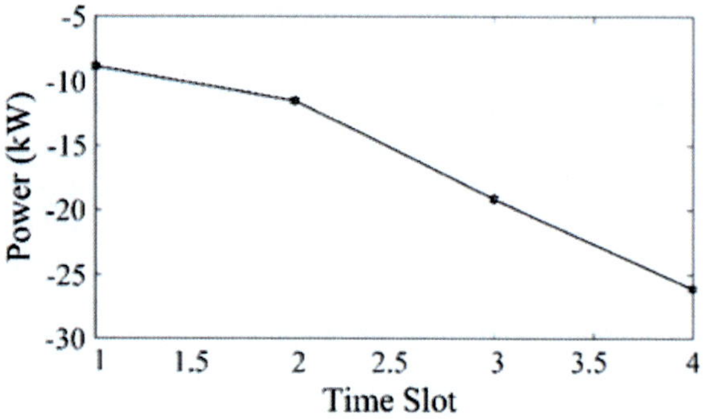

Fig. 10.10 Tie flow errors in the 13th hour

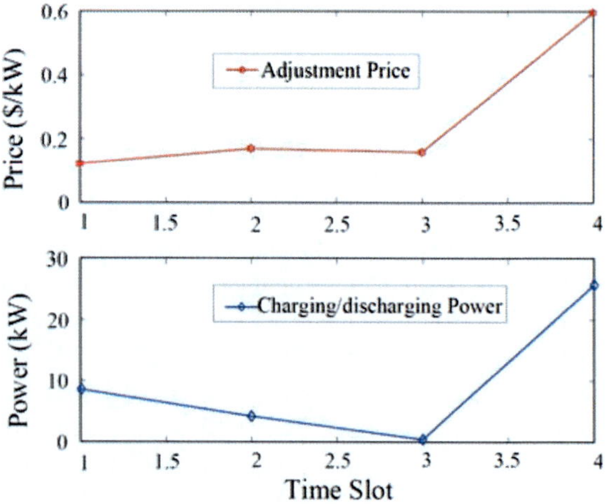

Fig. 10.11 Adjustment price and charging/discharging power of virtual batteries at 13th hour

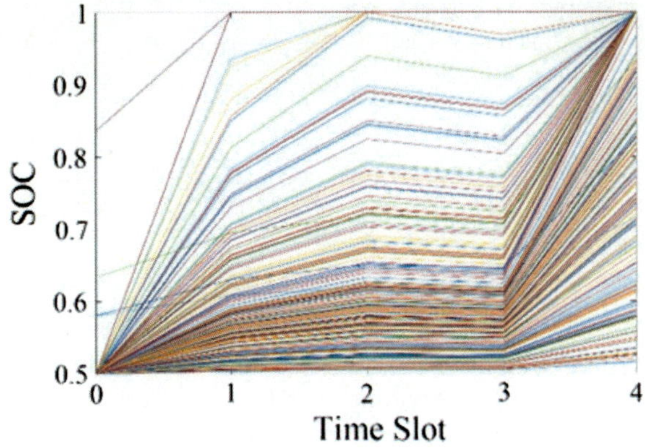

Fig. 10.12 SOC of virtual batteries at 13th hour

the SOC of virtual batteries becomes larger and virtual batteries will start to charge to address tie flow errors. Virtual batteries are always charging to absorb the excessive wind generation in the fourth time slot. These virtual batteries' operating behaviors are consistent with those of conventional batteries. However, virtual batteries continue charging 0.45kW even though ξ is decreased in the third time slot. This is because the charging/discharging coefficient β is within [0, 1], which makes virtual batteries charge even though SOC is reduced [17]. This is the difference between virtual batteries and conventional batteries.

In Fig. 10.12, the initial SOC is 0.5 for the majority of virtual batteries, while others are between 0.5 and 1. This is because most SOCs vary within $[SOC_0 - \Delta SOC^m, SOC_0 + \Delta SOC^m]$ according to the hour-time scale scheduling. However, when SOC is close to 1 and the initial SOC is larger than 0.5, the corresponding SOC varies within $[SOC_0 - \Delta SOC^m, 1]$.

Moreover, when only the day-ahead and hour-time scale scheduling are considered, the tie flow errors in third and fourth-time slots are -19.15 kW and -26.08 kW, respectively. The actual ξ are \$0.16/kW and \$0.6/kW, and the corresponding charging/discharging power of virtual batteries are 0.45 kW and 25.8kW. Considering the virtual battery's MC, ξ is \$0.47/kW and \$0.60/kW. If ξ is \$0.47/kW in the third time slot, virtual batteries will be charged to mitigate the tie flow errors (-19.15 kW). For a larger error (-26.08 kW) in the fourth time slot, ξ will be increased significantly to schedule more power. Due to SOC and MC limits, charging power may be much less than 26.08kW in the fourth time slot even though SOC is large. And virtual battery costs in the third and fourth-time slots will be large considering large SOCs. Therefore, virtual battery's SOCs in the third time slot is kept low and then increased to charge 25.8kW in the fourth time slot to mitigate the tie flow errors economically. This is why the tie flow errors in the 13th hour of Fig. 10.9 are not eliminated.

(3) Computation time analysis: The number of constraints in the minute-time scale optimization model will increase as we add more TCLs. Thus, the computation time is mainly determined by the number of TCLs. The computation time when TCLs are

Table 10.1 TCL number and computation time without the aggregation technique

TCL number	100	200	400
Time (s)	33.0	102.0	426.9

Table 10.2 TCL number and computation time using the aggregation technique

TCL number	1600	3200	6400
Time (s)	18.2	19.7	28.7

100, 200 and 400 is listed in Table 10.1. Here, the computation time increases sharply when the number of TCLs increases from 100 to 400. Obviously, the corresponding computation time will be even longer than the time step when the number of TCLs is large, which could be troublesome in microgrid scheduling. To address this issue, the aggregation technique [12, 13] is applied to integrate a certain number of TCLs as one entity for regulation. We assume 100 TCLs are aggregated as one virtual battery with one response curve provided to MO. The detailed TCL aggregation is not discussed here and will be considered in our future work. The computation times when TCLs are 1600, 3200 and 6400 are given in Table 10.2. The simulation results show that the computation time is reduced sharply when the aggregation technique is applied. Thus, the proposed scheduling scheme is suitable for microgrid minute-time scale scheduling.

10.6.4 Comparison of Single and Multi-time Scale Control of TCLs

Three scenarios are designed to show the impact of multi-time scale coordinated control and scheduling of inverter TCLs on mitigating the microgrid tie flow errors caused by wind generation variability.

Scenario 1 (S1): Inverter TCLs work as virtual generator and only participate in hour-time scale scheduling;

Scenario 2 (S2): Inverter TCLs work as virtual battery and only participate in minute-time scale scheduling;

Scenario 3 (S3): Inverter TCLs work as virtual generators and virtual batteries, and participate in both hour- and minute-time scale scheduling.

The hour-time scale tie flow errors in S1 and S2 without applying virtual generator are shown in the upper part of Fig. 10.13. The minute-time scale tie flow errors in S2 and S3 without employing virtual battery are given in the lower part of Fig. 10.13. Figure 10.14 shows the power of the virtual generator and battery in S1, S2 and S3. The tie flow errors in the three scenarios are compared in Fig. 10.15. Wind generation forecast error is decreased as the forecasting time scale becomes smaller [11]. In Fig. 10.13, the hour-time scale tie flow errors in S1 are larger than minute-time scale errors in S3 because of the larger forecast error in the former case. Meanwhile, minute-time scale tie flow errors in S2 are also larger than those in S3 because tie

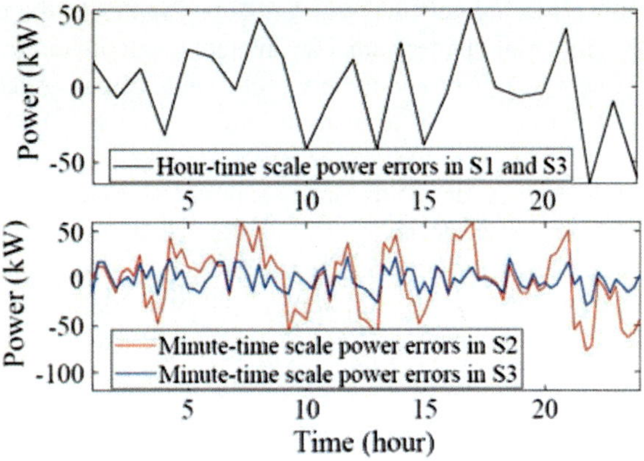

Fig. 10.13 Hour-and minute-time scale tie flow errors in S1, S2 and S3 without using virtual generator and battery to smooth out tie flow errors

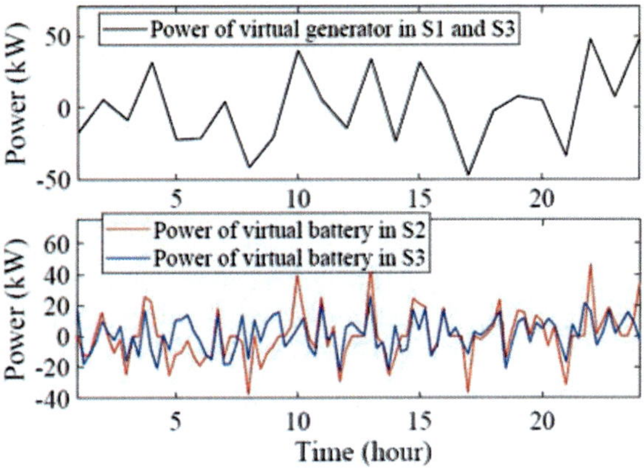

Fig. 10.14 Virtual generator and battery power in S1, S2 and S3

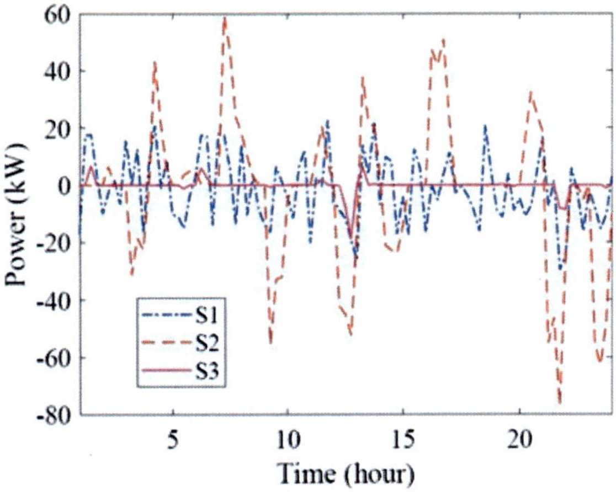

Fig. 10.15 Comparison of single and multi-time scale scheduling

flow errors at hour-time scale are not mitigated by the virtual generator. When virtual generator and battery are coordinated to smooth out tie flow errors, ΔSOC of virtual battery is constrained to be within a small range. Accordingly, the virtual generator will play a larger role than the virtual battery in Fig. 10.14 to smooth out tie flow errors.

Thus, smaller tie flow errors in minute-time scale cannot be mitigated in S1, while larger tie flow errors in hour-time scale cannot be addressed in S2 (see Fig. 10.15). In S3, larger tie flow errors are limited by virtual generators and smaller errors in minute-time scale are smoothed out by virtual battery (see Fig. 10.15). Moreover, MO costs in S1-S3 are $502.9, $1183.7 and $321.6, respectively. The coordination of multi-time scale TCL scheduling can effectively address tie flow errors caused by wind generation variability. Meanwhile, adjustment fees for inverter TCLs in S1-S3 are $74.9, $62.3 and $123.2, respectively. It is demonstrated that multi-time scale coordinated control can utilize DR potentials of inverter TCLs to gain higher end-user revenues.

10.7 Conclusions

This chapter proposes a multi-time scale coordinated control and scheduling of inverter TCLs to mitigate the microgrid tie flow errors caused by wind generation variability. Simulation results demonstrate that (1) the proposed response curve-based control of inverter TCLs can effectively address the tie flow errors in hour- and minute-time scale; (2) inverter TCLs will be scheduled to mitigate the tie flow errors only when marginal costs of TCLs are less than penalty factor; (3) larger penalty factors will be helpful to smooth out the tie flow errors; (4) compared to single-time scale scheduling, the multi-time scale coordinated scheduling of inverter TCLs will address the tie flow errors more effectively; (5) multi-time scale coordinated control will employ DR potentials of inverter TCLs to bring more revenues to end-users. In addition to introducing the wind generation uncertainty, our future work will consider uncertain factors embedded in loads, TCL responses, and communication delays. Moreover, we will enhance the robustness of power system operation when multiple uncertainties are considered.

References

1. Song, M., Sun, W., Shahidehpour, M., Yan, M., & Gao, C. (2020). Multi-time scale coordinated control and scheduling of inverter-based TCLs with variable wind generation. *IEEE Transactions on Sustainable Energy, 12*(1), 46–57.
2. Wang, X., Li, Z., Shahidehpour, M., & Jiang, C. (2017). Robust line hardening strategies for improving the resilience of distribution systems with variable renewable resources. *IEEE Transactions on Sustainable Energy, 10*(1), 386–395.

3. Trovato, V., Teng, F., & Strbac, G. (2017). Role and benefits of flexible thermostatically controlled loads in future low-carbon systems. *IEEE Transactions on Smart Grid, 9*(5), 5067–5079.
4. Radaideh, A., Vaidya, U., & Ajjarapu, V. (2017). Sequential set-point control for heterogeneous thermostatically controlled loads through an extended Markov chain abstraction. *IEEE Transactions on Smart Grid, 10*(1), 116–127.
5. Wai, C. H., Beaudin, M., Zareipour, H., Schellenberg, A., & Lu, N. (2014). Cooling devices in demand response: A comparison of control methods. *IEEE Transactions on Smart Grid, 6*(1), 249–260.
6. Bashash, S., & Fathy, H. K. (2012). Modeling and control of aggregate air conditioning loads for robust renewable power management. *IEEE Transactions on Control Systems Technology, 21*(4), 1318–1327.
7. Hu, J., Cao, J., Chen, M. Z., Yu, J., Yao, J., Yang, S., & Yong, T. (2016). Load following of multiple heterogeneous TCL aggregators by centralized control. *IEEE Transactions on Power Systems, 32*(4), 3157–3167.
8. Kandasamy, N. K., Tseng, K. J., & Boon-Hee, S. (2017). Virtual storage capacity using demand response management to overcome intermittency of solar PV generation. *IET Renewable Power Generation, 11*(14), 1741–1748.
9. Blum, D. H., Zakula, T., & Norford, L. K. (2016). Opportunity cost quantification for ancillary services provided by heating, ventilating, and air-conditioning systems. *IEEE Transactions on Smart Grid, 8*(3), 1264–1273.
10. Luo, F., Zhao, J., Dong, Z. Y., Tong, X., Chen, Y., Yang, H., & Zhang, H. (2015). Optimal dispatch of air conditioner loads in southern China region by direct load control. *IEEE Transactions on Smart Grid, 7*(1), 439–450.
11. Xu, Y., & Singh, C. (2012). Adequacy and economy analysis of distribution systems integrated with electric energy storage and renewable energy resources. *IEEE Transactions on power systems, 27*(4), 2332–2341.
12. Zhang, W., Lian, J., Chang, C. Y., & Kalsi, K. (2013). Aggregated modeling and control of air conditioning loads for demand response. *IEEE Transactions on Power Systems, 28*(4), 4655–4664.
13. Hao, H., Sanandaji, B. M., Poolla, K., & Vincent, T. L. (2014). Aggregate flexibility of thermostatically controlled loads. *IEEE Transactions on Power Systems, 30*(1), 189–198.
14. Hussain, A., Bui, V. H., & Kim, H. M. (2016). A resilient and privacy-preserving energy management strategy for networked microgrids. *IEEE Transactions on Smart Grid, 9*(3), 2127–2139.
15. Tindemans, S. H., Trovato, V., & Strbac, G. (2015). Decentralized control of thermostatic loads for flexible demand response. *IEEE Transactions on Control Systems Technology, 23*(5), 1685–1700.
16. Vrettos, E., Ziras, C., & Andersson, G. (2016). Fast and reliable primary frequency reserves from refrigerators with decentralized stochastic control. *IEEE Transactions on Power Systems, 32*(4), 2924–2941.
17. Song, M., Gao, C., Yan, H., & Yang, J. (2017). Thermal battery modeling of inverter air conditioning for demand response. *IEEE Transactions on Smart Grid, 9*(6), 5522–5534.
18. Shah, N., Waide, P., & Phadke, A. (2013). Cooling the planet: Opportunities for deployment of superefficient room air conditioners (No. LBNL-6164E). Lawrence Berkeley National Lab.(LBNL), Berkeley, CA, USA.
19. Wang, D., Ge, S., Jia, H., Wang, C., Zhou, Y., Lu, N., & Kong, X. (2014). A demand response and battery storage coordination algorithm for providing microgrid tie-line smoothing services. *IEEE Transactions on Sustainable Energy, 5*(2), 476–486.
20. Wu, Q., Shahidehpour, M., Li, C., Huang, S., & Wei, W. (2018). Transactive real-time electric vehicle charging management for commercial buildings with PV on-site generation. *IEEE Transactions on Smart Grid, 10*(5), 4939–4950.
21. Kim, Y. J., Norford, L. K., & Kirtley, J. L. (2014). Modeling and analysis of a variable speed heat pump for frequency regulation through direct load control. *IEEE Transactions on Power Systems, 30*(1), 397–408.

22. Song, M., Gao, C., Yang, J., & Yan, H. (2018). Energy storage modeling of inverter air conditioning for output optimizing of wind generation in the electricity market. *CSEE Journal of Power and Energy Systems, 4*(3), 305–315.
23. Song, M., Gao, C., Shahidehpour, M., Li, Z., Lu, S., & Lin, G. (2018). Multi-time-scale modeling and parameter estimation of TCLs for smoothing out wind power generation variability. *IEEE Transactions on Sustainable Energy, 10*(1), 105–118.
24. Luo, F., Dong, Z. Y., Meng, K., Wen, J., Wang, H., & Zhao, J. (2016). An operational planning framework for large-scale thermostatically controlled load dispatch. *IEEE Transactions on Industrial Informatics, 13*(1), 217–227.
25. Kang, M. S., Chen, C. S., Ke, Y. L., Lin, C. H., & Huang, C. W. (2007). Load profile synthesis and wind-power-generation prediction for an isolated power system. *IEEE Transactions on Industry Applications, 43*(6), 1459–1464.
26. Xu, Y., & Singh, C. (2012, October). Operation strategies of the load aggregator with electric energy storage. In *2012 IEEE International Conference on Power System Technology (POWERCON)* (pp. 1–6).